AF686139

# New Fuels and Advanced Combustion Modes for Innovative Internal Combustion Engines

# New Fuels and Advanced Combustion Modes for Innovative Internal Combustion Engines

Guest Editors

**Enzo Galloni**
**Davide Lanni**

MDPI

Basel • Beijing • Wuhan • Barcelona • Belgrade • Novi Sad • Cluj • Manchester

*Guest Editors*
Enzo Galloni
Department of Civil and Mechanical Engineering
University of Cassino and Southern Latium
Cassino
Italy

Davide Lanni
Department of Civil and Mechanical Engineering
University of Cassino and Southern Latium
Cassino
Italy

*Editorial Office*
MDPI AG
Grosspeteranlage 5
4052 Basel, Switzerland

This is a reprint of the Special Issue, published open access by the journal *Energies* (ISSN 1996-1073), freely accessible at: https://www.mdpi.com/journal/energies/special_issues/RMC84E3Z96.

For citation purposes, cite each article independently as indicated on the article page online and as indicated below:

| Lastname, A.A.; Lastname, B.B. Article Title. *Journal Name* **Year**, *Volume Number*, Page Range. |
|---|

**ISBN 978-3-7258-3209-5 (Hbk)**
**ISBN 978-3-7258-3210-1 (PDF)**
**https://doi.org/10.3390/books978-3-7258-3210-1**

© 2025 by the authors. Articles in this book are Open Access and distributed under the Creative Commons Attribution (CC BY) license. The book as a whole is distributed by MDPI under the terms and conditions of the Creative Commons Attribution-NonCommercial-NoDerivs (CC BY-NC-ND) license (https://creativecommons.org/licenses/by-nc-nd/4.0/).

# Contents

# About the Editors

**Enzo Galloni**

Enzo Galloni is an Associate Professor at the Department of Civil and Mechanical Engineering of the University of Cassino and Southern Latium (Italy). He is the author of several publications in peer-reviewed scientific journals and conference proceedings on topics related to Fluid Machines and Heat Engines. His research activities, carried out through both numerical and experimental approaches, are mainly focused on the development of more sustainable internal combustion engines. Currently, he is in charge of the following courses: Fluid Machines Design, a Master's Degree Course in Mechanical Engineering and Fluid Dynamics, and a Degree Course in Industrial Engineering.

**Davide Lanni**

Davide Lanni (Ph.D) is currently an Assistant Professor at the Department of Civil and Mechanical Engineering of the University of Cassino and Southern Latium. His research activities mainly concern experimental testing, modelling, and simulation of internal combustion engines and techno-economic analyses related to power-to-gas plants. In particular, he has carried out numerical simulations of internal combustion engines fed with both biofuels and carbon-free fuels and both experimental and numerical analyses on the use of the water injection technique in spark-ignition engines. Moreover, he has conducted techno-economic analyses focused on the optimal sizing of power-to-gas plants for the production of hydrogen and biomethane. He currently teaches "Advanced Power Systems" in the master's degree program in Mechanical Engineering at the University of Cassino and Southern Latium and "Sustainable Energy Technologies" in the master's degree program in Management Engineering at the University of Naples Parthenope. He also teaches "Hydrogen Production and Utilization Technologies" in the Ph.D. program in Methods, Models and Technologies for Engineering at the University of Cassino and Southern Latium.

*energies*

MDPI

*Editorial*

# New Fuels and Advanced Combustion Modes for Innovative Internal Combustion Engines: An Overview

**Davide Lanni * and Enzo Galloni**

Department of Civil and Mechanical Engineering, University of Cassino and Southern Lazio, 03043 Cassino, Italy
* Correspondence: davide.lanni@unicas.it

**Abstract:** Internal combustion engines (ICEs) currently account for approximately 25% of global power generation. Notably, this technology still plays a crucial role in a large segment of the transportation sector. In this editorial, a short overview of the latest developments and current research trends related to internal combustion engines is presented. Furthermore, the 11 contributions of this Special Issue are introduced. They cover three main topics: the use of new fuels for internal combustion engines for both automotive and railway applications; testing of additives for ICEs fed with conventional fuels; and CFD simulation applied to the analysis and design of ICE components.

**Keywords:** internal combustion engines; carbon-free fuels; e-fuels; biofuels; fuel additives; CFD engine modeling; advanced combustion modes

**Citation:** Lanni, D.; Galloni, E. New Fuels and Advanced Combustion Modes for Innovative Internal Combustion Engines: An Overview. *Energies* **2024**, *17*, 6228. https://doi.org/10.3390/en17246228

Received: 5 December 2024
Revised: 8 December 2024
Accepted: 9 December 2024
Published: 11 December 2024

**Copyright:** © 2024 by the authors. Licensee MDPI, Basel, Switzerland. This article is an open access article distributed under the terms and conditions of the Creative Commons Attribution (CC BY) license (https://creativecommons.org/licenses/by/4.0/).

## 1. Introduction

Today, the energy sector faces the pressing challenge of mitigating greenhouse gas and pollutant emissions. This situation necessitates the adoption of innovative technologies alongside enhancements to conventional ones, such as internal combustion engines. Transitioning from ICEs to technologies that do not rely on combustion presents significant technical and economic hurdles. Achieving this transition on a global scale within a short timeframe appears to be a tough challenge. Additionally, it is essential to conduct a thorough evaluation of the impact of all technologies, taking into account their complete life and usage cycles.

The pollutant emissions from ICEs that comply with current regulations have shown a notable decrease in recent years [1]. However, the greenhouse gas emissions produced by burning traditional fossil fuels still require substantial examination. Thus, in order to reduce the environmental impact of ICEs, several solutions are currently under study.

The use of carbon-free fuels (hydrogen and ammonia [2]), low-carbon fuels (methane, methanol, etc.) [3], e-fuels [4], and biofuels [5] can strongly contribute to reducing both the greenhouse gas and pollutant emissions of ICEs.

In particular, in recent years, both hydrogen and ammonia have been receiving particular attention from researchers, since, if produced starting from renewable energy sources, they drastically reduce the environmental impact of engines if compared to traditional fuels. Hydrogen is suitable for both spark-ignition [6–9] and compression-ignition engines [10]. The expected improvements in the next generation of hydrogen internal combustion engines suggest that their thermodynamic efficiency will be similar to that of current fuel cell powertrains [11]. Ammonia, which is also a hydrogen carrier [12], presents advantages in terms of storage over hydrogen but also issues when burned in internal combustion engines. Several studies assessed both its combustion characteristics and emissions in spark-ignition engines, highlighting the critical issues of using neat ammonia as a single fuel [13–15]. Different techniques to improve ammonia combustion in spark-ignition engines have been proposed, such as hydrogen enrichment [16–18], multiple spark strategies [19,20] and plasma ignition [21]. Ammonia can also be used in compression ignition engines in dual-fuel mode, even if advanced injection strategies and aftertreatment systems are needed to improve both engine performance and emissions [22].

The use of fuel additives is a way to improve performance and reduce the emissions of ICEs fed with conventional fuels, particularly in regard to existing vehicle fleets. If coupled with alternative fuels, fuel additives can play a significant role in promoting the shift toward cleaner and more sustainable energy sources in the transportation sector [23]. However, achieving a balance between cost and performance is essential to ensure their widespread adoption [23].

Considering both conventional and non-conventional fuels, advanced combustion strategies can improve the energy conversion efficiency in internal combustion engines.

Homogeneous charge compression ignition (HCCI) has gained recognition and undergone extensive testing, research, and analysis as a promising "green" engine technology in contemporary times [24]. Nevertheless, challenges such as ignition control, difficulties with cold starts, and the need for consistent fuel/air mixture preparation, among other limitations, continue to pose significant obstacles [24].

Reactivity-controlled compression ignition (RCCI) demonstrated significant commercial potential for automotive applications in the short to medium term [25]. Nevertheless, the implementation of accurate feedback control systems is essential to regulate engine parameters, thereby reducing cycle-to-cycle variability and ensuring reliable and safe operation [25].

Pre-chamber ignition, also known as turbulent jet ignition (TJI), allows the use of both stoichiometric and lean air-fuel mixtures in ICEs. The passive TJI system provides low-cost technology and stoichiometric operation, enhancing combustion and potentially improving fuel consumption by 2–3% [26]. In contrast, the active TJI system operates under lean-burn conditions with an air-excess value up to 2, offering greater efficiency and lower $NO_x$ emissions, but at higher costs and complexity [26]. Several significant challenges remain to be thoroughly resolved, including the design and integration of hardware in current cylinder heads, ensuring satisfactory combustion stability at low loads such as idle and catalyst light-off, and implementing lean aftertreatment for the active configuration [26].

Water injection (WI) has garnered significant interest in recent years across various engine types as a promising method to lower in-cylinder and exhaust temperatures, reduce knock events, enhance combustion timing, and reduce $NO_x$ emissions [27]. In particular, WI has significant potential to enhance the thermal efficiency of spark-ignition engines, mainly because of its ability to effectively suppress knock and provide cooling benefits [28]. Water can be injected into the intake ports (port water injection, PWI) [29,30] or directly into the combustion chamber (direct water injection, DWI) [31,32]. DWI is capable of removing a larger quantity of heat in comparison to PWI; however, the latter facilitates a more uniform distribution of water near the cylinder walls [33]. The fundamental thermophysical and chemical kinetic effects of adding water on combustion processes and emissions require further exploration, particularly concerning various methods of WI and different engine configurations. Moreover, there is a scarcity of research focused on the long-term effects of water injection in ICEs [27].

The presented short overview shows that ICEs are still widely studied and developed. They may still play an important role in the near future in order to increase the transport sector's sustainability.

In this scenario, this Special Issue presents seven research articles assessing the use of new fuels for internal combustion engines (Section 2), two studies related to experimental investigations on additives for ICEs fed with conventional fuels (Section 3), and two contributions (a research article and a review) focused on the CFD simulation applied to the analysis and design of ICE components (Section 4).

## 2. New Fuels for ICEs

Lanni et al. [34] employed a one-dimensional (1D) numerical method to assess the performance and operational limits of a downsized port fuel injection spark ignition

(PFI SI) engine running on pure ammonia. The investigation covered both throttled and unthrottled operation and also evaluated the impact of different geometrical compression ratios. The findings revealed that ammonia's low laminar flame speed leads to longer combustion durations and optimal spark timing that is more advanced than in conventional SI engines, with knock consistently mitigated. Long combustion durations and high exhaust gas temperatures limited both the maximum allowed engine speed and the lowest achievable load. However, higher compression ratios allowed for an increase in the engine's maximum speed.

The work of Tutak et al. [35] examined the co-combustion of hydrogen with diesel and biodiesel (RME) in a compression-ignition engine under maximum load, with hydrogen content up to 34%. Hydrogen minimally affected ignition delay but reduced combustion duration, especially with biodiesel. A hydrogen content of 12% provided optimal energy efficiency, with better results when using RME. However, the hydrogen–RME engine showed slightly less stability. Emissions analysis revealed reductions in CO, $CO_2$, and soot, but increases in NO and HC emissions.

Tutak et al. [36] also evaluated the operation of a cooperative fuel research spark-ignition engine using ammonia and dimethyl ether (DME) as fuels. DME facilitated ignition and improved combustion, with just a 10% energy share of DME contributing to proper combustion. The addition of DME reduced both ignition delay and combustion duration. Even minimal DME inclusion ensured high repeatability of IMEP (indicated mean effective pressure), staying below 5%. With an 18% energy share of ammonia, the engine achieved the maximum efficiency.

Arsie et al. [37] presented preliminary findings from the H2-ICE project, focusing on hydrogen fuel utilization and hybrid powertrain integration for urban buses. A robust methodology has been developed, including validated high-fidelity CFD models to assess various injection methods (port fuel injection and direct injection) and optimize combustion efficiency. Two Waste Heat Recovery (WHR) systems were tested. The development of catalytic systems for the Selective Catalytic Reduction (SCR) of nitrogen oxides ($NO_x$) using hydrogen has been completed, along with a control strategy utilizing an Artificial Neural Network (ANN) model. Two Energy Management System (EMS) methodologies were evaluated to meet a target fuel consumption of 0.1 kg/km.

Kolahchian et al. [38] employed a 1D engine modeling approach to evaluate hydrogen as a fuel for railways, using a turbocharged diesel engine as a reference. Modifications to the turbocharger and injection systems allow the engine to run on hydrogen while retaining original performance. Results show a reduction in traction power from 600 kW to 400 kW and nearly double the energy consumption during typical missions compared to diesel. However, a Life Cycle Assessment indicates a 56% reduction in equivalent $CO_2$ emissions when using photovoltaic-based green hydrogen, lowering emissions from 4.27 to below 2 kg $CO_2$ equivalent per kilometer. This suggests that engines with moderate modifications can effectively reduce carbon emissions in non-electrified railway segments.

The research of Capatano et al. [39] experimentally assessed the potential of low- and zero-carbon fuels, such as methanol, methane, and hydrogen, in a spark-ignition engine. Results showed that alternative fuels reduced CO and $CO_2$ emissions, with the reduction varying by fuel type. Methanol had higher total hydrocarbon (THC) emissions but lower nitrogen oxides ($NO_x$) emissions compared to gasoline. Methane and hydrogen further lowered THC emissions, while $NO_x$ emissions were influenced by operational conditions. Particle emissions were affected by fuel properties, engine conditions, and lubricating oil type, particularly with hydrogen, which produced significant particle emissions despite lacking carbon atoms.

The safety aspects related to the use of hydrogen in an engine-based combustion system were evaluated by Gill et al. [40]. Their methodology utilizes the FMESA (Failure Mode and Effects Severity Analysis) framework with specialized tabular scales to evaluate failure severity, aiming to reduce epistemic uncertainty in hazard severity and risk models. This approach modifies traditional methods like FMEA/FMECA (Failure Mode and Effect

Analysis/Failure Mode, Effects and Criticality Analysis) and includes the development of new failure severity scales specific to hydrogen systems. The article reviews relevant literature, discusses the FMESA method and its mathematical framework, presents failure severity scales, shares experimental findings, and quantitatively analyzes the severity of different failure modes in the hydrogen combustion system.

## 3. Additives for ICEs Fed with Conventional Fuels

Chivu et al. [41] investigated the performance, emissions, and fuel consumption of a 1.6 L four-cylinder direct-injection diesel engine using blends of commercial diesel and pine tree turpentine (up to 30% of turpentine). Key findings indicate that the biofuel had minimal overall performance impact, but a blend with 15% turpentine increased torque by up to 7.9% at low load and 6.8% at high load, with power output rising by 9% at low speeds and 5% at high speeds compared to baseline diesel. While efficiency and greenhouse gas emissions improved, pollutant emissions varied; hydrocarbons and particulate matter responses differed, but $NO_x$ emissions increased by 30% at high loads and 20% at low loads, mainly due to enhanced engine performance rather than turpentine's higher oxygen content.

The study of Marchitto et al. [42] assessed the effects of two performance packages on exhaust emissions and fuel efficiency in five vehicles from the current Italian fleet, including three Euro 4 vehicles (two passenger cars and one light commercial vehicle) and two Euro 6 diesel vehicles. Considering the worldwide harmonized light vehicles test cycle (WLTC), the results indicated a 1.2% reduction in fuel consumption for the Euro 6 diesel passenger car and 8.1% for the Euro 4 diesel passenger car, along with similar trends in $CO_2$ emissions and significant reductions in total hydrocarbons, carbon monoxide, and particulate matter across all vehicles.

## 4. CFD Simulation for the Analysis and Design of ICE Components

Martos et al. [43] developed a CFD model to simulate throttle valve operation and identify optimal exhaust backpressure for effective low-pressure exhaust gas recirculation in Euro 6 compliant engines. The model examines flow control dynamics for integrating these valves with thermoelectric generators that convert residual thermal energy into electricity. The results indicate consistent pressure drop values across different scenarios, with over 90% agreement between model and experimental results, leading to a correlation for estimating exhaust gas mass flow rates based on easily measurable parameters.

The perspective review of Jeong et al. [44] highlighted that CFD technology is widely used in industry and academia to improve pre-chamber designs, but it faces challenges in predictive accuracy, particularly in turbulence modeling, which impacts mixing, combustion, and wall heat transfer. They concluded that, to unlock the full potential of CFD as a sophisticated design tool rather than merely a predictive one, it is crucial to enhance CFD methodologies with specifically tailored, physics-based numerical models. Furthermore, to acheive high-fidelity CFD modeling of pre-chamber engines, it is imperative to identify and rectify the primary sources of uncertainty prior to implementing refinements and enhancements related to both turbulence and combustion models.

**Conflicts of Interest:** The authors declare no conflicts of interest.

## References

1. Reitz, R.D.; Ogawa, H.; Payri, R.; Fansler, T.; Kokjohn, S.; Moriyoshi, Y.; Agarwal, A.; Arcoumanis, D.; Assanis, D.; Bae, C.; et al. IJER editorial: The future of the internal combustion engine. *Int. J. Engine Res.* **2020**, *21*, 3–10. [CrossRef]
2. El-Adawy, M.; Nemitallah, M.A.; Abdelhafez, A. Towards sustainable hydrogen and ammonia internal combustion engines: Challenges and opportunities. *Fuel* **2024**, *364*, 131090. [CrossRef]
3. Benajes, J.; García, A.; Monsalve-Serrano, J.; Guzmán-Mendoza, M. A review on low carbon fuels for road vehicles: The good, the bad and the energy potential for the transport sector. *Fuel* **2024**, *361*, 130647. [CrossRef]
4. Dell'aversano, S.; Villante, C.; Gallucci, K.; Vanga, G.; Di Giuliano, A. E-Fuels: A Comprehensive Review of the Most Promising Technological Alternatives towards an Energy Transition. *Energies* **2024**, *17*, 3995. [CrossRef]

5. Caliskan, H.; Yildiz, I.; Mori, K. Biofuels combustion in internal combustion engines. In *Advances in Biofuels Production, Optimization and Applications*; Elsevier: Amsterdam, The Netherlands, 2024; pp. 185–205.
6. Manzoor, M.U.; Dou, X.; Yosri, M.; Talei, M.; Yang, Y. Capturing different modes of hydrogen combustion in a spark-ignition engine using numerical simulations. *Fuel* **2024**, *375*, 132343. [CrossRef]
7. Galloni, E.; Lanni, D.; Fontana, G.; D'antuono, G.; Stabile, S. Performance Estimation of a Downsized SI Engine Running with Hydrogen. *Energies* **2022**, *15*, 4744. [CrossRef]
8. Li, X.; Zhuang, Y.; Wang, Y.; Zhu, Z.; Qian, Y.; Zhai, R. An experimental investigation on the lean-burn characteristics of a novel hydrogen fueled spark ignition engine: Hydrogen injection via a micro-hole on the spark plug. *Int. J. Hydrogen Energy* **2024**, *57*, 990–999. [CrossRef]
9. Matla, J.; Kaźmierczak, A.R.; Haller, P.; Trocki, M. Hydrogen as a fuel for spark ignition combustion engines-state of knowledge and concept. *Combust. Engines* **2024**, *196*, 73–79. [CrossRef]
10. Rueda-Vázquez, J.M.; Serrano, J.; Pinzi, S.; Jiménez-Espadafor, F.J.; Dorado, M.P. A Review of the Use of Hydrogen in Compression Ignition Engines with Dual-Fuel Technology and Techniques for Reducing $NO_x$ Emissions. *Sustainability* **2024**, *16*, 3462. [CrossRef]
11. Onorati, A.; Payri, R.; Vaglieco, B.; Agarwal, A.; Bae, C.; Bruneaux, G.; Canakci, M.; Gavaises, M.; Günthner, M.; Hasse, C.; et al. The role of hydrogen for future internal combustion engines. *Int. J. Engine Res.* **2022**, *23*, 529–540. [CrossRef]
12. Pinzón, M.; García-Carpintero, R.; de la Osa, A.; Romero, A.; Abad-Correa, D.; Sánchez, P. Ammonia as a hydrogen carrier: An energy approach. *Energy Convers. Manag.* **2024**, *321*, 118998. [CrossRef]
13. Lhuillier, C.; Brequigny, P.; Contino, F.; Rousselle, C. *Combustion Characteristics of Ammonia in a Modern Spark-Ignition Engine*; SAE Technical Paper No. 2019-24-0237; SAE International: Warrendale, PA, USA, 2019.
14. Lhuillier, C.; Brequigny, P.; Contino, F.; Mounaïm-Rousselle, C. Experimental investigation on ammonia combustion behavior in a spark-ignition engine by means of laminar and turbulent expanding flames. *Proc. Combust. Inst.* **2021**, *38*, 5859–5868. [CrossRef]
15. D'Antuono, G.; Galloni, E.; Lanni, D.; Contino, F.; Brequigny, P.; Mounaïm-Rousselle, C. Assessment of combustion development and pollutant emissions of a spark ignition engine fueled by ammonia and ammonia-hydrogen blends. *Int. J. Hydrogen Energy* **2024**, *85*, 191–199. [CrossRef]
16. Lhuillier, C.; Brequigny, P.; Contino, F.; Rousselle, C. *Performance and Emissions of an Ammonia-Fueled SI Engine with Hydrogen Enrichment*; SAE Technical Paper No. 2019-24-0137; SAE International: Warrendale, PA, USA, 2019.
17. Lee, J.; Jang, Y.; Park, C.; Kim, Y.; Choi, Y. Operable range extension of ammonia direct injection spark ignition engine by hydrogen addition. *Int. J. Hydrogen Energy* **2024**, *58*, 1631–1639. [CrossRef]
18. D'Antuono, G.; Lanni, D.; Galloni, E.; Fontana, G. Numerical Modeling and Simulation of a Spark-Ignition Engine Fueled with Ammonia-Hydrogen Blends. *Energies* **2023**, *16*, 2543. [CrossRef]
19. Uddeen, K.; Tang, Q.; Shi, H.; Magnotti, G.; Turner, J. A novel multiple spark ignition strategy to achieve pure ammonia combustion in an optical spark-ignition engine. *Fuel* **2023**, *349*, 128741. [CrossRef]
20. Huang, Q.; Liu, J. Preliminary assessment of the potential for rapid combustion of pure ammonia in engine cylinders using the multiple spark ignition strategy. *Int. J. Hydrogen Energy* **2024**, *55*, 375–385. [CrossRef]
21. Zhang, B.; Rubio, M.; Egolfopoulos, F.; Cronin, S.B. Stable combustion of ammonia in an internal combustion engine: A single fuel approach enabled by multi-pulse transient plasma ignition. *Fuel* **2025**, *381*, 133502. [CrossRef]
22. Dimitriou, P.; Javaid, R. A review of ammonia as a compression ignition engine fuel. *Int. J. Hydrogen Energy* **2020**, *45*, 7098–7118. [CrossRef]
23. Malik, M.A.I.; Kalam, M.; Abbas, M.M.; Silitonga, A.S.; Ikram, A. Recent advancements, applications, and technical challenges in fuel additives-assisted engine operations. *Energy Convers. Manag.* **2024**, *313*, 118643. [CrossRef]
24. Zhou, W. HCCI Technology: Operating Principles, Advantages, Challenges, and Future Prospects. *Highlights Sci. Eng. Technol.* **2024**, *88*, 585–592. [CrossRef]
25. Paykani, A.; Garcia, A.; Shahbakhti, M.; Rahnama, P.; Reitz, R.D. Reactivity controlled compression ignition engine: Pathways towards commercial viability. *Appl. Energy* **2021**, *282*, 116174. [CrossRef]
26. Zhu, S.; Akehurst, S.; Lewis, A.; Yuan, H. A review of the pre-chamber ignition system applied on future low-carbon spark ignition engines. *Renew. Sustain. Energy Rev.* **2022**, *154*, 111872. [CrossRef]
27. Zhu, S.; Hu, B.; Akehurst, S.; Copeland, C.; Lewis, A.; Yuan, H.; Kennedy, I.; Bernards, J.; Branney, C. A review of water injection applied on the internal combustion engine. *Energy Convers. Manag.* **2019**, *184*, 139–158. [CrossRef]
28. Wan, J.; Zhuang, Y.; Huang, Y.; Qian, Y.; Qian, L. A review of water injection application on spark-ignition engines. *Fuel Process. Technol.* **2021**, *221*, 106956. [CrossRef]
29. Lanni, D.; Galloni, E.; Fontana, G. Numerical analysis of the effects of port water injection in a downsized SI engine at partial and full load operation. *Appl. Therm. Eng.* **2022**, *205*, 118060. [CrossRef]
30. Fan, Y.; Wu, T.; Xiao, D.; Xu, H.; Li, X.; Xu, M. Effect of port water injection on the characteristics of combustion and emissions in a spark ignition direct injection engine. *Fuel* **2021**, *283*, 119271. [CrossRef]
31. Fratita, M.; Popescu, F.; Martins, J.; Brito, F.; Costa, T. Direct water injection and combustion time in SI engines. *Energy Rep.* **2021**, *7*, 798–803. [CrossRef]
32. Lanni, D.; Galloni, E. *Direct Water Injection Strategies for Performance Improvement of a Turbocharged Spark-Ignition Engine at High Load Operation*; SAE Technical Paper No. 2022-37-0007; SAE International: Warrendale, PA, USA, 2022.

33. Lanni, D.; Galloni, E.; Fontana, G.; Erme, G. *Experimental and Numerical Analyses of Direct and Port Water Injection in a Turbocharged Spark-Ignition Engine*; SAE Technical Paper 2021-24-0035; SAE International: Warrendale, PA, USA, 2021. [CrossRef]
34. Lanni, D.; Galloni, E.; Fontana, G.; D'Antuono, G. Assessment of the Operation of an SI Engine Fueled with Ammonia. *Energies* **2022**, *15*, 8583. [CrossRef]
35. Tutak, W.; Jamrozik, A.; Grab-Rogaliński, K. Co-Combustion of Hydrogen with Diesel and Biodiesel (RME) in a Dual-Fuel Compression-Ignition Engine. *Energies* **2023**, *16*, 4892. [CrossRef]
36. Tutak, W.; Pyrc, M.; Gruca, M.; Jamrozik, A. Ammonia Combustion in a Spark-Ignition Engine Supported with Dimethyl Ether. *Energies* **2023**, *16*, 7283. [CrossRef]
37. Arsie, I.; Battistoni, M.; Brancaleoni, P.P.; Cipollone, R.; Corti, E.; Di Battista, D.; Millo, F.; Occhicone, A.; Peiretti Paradisi, B.; Rolando, L.; et al. A New Generation of Hydrogen-Fueled Hybrid Propulsion Systems for the Urban Mobility of the Future. *Energies* **2024**, *17*, 34. [CrossRef]
38. Kolahchian Tabrizi, M.; Cerri, T.; Bonalumi, D.; Lucchini, T.; Brenna, M. Retrofit of Diesel Engines with H2 for Potential Decarbonization of Non-Electrified Railways: Assessment with Lifecycle Analysis and Advanced Numerical Modeling. *Energies* **2024**, *17*, 996. [CrossRef]
39. Catapano, F.; Di Iorio, S.; Magno, A.; Sementa, P.; Vaglieco, B.M. A Complete Assessment of the Emission Performance of an SI Engine Fueled with Methanol, Methane and Hydrogen. *Energies* **2024**, *17*, 1026. [CrossRef]
40. Gill, A.; Pielecha, I.; Szwajca, F. A New Method of Failure Mode and Severity Effects Analysis for Hydrogen-Fueled Combustion Systems. *Energies* **2024**, *17*, 4802. [CrossRef]
41. Chivu, R.M.; Martins, J.; Popescu, F.; Uzuneanu, K.; Ion, I.V.; Goncalves, M.; Codău, T.-C.; Onofrei, E.; Brito, F.P. Turpentine as an Additive for Diesel Engines: Experimental Study on Pollutant Emissions and Engine Performance. *Energies* **2023**, *16*, 5150. [CrossRef]
42. Marchitto, L.; Costagliola, M.A.; Berra, A. Influence of Performance Packages on Fuel Consumption and Exhaust Emissions of Passenger Cars and Commercial Vehicles under WLTP. *Energies* **2024**, *17*, 3356. [CrossRef]
43. Martos, F.J.; Soriano, J.A.; Braic, A.; Fernández-Yáñez, P.; Armas, O. A CFD Modelling Approach for the Operation Analysis of an Exhaust Backpressure Valve Used in a Euro 6 Diesel Engine. *Energies* **2023**, *16*, 4112. [CrossRef]
44. Jeong, S.-J. CFD Simulation of Pre-Chamber Spark-Ignition Engines—A Perspective Review. *Energies* **2024**, *17*, 4696. [CrossRef]


**Disclaimer/Publisher's Note:** The statements, opinions and data contained in all publications are solely those of the individual author(s) and contributor(s) and not of MDPI and/or the editor(s). MDPI and/or the editor(s) disclaim responsibility for any injury to people or property resulting from any ideas, methods, instructions or products referred to in the content.

energies

MDPI

Article

# Assessment of the Operation of an SI Engine Fueled with Ammonia

**Davide Lanni *, Enzo Galloni *, Gustavo Fontana and Gabriele D'Antuono**

Department of Civil and Mechanical Engineering, University of Cassino and Southern Latium, 03043 Cassino, Italy
* Correspondence: davide.lanni@unicas.it (D.L.); galloni@unicas.it (E.G.)

**Abstract:** Recently, the research interest regarding ammonia applications in energy systems has been increasing. Ammonia is an important hydrogen carrier that can also be obtained starting from renewable energy sources. Furthermore, ammonia can be used as a carbon-free fuel in combustion systems. In particular, the behavior of internal combustion engines (ICEs), fueled by ammonia, needs to be further investigated. The main disadvantage of this kind of fuel is its low laminar flame speed when it is oxidized with air. On the other hand, considering a spark-ignition (SI) engine, the absence of knock phenomena could allow a performance improvement. In this work, a 1D numerical approach was used in order to assess the performance and the operating limits of a downsized PFI SI engine fueled with pure ammonia. Furthermore, the reliability of the 1D model was verified by means of a 3D approach. Both throttled and unthrottled engine operation was investigated. In particular, different boost levels were analyzed under WOT (wide-open throttle) conditions. The potential of the 1D approach was also exploited to evaluate the effect of different geometrical compression ratio on the ammonia engine behavior. The results show that the low laminar flame speed of ammonia–air mixtures leads to increased combustion durations and optimal spark timings more advanced than the typical ones of SI engines. On the other hand, knock phenomena are always avoided. Due to the engine operating limits, the maximum rotational speed guaranteeing proper engine operation is 3000 rpm, except for at the highest boost level. At this regime, the load regulation can be critical in terms of unburned fuel emissions. Considering increased compression ratios and no boost conditions, even the 4000 rpm operating point guarantees proper engine operation.

**Keywords:** ammonia; carbon-free fuels; e-fuels; spark-ignition engines; downsizing

**Citation:** Lanni, D.; Galloni, E.; Fontana, G.; D'Antuono, G. Assessment of the Operation of an SI Engine Fueled with Ammonia. *Energies* **2022**, *15*, 8583. https://doi.org/10.3390/en15228583

Academic Editor: Jamie W.G. Turner

Received: 29 October 2022
Accepted: 13 November 2022
Published: 16 November 2022

**Publisher's Note:** MDPI stays neutral with regard to jurisdictional claims in published maps and institutional affiliations.

**Copyright:** © 2022 by the authors. Licensee MDPI, Basel, Switzerland. This article is an open access article distributed under the terms and conditions of the Creative Commons Attribution (CC BY) license (https://creativecommons.org/licenses/by/4.0/).

## 1. Introduction

Pollution and greenhouse gases are the main problems related to the combustion of fossil fuels in internal combustion engines (ICEs). Thus, it is crucial to find new fuels characterized by low emissions associated with the burning process. Ammonia is an important energy vector that can be used as a hydrogen carrier or as a fuel in combustion processes [1]. The energy required from the ammonia production process can be obtained via conventional processes, but also from solar or wind energy [2]. Therefore, the type of process used to produce ammonia determines the classification in brown ammonia, blue ammonia, and green ammonia [3].

The interest in ammonia as a power source started in 1878 [4] and continued during the Second World War in Belgium due to the lack of conventional fuel for heavy duty applications. Then, the use of ammonia as a fuel lost attractiveness due to its bad combustion properties. Nowadays, the interest in ammonia as an energy carrier is increasing because of having zero carbon emissions and low storage and transport costs compared to those of hydrogen. There are also some disadvantages, like low energy density, corrosive behavior, and worse combustion characteristics compared to conventional fuels. It is worth underlining that ammonia can have dangerous effects on human health, such as irritation of the eyes and nose. Furthermore, ammonia may also be dangerous for the environment, in particular in dry, windy, and warm conditions, but its high volatility promotes dispersion in the atmosphere [1].

Table 1 summarizes the main characteristics of ammonia and compares them to those of gasoline and hydrogen. Ammonia shows clearly worse combustion characteristics with respect to gasoline. The laminar flame speed and the LHV are 87% less and 57% less than those of gasoline respectively. Nevertheless, ammonia shows a higher octane number, and therefore the knock resistance is better than that of gasoline. This allows the use of a high compression ratio in order to increase the engine efficiency and/or high boost pressure values in a supercharged engine to increase the specific power. Regarding the allowable mixture strength of ammonia–air mixtures, the flame can propagate in mixture characterized by an equivalence ratio ranging between 0.63 and 1.4 [5]; therefore, ammonia engines will be characterized by a substantially stoichiometric operation, similar to that of gasoline engines.

**Table 1.** Ammonia properties vs. gasoline and hydrogen properties.

| | Ammonia | Gasoline | Hydrogen |
|---|---|---|---|
| Storage [1] | Liquid [300 K–11 bar] | Liquid [300 K–1 bar] | Compressed [300 K, 700 bar] |
| LHV [MJ/kg] [6] | 18.8 | 44.5 | 120 |
| Octane number [6,7] | 130 | 92–98 | >130 |
| Laminar burn velocity [m/s] [6] | 0.015 | 0.58 | 3.51 |
| Auto-ignition temperature [K] [6] | 924 | 550 | 858 |
| Explosion limit [% volume ratio] [8] | 16–28 | 1.4–7.6 | 4.5–75 |
| Density [kg/m$^3$] [6] | 0.703 | 740 | 0.082 |

Ammonia may be useful in ICEs [8] because it is a carbon-free fuel, it does not emit carbon dioxide, and it can be used in both spark-ignition (SI) engines and compression-ignition (CI) engines. Furthermore, the possibility of using ammonia in the marine sector is also under investigation [9,10]. ICEs could be fired using ammonia as a neat fuel [11], mixed with hydrogen [12], or in blends with conventional fuels like gasoline [13] or diesel [14,15]. In particular, hydrogen enrichment can significantly improve the combustion process of gaseous fuels in SI engines, as already shown in the work of Sun et al. [16] and Duan et al. [17].

However, the potential of pure ammonia as a fuel in internal combustion engines has been little investigated until now. As an example, Lhuillier et al. [18,19] experimentally investigated the combustion characteristics of ammonia in a modified GDI engine at 1500 rpm. They compared the combustion behavior of ammonia with hydrogen-enriched ammonia and pure methane. The results showed that ammonia can be used in an SI engine, but, in order to achieve proper engine operation, it was necessary to advance the spark timing. The efficiency was similar for each case studied, with pressure peaks lower for ammonia compared to those of other fuels. Mounaim-Rousselle et al. [20] focused their attention on the operating limits of a small GDI engine modified to run with premixed ammonia–hydrogen–air mixtures (hydrogen content ranging from 0 to 10% by volume). Keeping the original spark plug device, they analyzed engine cycle variation and exhaust gas emissions when varying the intake air pressure (from about 0.55 bar up to 1 bar) at different engine speeds (from idle speed to 2000 rpm) for lean, stoichiometric and rich air–fuel mixtures. Running with pure ammonia, they found the engine could stably operate at up to 1500 rpm when the intake air pressure was equal to 1 bar, while the engine cycle variation limited low load operation. In particular, they indicated a coefficient of variation of IMEP higher than 5% when the intake air pressure was less than 0.75 bar. The engine was not able to operate at 2000 rpm for any of the air intake pressures. When the hydrogen percentage in the fuel mixture was 10%, the authors achieved proper engine operation for all considered cases.

In [21], the authors investigated the combustion of lean gasoline–ammonia–air mixtures in an SI engine derived from a CI engine characterized by a glow plug and a sub-chamber. They found that increasing the ammonia concentration reduces the pressure peak, while $NO_x$ increases. The influence of the glow plug on IMEP and thermal efficiency was negligible in almost all conditions.

El Fattah et al. [22] focused on the exhaust emissions and performance of an SI engine fueled with water–ammonia–gasoline blends. They found that when adding water–ammonia to gasoline, the engine thermal efficiency increased, but the CO formation also increased due to the occurrence of incomplete combustion. Additionally, $NO_x$ emissions increased with the volume percentage of the water–ammonia solution.

In [23] Sahin et al. investigated the behavior of a small dual fuel CI engine running with premixed water–ammonia solutions. The addition of ammonia determined an increase in engine efficiency at all the investigated speed and load values. Even in this case, they found an increase in $NO_x$ related to the nitrogen present in $NH_3$. The authors also pointed the attention on the damages of some elements of the carburetor used to add the fuel.

A literature review shows that fueling ICEs with pure ammonia leads to increased combustion durations, which could limit the operation of light-duty engines. The aim of this work is to explore the operating limits of a light-duty turbocharged SI engine fueled with pure ammonia considering different engine speeds, boost pressure levels and throttle openings. In particular, the behavior of a small ammonia engine is investigated by means of a numerical approach able to reproduce the whole engine layout. This approach also allows investigating the influence of increased geometrical compression ratios on both the performances and the operating range of the engine.

## 2. Materials and Methods

To investigate the potential of ammonia as a fuel for internal combustion engines, the conversion of a typical downsized gasoline engine, developed for light duty vehicles, was considered (Section 2.1). The performance of the converted engine was estimated by means of a 1D model developed to reproduce the engine operation when the fuel is varied (Section 2.2). The reliability of 1D results mainly depends on a proper prediction of the in-cylinder pressure development. Thus, 3D calculations were also considered in order to model the flame propagation within the combustion chamber of the ammonia engine (Section 2.3). The comparison between 1D and 3D results, with respect to some operating points, gives an idea of the reliability of both of the calculation models.

### *2.1. The Engine*

The engine analyzed in this work is a typical port fuel injection (PFI) turbocharged SI engine for automotive use. It was developed to run with straight regular grade gasoline fueling. Its main characteristics are shown in Table 2.

**Table 2.** Engine main characteristics.

| Model | 4 Cylinders, 4 Valves/cyl |
|---|---|
| Displacement [$cm^3$] | 1368 |
| Bore/Stroke/Con. Rod [mm] | 72/84/129 |
| Compression Ratio [-] | 9.8 |
| Combustion Chamber Shape | Pent Roof |
| Max Power (ISO Conditions) [kW] | 110 @ 5500 rpm |
| Max Torque (ISO Conditions) [Nm] | 230 @ 3000 rpm |
| Turbocharger group | IHI RHF3 |

### *2.2. 1D Numerical Approach*

A 1D approach was used to assess the operation of the engine running with ammonia. The model reproduces the whole engine layout (Figure 1) in order to perform full engine cycle simulations. It was described in detail in [24,25]. Briefly, in-pipe flow was estimated using a finite volume approach, while the turbocharger was modeled using the steady flow maps provided by the manufacturer. To allow boost pressure control, a part of the exhaust gas can bypass the turbine simulating the wastegate valve operation. The heat transfer rate through the walls of the combustion chamber was estimated by means of the well-known Woshni correlation, while the friction losses were calculated as a function of the engine speed.

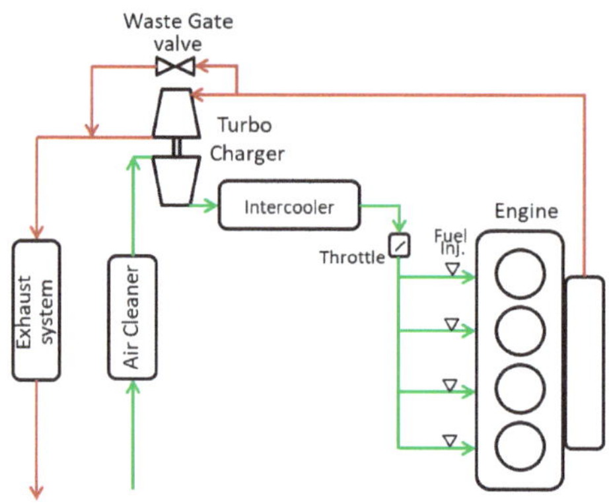


**Figure 1.** Schematic layout of the engine [25].

Ammonia oxidizes according to Equation (1). The dissociation of the combustion products is not considered, so the ammonia molecules can be oxidized completely or not depending on the charge conditions.

$$4NH_3 + 3O_2 \rightarrow 2N_2 + 6H_2O \tag{1}$$

The model used predicts the burning rate for homogeneous charge spark-ignition engines based on the work of Keck and Tabaczynski [26,27]. This approach takes into account the cylinder geometry, the spark location and timing, the air motion and the fuel properties. The calculated turbulent flame speed depends on both laminar flame speed and turbulent flow indices according to an entrainment model. A spherical flame propagates from the sparkplug, while a turbulent entrainment process controls the combustion rate, as shown in Equations (2) and (3).

$$\frac{dM_e}{dt} = \rho_e A_f (u' + S_L) \tag{2}$$

$$\frac{dM_b}{dt} = \frac{(M_e - M_b)}{\tau} \tag{3}$$

where $M_e$ is the mass of the fresh gas mixture entering the flame front and $M_b$ is the burned mass. $A_f$ is the surface area at the edge of the flame front, $\rho_e$ is the fresh gas density, $u'$ is the turbulent intensity and $S_L$ is the laminar flame speed. $\tau$ is a time constant depending on the ratio between the Taylor microscale of turbulence and the laminar flame speed. The Taylor scale is calculated as a function of the integral length of turbulence and the kinematic viscosity of the unburned mixture. Following the hierarchical approach described in [28], the mean turbulent length scale and the mean turbulent intensity were imposed, exploiting the flow field details provided by 3D calculations.

The knock occurrence was predicted taking into account the empirical induction time ($\tau_{id}$) of the air–fuel mixture. It was assumed that knock occurs when $\int_0^t dt/\tau_{id}(t) = 1$ [29]. When knock occurs, the knock intensity is estimated as in Equation (4) [30].

$$KI = (1 - x_b)(cr - 1)\sqrt{1 - \frac{\theta_k}{\theta_{ref}} \frac{N}{N_{ref}}} \tag{4}$$

where $x_b$ is the mass of burned fuel, $cr$ is the compression ratio, $\theta_k$ is the knock onset crank angle, and $N$ the engine speed, $\theta_{ref}$ the maximum crank angle for which knock is still audible and is set to 50 CAD. $N_{ref}$ is a tuning parameter set equal to 3500 rpm. As in [24] and [25], it was assumed that tolerable knock corresponds to a $KI$ level less than or equal to 0.5.

The model was previously validated considering the engine running with pure gasoline both at full and partial load [24]. Furthermore, it correctly reproduced the behavior of the engine running with different alcohol–gasoline blends [31] and was also used to predict the performance of the same engine in case of hydrogen fueling [25]. This encouraged the authors to use the model also to predict the behavior of the analyzed engine running with pure ammonia. Of course, the thermochemical properties of ammonia–air mixture were considered, while the laminar flame speed was calculated using the correlations proposed by Goldmann and Dinkelacker [32]:

$$S_L = S_{L0}\left(\frac{T_u}{T_0}\right)^{\alpha}\left(\frac{p}{p_0}\right)^{\beta} k \tag{5}$$

$$S_{L,0} = (a_1 + b_1\lambda)/\left(1 + c_1\lambda + d_1\lambda^2\right) \tag{6}$$

$$\alpha = a_2 + b_2\lambda + c_2\lambda^2 + d_2\lambda^3 + e_2\lambda^4 + f_2\lambda^5 \tag{7}$$

$$\beta = a_3 + b_3\lambda + c_3\lambda^2 + d_3\lambda^3 + e_3\lambda^4 + f_3\lambda^5 \tag{8}$$

$$k = \begin{cases} k_1 k_2^{\frac{p}{p_0}}\left(\frac{p}{p_0}\right)^{k_3}\lambda^{k_4} \; \forall T_u \le T_0 \wedge p \le 0.5\,\text{MPa} \\ 1.0 \qquad\qquad \forall\, T_u > T_0 \wedge p > 0.5\,\text{MPa} \end{cases} \tag{9}$$

where $T_u$ and $p$ are the temperature and pressure of the unburned mixture, respectively. $T_0$ and $p_0$ are the reference conditions of temperature and pressure equal to 300 K and 0.1 MPa, respectively. $\lambda$ is the air–fuel equivalence ratio and is equal to $1/\phi$. Coefficients $a, b, c, d, e, f$ and $k$ are tabulated as a function of pressure and temperature ranges [32].

Figure 2 shows the laminar flame speed at ambient condition for ammonia–air mixtures calculated by the relationships mentioned above and compares them with those calculated for gasoline–air mixtures according to [33]. As expected, gasoline shows a faster laminar flame speed. Both ammonia and gasoline exhibit a peak at an equivalence ratio of 1.1. Considering stoichiometric conditions for both mixtures, the laminar flame speed for ammonia is 0.06 m/s, approximately 82% less than that of gasoline.

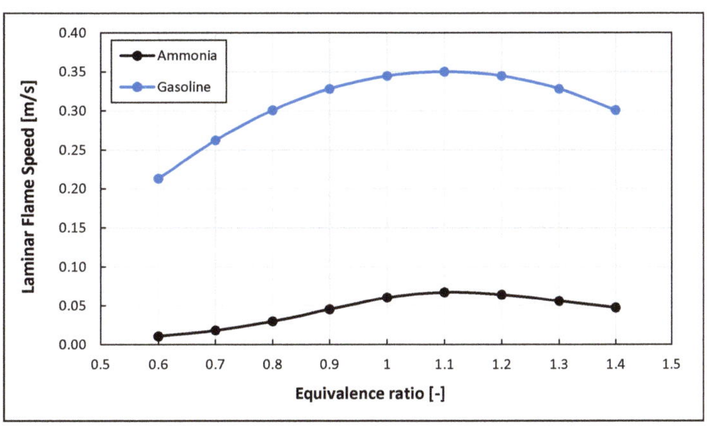


**Figure 2.** Laminar flame speed under ambient conditions calculated for gasoline–air mixtures and ammonia–air mixtures.

The induction time of the ammonia–air mixture is calculated by means of Equation (10) [34]:

$$\tau_{id} = Ap^{n} \cdot e^{\frac{E_a}{RT}} \tag{10}$$

where $A$ is equal to 0.0658 ms, $n$ is equal to $-0.779$, $E_a$ represents the global activation energy and is equal to 35.574 kcal/mol, and $R$ is the universal gas constant.

*2.3. 3D Numerical Approach*

A 3D computational model was used to reproduce some of the operating points investigated by means of the 1D model. The model was built using the commercial code AVL Fire. The partial differential transport equations were discretized on the basis of a finite volume method. A first-order (Euler) implicit differencing scheme was used for time discretization. Second-order schemes were used for continuity and momentum equations. An upwind scheme was used for turbulence and energy equations. Turbulent effects were described by means of the standard k-ε approach. The ammonia oxidation process was simulated implementing the chemical-kinetics mechanism proposed by Nakamura et al. [35] within the 3D model. This mechanism consists of 33 species and 232 reactions.

Due to the symmetry of the engine geometry, only half of the fluid domain was modeled. The combustion chamber, the intake valve, and the intake port were discretized by means of polyhedral elements in order to obtain a dynamic unstructured grid (Figure 3) which reproduced the intake stroke, the compression stroke and the following expansion stroke of the engine. The minimum number of cells in the grid is about 50,000 (firing TDC), while the maximum number is about 350,000 (BDC during the intake stroke). Boundary and initial conditions were set by means of calculations carried out using the 1D model described above.

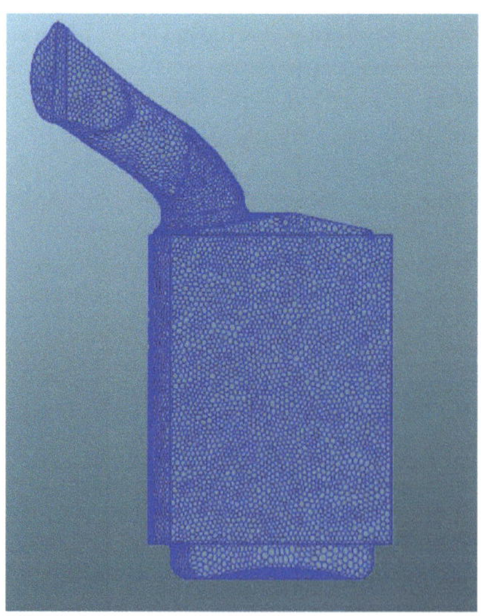

**Figure 3.** Engine grid.

## 3. Results

To assess the behavior of the analyzed engine with ammonia fueling, steady-state engine operating points were investigated. In each case, wide open throttle (WOT) operation was imposed, except for the analysis referring to the effect of throttling on engine performance. First, 3D calculations were carried out to reproduce some operating points and to compare the 3D results with those obtained from the 1D calculations. Then, several analyses were performed using the 1D approach in order to assess both the performance and the operating limits of the ammonia engine. In the following sub-sections, the results of each analysis are described in detail.

### *3.1. Comparison between 3D and 1D Results*

The engine operating points shown in Table 3 were simulated using both 3D and 1D approaches. No boost indicates fully open wastegate valve conditions. In particular, the optimal spark timing for maximum torque was identified by means of 1D calculations. The same operating point was reproduced using the 3D approach. It is worth highlighting that spark timings more advanced than typical ones of SI engines are needed to obtain a proper combustion with ammonia fueling due to its low laminar flame speed. This is confirmed by the results of the experimental analysis carried out in [18]. Naturally, the more the engine speed increases, the more the optimal spark timing will advance (Table 3).

**Table 3.** Test cases, WOT, no boost. Optimal spark timing.

| Engine Speed [rpm] | 2000 | 3000 |
|---|---|---|
| Spark advance [°bTDC] | 82.9 | 102.1 |
| Equivalence ratio [–] | 1 | 1 |

Figure 4 shows the in-cylinder pressure evolution and the heat release rate predicted by both 3D and 1D models. The good agreement between the results of both approaches further encouraged the authors to use the 1D approach to perform an analysis of both the performance and the operating limits of the ammonia engine. Furthermore, the calculated burned fuel fraction at the exhaust valve opening (EVO) is shown in Table 4. The 1D model overestimates the amount of unburned ammonia because it does not take into account the partial oxidation of ammonia. The result is an underestimation in terms of burned fuel fraction less than 2%.

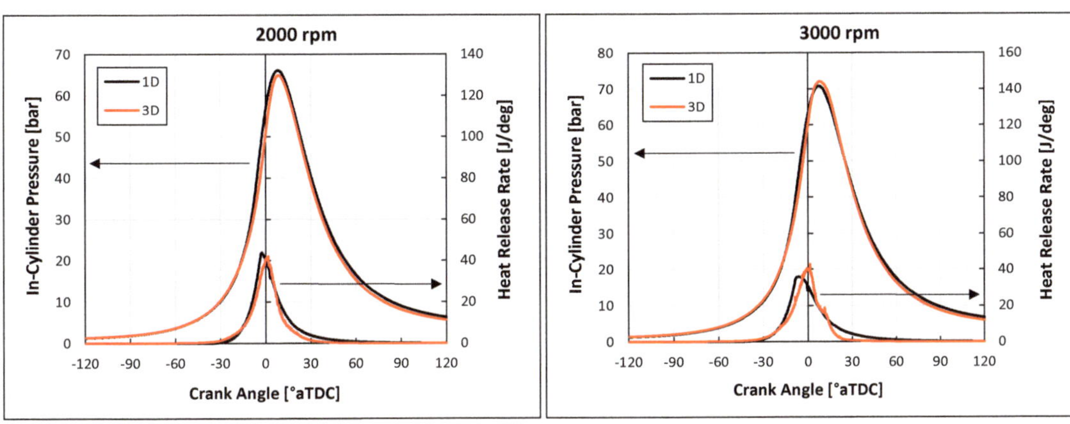


**Figure 4.** In-cylinder pressure and heat release rate for both 2000 and 3000 rpm operating points, WOT. Comparison between 1D and 3D results. $\phi = 1$, no boost.

**Table 4.** Burned fuel fraction at EVO calculated by both 3D and 1D model. WOT, no boost, optimal spark timing.

| Engine Speed [rpm] | 2000 | | 3000 | |
|---|---|---|---|---|
| | 3D | 1D | 3D | 1D |
| Burned fuel fraction at EVO [%] | 99.95 | 98.84 | 99.99 | 98.42 |

### 3.2. Engine Performance and Operating Limits

First, parametric analyses of the effects of both spark timing and equivalence ratio on engine performance and combustion development were performed at 1500 rpm. Then, in subsequent analyses, the spark angle was set for maximum torque and stoichiometric combustion was imposed. Throttled and unthrottled engine operating points were analyzed. In particular, WOT operation was investigated both for increasing boost pressures (BP) and increasing geometrical compression ratio.

Analyses have been carried out taking into account the engine operating limits reported in Table 5. Unfortunately, the 1D model is not able to predict the cycle-to-cycle variation. Thus, this important constraint was not considered.

**Table 5.** Engine operating limits.

| | |
|---|---|
| Peak pressure [bar] | ≤100 |
| Exhaust gas temperature [K] | ≤1173 |
| Spark advance [°bTDC] | ≤110 |
| Knock intensity [–] | ≤0.5 |

It is worth underlining that no knocking condition was detected in any of the analyzed cases.

Finally, it was considered that unburned ammonia cannot be released through the engine exhaust. As shown in Table 4, the 1D model overestimates the amount of unburned ammonia calculated in the various cases. For this reason, in the subsequent analysis a fuel burned mass fraction equal to 98% was considered as the minimum acceptable value.

#### 3.2.1. Effect of Spark Timing

The effect of different spark advances (SA) on engine performance and combustion characteristics was evaluated at 1500 rpm setting the equivalence ratio equal to 1 and no boost WOT conditions. The results are shown in Figure 5 and Table 6. Advanced spark timings lead to an increase in combustion delay (CA0-2) and a reduction of combustion duration (CA10-90), as reported in Table 6. The CA10-90 reduction prevails over the CA0-2 increase. Thus, when the spark advance increases, both the pressure peak (Figure 5, left) and the burned fuel fraction (Table 6) at EVO increases. The spark advance value which maximizes both engine power (Figure 5, right) and efficiency (Table 6) is 69 °bTDC. Setting the latter value, the brake specific fuel consumption (BSFC) is about 420 g/kWh (Table 6).

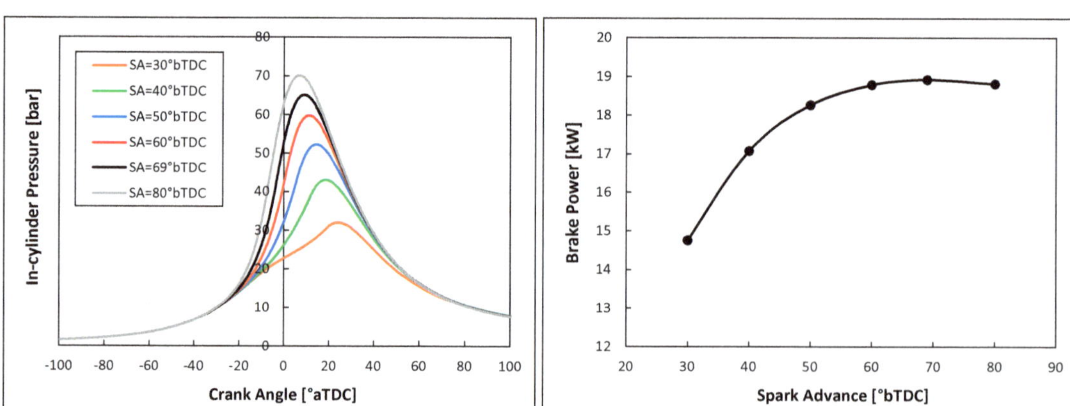


**Figure 5.** In-cylinder pressure (**left**) and brake power (**right**) for different spark advances; 1500 rpm, $\phi = 1$. WOT, no boost.

**Table 6.** Combustion characteristics and engine performance; 1500 rpm, $\phi$ = 1, WOT, no boost.

| Spark Advance [°bTDC] | 30 | 40 | 50 | 60 | 69 | 80 |
|---|---|---|---|---|---|---|
| CA0-2 [CAD] | 29.3 | 33.3 | 38.5 | 44.2 | 49.7 | 56.5 |
| CA10-90 [CAD] | 101.9 | 50.1 | 35.8 | 30.5 | 28.1 | 26.8 |
| Burned Fuel Fraction at EVO [%] | 91.3 | 95.7 | 97.7 | 98.6 | 99.1 | 99.4 |
| Brake Efficiency [%] | 29.7 | 34.4 | 36.8 | 37.9 | 38.2 | 38.0 |
| BSFC [g/kWh] | 539.9 | 465.8 | 435.1 | 422.6 | 419.5 | 421.7 |

### 3.2.2. Effect of Equivalence Ratio

The effect of the mixture composition was investigated under 1500 rpm, wide open throttle, no boost conditions. The spark advance was set to be equal to 69 °bTDC, i.e., the optimal spark advance under stoichiometric conditions for $\phi$ = 1. Figure 6 and Table 7 show the obtained results. Both combustion delay and combustion duration decrease for increasing values of the equivalence ratio leading to higher values of the burned fuel fraction at EVO (Table 7). In particular, the maximum mass fraction burnt (99.1%) is obtained for the stoichiometric mixture, while for $\phi < 0.9$ the 90% mass fraction burnt is not reached. As a result, of the faster combustion, the peak pressure increases for increasing values of the equivalence ratio (Figure 6, left) leading to higher engine power (Figure 6, right). The maximum engine efficiency and the minimum BSFC are reached for $\phi$ = 0.9, but the overall combustion duration is higher with respect to stoichiometric conditions (Table 7). For this reason, in the following analyses, stoichiometric combustion was imposed in order to obtain a good compromise between engine efficiency and combustion duration. This choice was made since, at higher engine speeds, a too slow combustion could not allow a proper engine operation. On the other hand, rich mixtures were avoided since they lead to ammonia emissions which are not acceptable from an environmental point of view.

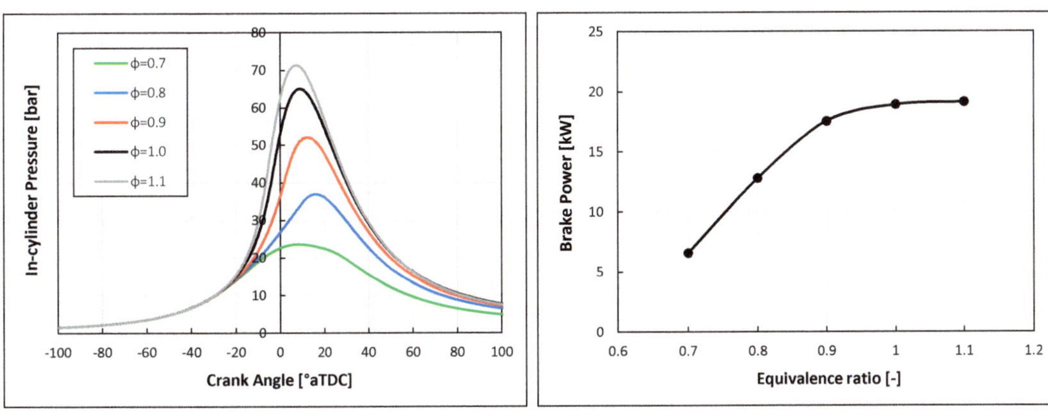


**Figure 6.** In-cylinder pressure (**left**) and brake power (**right**) for different equivalence ratios; 1500 rpm, SA = 69 °bTDC, WOT, no boost.

**Table 7.** Combustion characteristics and engine performance; 1500 rpm, SA = 69 °bTDC, WOT, no boost.

| Equivalence Ratio [-] | 0.7 | 0.8 | 0.9 | 1.0 | 1.1 |
|---|---|---|---|---|---|
| CA0-2 [CAD] | 66.3 | 59.3 | 53.9 | 49.7 | 47.7 |
| CA10-90 [CAD] | / | / | 41.2 | 28.1 | 24.6 |
| Burned Fuel Fraction at EVO [%] | 61.7 | 87.4 | 96.5 | 99.1 | 91.0 |
| Brake Efficiency [%] | 18.0 | 31.1 | 38.6 | 38.2 | 35.9 |
| BSFC [g/kWh] | 889 | 516 | 415 | 420 | 446 |

### 3.2.3. Effect of Boost Pressure

The performances of the ammonia engine were investigated by varying the engine speed and considering different boost levels (Figures 7–10). In each operating point, the mixture was set to be stoichiometric and the spark angle was adjusted to the optimal value complying with the limit SA $\leq$ 110 °bTDC.

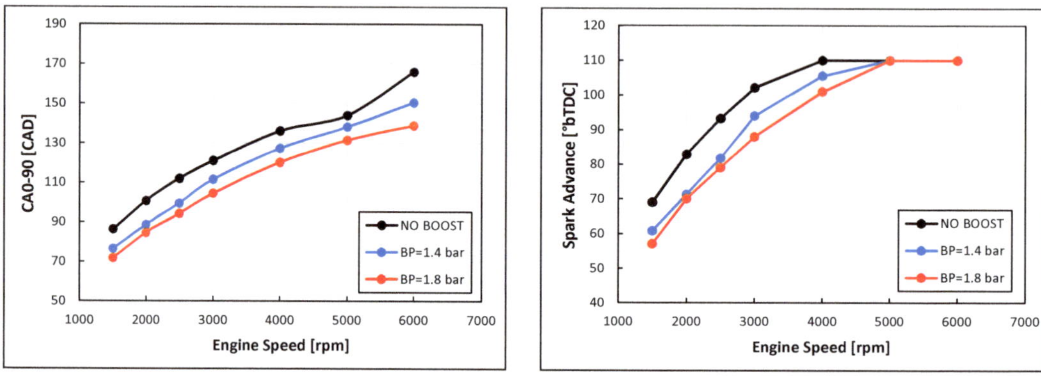


**Figure 7.** Combustion duration (0–90%) (**left**) and optimal spark advance (**right**) for different boost levels. $\phi$ = 1. WOT.

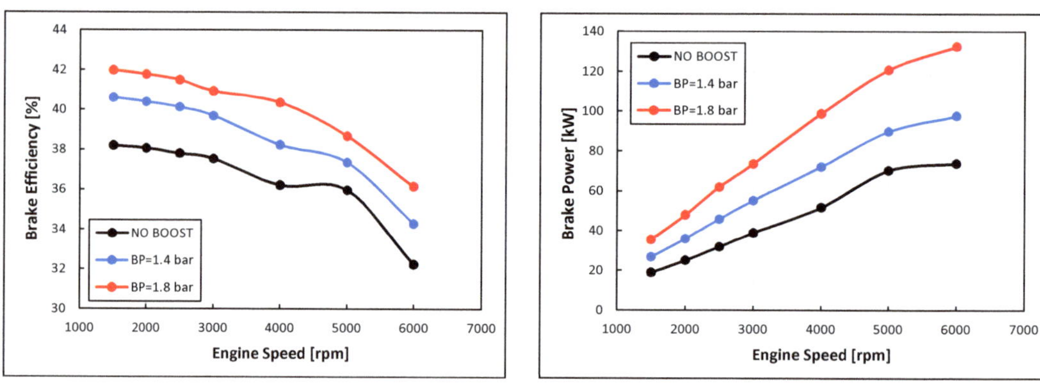


**Figure 8.** Brake efficiency (**left**) and brake power (**right**) for different boost levels. $\phi$ = 1. WOT.

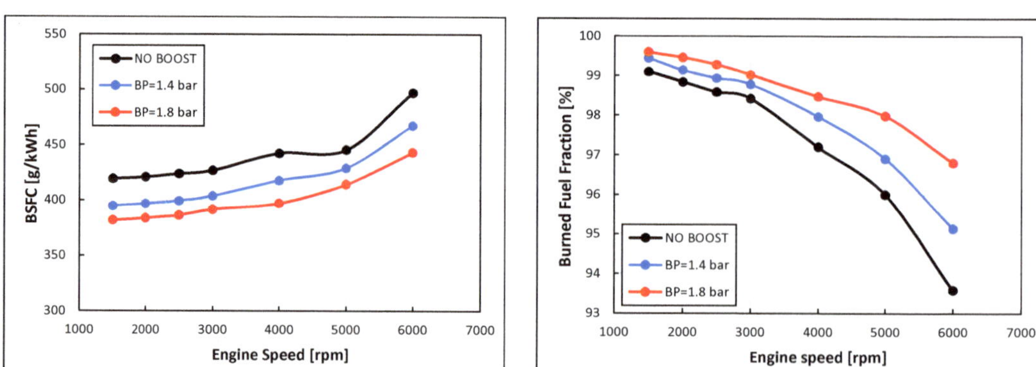


**Figure 9.** BSFC (**left**) and burned fuel fraction (**right**) for different boost levels. $\phi$ = 1. WOT.

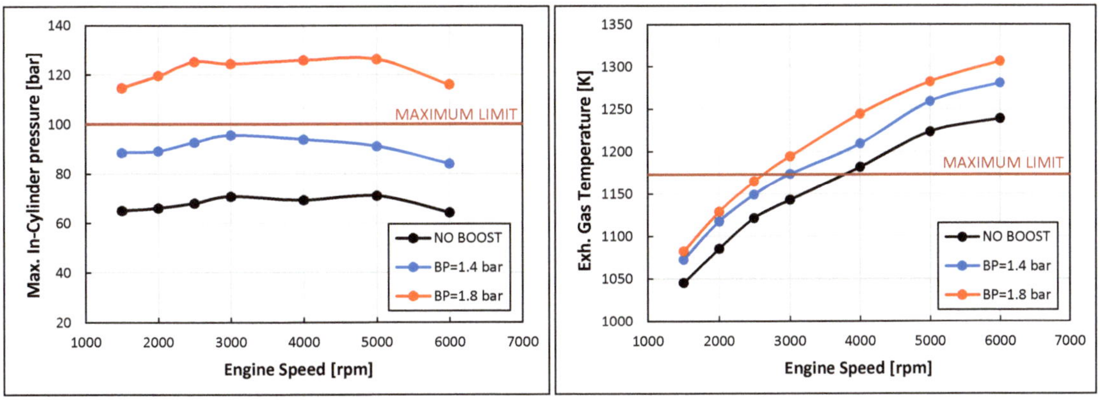


**Figure 10.** Maximum in-cylinder pressure (**left**) and exhaust gas temperature (**right**) for different boost levels. $\phi = 1$. WOT.

Results related to engine speeds ranging from 1500 rpm to 6000 rpm are shown. The combustion duration increases with increasing engine speed, while it decreases with increasing boost pressure (Figure 7, left). The optimal spark advance consequently varies, reaching the imposed constraint at the highest engine speeds (Figure 7, right). It is worth underlining that the increase in the angular combustion duration (CA0-90) with increasing engine speed is typical of SI engines, but it can be critical in the case of ammonia fueling due to the low laminar flame speed of ammonia–air mixtures (Figure 2). At medium and high rotation speeds, the engine could not properly operate [20]. Combustion durations comparable to those obtained in this work were experimentally measured in [18].

The engine efficiency grows with the boost pressure (Figure 8, left) showing very high values at low engine speeds (from 38% to 42% at 1500 rpm). It decreases when the engine speed increases because the combustion slows down. However, the engine efficiency always remains over 32%. The BSFC consequently varies from a minimum value of about 380 g/kWh to a maximum value of about 500 g/kWh (Figure 9, left). The engine power also reaches very interesting values (Figure 8, right), like those of gasoline engines. At 6000 rpm, with a boost pressure equal to 1.4, the engine can deliver about 100 kW.

Unfortunately, higher engine speeds appear unsustainable considering the trends of both the unburned ammonia content and the temperature of the exhaust gas leaving the cylinders. Indeed, increasing combustion durations lead to an incomplete fuel combustion (Figure 9, right) and to higher temperatures of the exhaust gases which leave the cylinders (Figure 10, right). It is worth noting that when increasing the boost level, the burned ammonia fuel fraction improves, while the exhaust gas temperature increases at a given engine speed.

In the end, the maximum allowable engine speed is about 3000 rpm both at no boost operation (due to the unburned fuel constraint) and at BP = 1.4 bar (due to the exhaust gas temperature constraint). The boost level 1.8 is unacceptable, since the maximum in-cylinder pressure exceeds the maximum allowable value in each operating point (Figure 10, left).

#### 3.2.4. Effect of Compression Ratio

The effects of increasing geometrical compression ratios (from 9.8 to 11.0) on the engine behavior were evaluated at wide throttle operation, no boost condition. As in the previous analysis, the spark angle was adjusted to the optimal value complying with the limit SA $\leq$ 110 °bTDC.

For each engine speed, the increase in the compression ratio leads to a reduction in the combustion duration and to a consequent decrease in the optimal spark advance (Figure 11). This is due to the higher pressure in the cylinder during the compression stroke, which increases the charge temperature and, consequently, the flame speed. The favorable thermodynamic effect due to both the increase in the compression ratio and the faster combustion determines a not negligible improvement in the engine efficiency and a slight increase in the engine power (Figure 12).

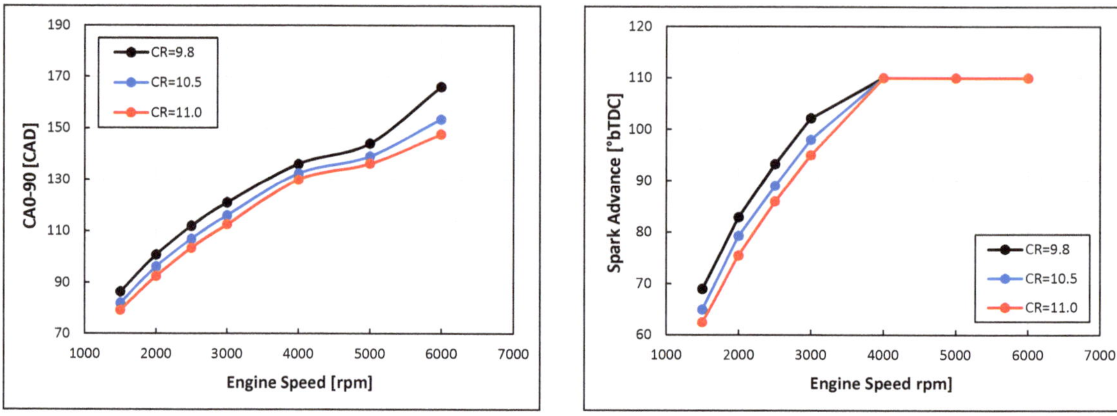


**Figure 11.** Combustion duration (**left**) and optimal spark advance (**right**) for different compression ratios. $\phi = 1$, WOT, no boost.

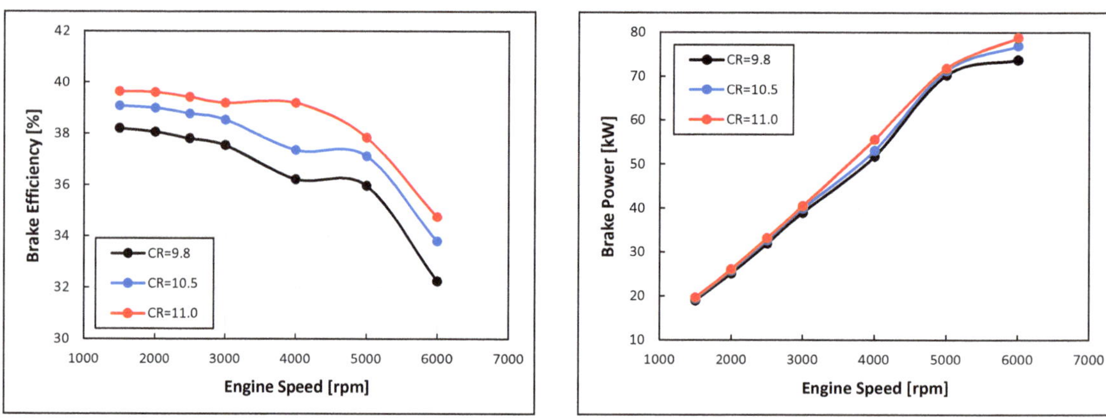


**Figure 12.** Brake efficiency (**left**) and brake power (**right**) for different compression ratios. $\phi = 1$, WOT, no boost.

Higher geometrical compression ratios are beneficial for both the quantity of unburned fuel at EVO (Figure 13, right) and the exhaust gas temperature (Figure 14, right). Indeed, faster combustion makes the combustion process more complete. Furthermore, the temperature of the exhaust gas leaving the cylinder decreases due to the increased effective expansion phase. On the other hand, the peak pressures increase while remaining below the maximum limit (Figure 14, left). At the end, for no boost conditions, with increased compression ratios, even the 4000 rpm operating point complies with the operating limits, guaranteeing proper engine operation.

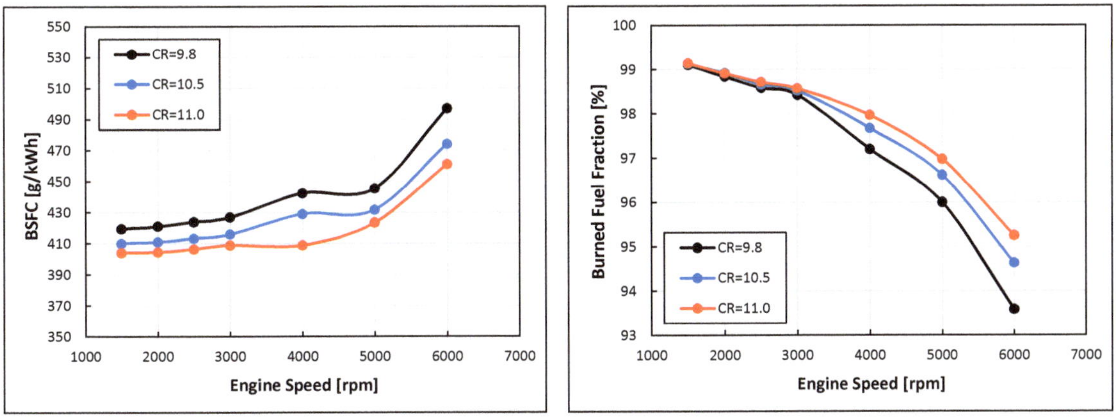


**Figure 13.** BSFC (**left**) and burned fuel fraction (**right**) for different compression ratios. $\phi$ = 1, WOT, no boost.

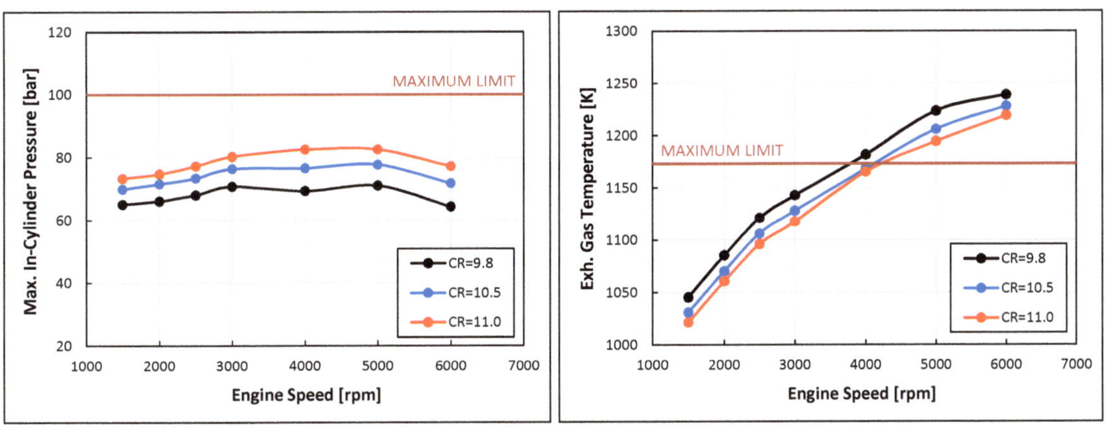


**Figure 14.** Maximum in-cylinder pressure (**left**) and exhaust gas temperature (**right**) for different compression ratios. $\phi$ = 1, WOT, no boost.

#### 3.2.5. Effect of Throttle Opening

Considering no boost conditions, stoichiometric air–ammonia mixture and optimal spark timing, the effect of different throttle openings on engine behavior was evaluated. The analysis was carried out at 1500 and 3000 rpm operating points. The load regulation curves are shown in Figure 15. As is typical in PFI SI engines, when throttling the engine, the brake efficiency decreases due to the growing of the pumping losses. Furthermore, the combustion duration increases due to the higher dilution of the charge (Figure 16, left) which reduces the laminar flame speed Equations (3)–(7). Higher CA0-90 values at decreasing engine loads lead to a reduction of the burned fuel fraction which is amplified for the highest engine speed (Figure 16, right). These results clearly show that, at 3000 rpm, the load regulation can be critical in terms of unburned fuel emissions. It is worth noting that using a numerical approach it has not been possible to evaluate the engine cycle variation. Thus, the operating range could be further reduced due to an excessive value of the coefficient of variation of IMEP [20].

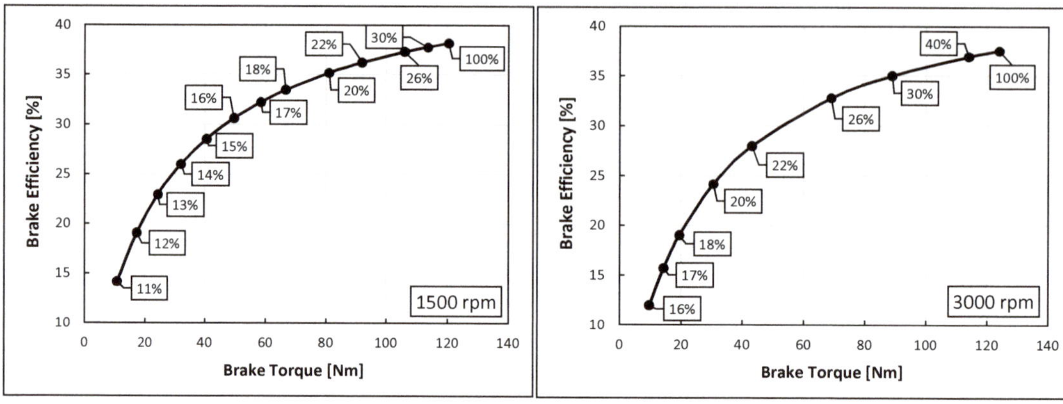


**Figure 15.** Brake efficiency as a function of the brake torque. The labels highlight the different throttle opening values. $\phi = 1$, no boost.

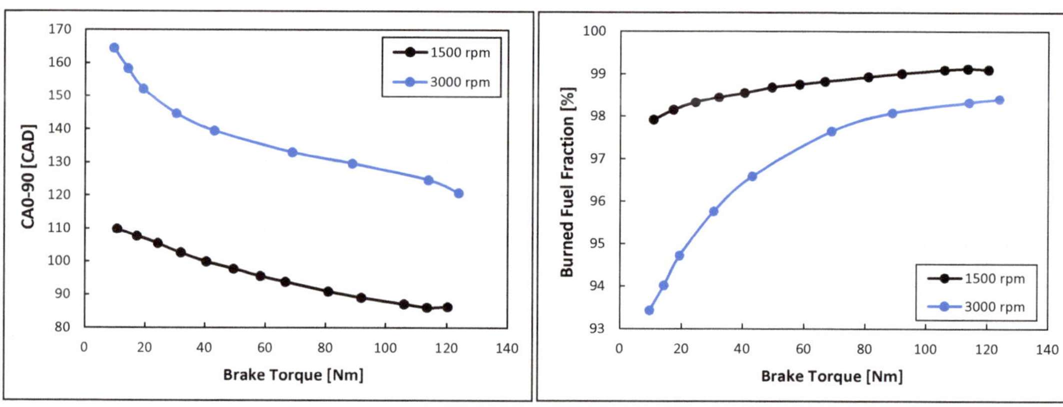


**Figure 16.** CA0-90 (**left**) and burned fuel fraction (**right**) vs. engine torque for both 1500 and 3000 rpm operating points. $\phi = 1$, no boost.

## 4. Conclusions

In this paper, the performance and the operating limits of a downsized PFI SI engine fueled with pure ammonia were estimated by means of a 1D predictive model. 3D calculations were performed to verify the reliability of the results provided by the 1D approach. The main results can be summarized as follows:

- The low laminar flame speed of ammonia–air mixtures leads to increased combustion durations and to optimal spark timings more advanced than typical ones of SI engines. On the other hand, no knock occurrence was detected.
- At 1500 rpm, no boost WOT condition and fixed spark advance, the ammonia burned fuel fraction decreases with decreasing equivalence ratios. Despite the maximum engine efficiency (38.6%) and the minimum BSFC (415 g/kWh) being reached for $\phi = 0.9$, both the lowest combustion duration and the minimum ammonia content in the exhaust gas are obtained for $\phi = 1$.
- Considering unthrottled engine operation and stoichiometric conditions, the combustion duration decreases for increasing boost pressure. Due to the operating limits of both in-cylinder pressure and exhaust gas temperature, the maximum regime which guarantees a proper engine operation is 3000 rpm, except for the highest boost level (BP = 1.8 bar).

- Imposing no boost WOT conditions and optimal spark timing, increasing compression ratios (from 9.8 to 11.0) lead to a reduction in the combustion duration (up to 11% at the maximum regime) and to a consequent increase in the burned fuel fraction (up to about 1.7% at the maximum regime). Of course, engine power slightly increases and BSFC decreases. The exhaust gas temperature also decreases, allowing the engine to properly operate at up to 4000 rpm.
- Considering no boost conditions, stoichiometric air–ammonia mixture and optimal spark timing, the more the engine is throttled, the more the burned fuel fraction decreases. This reduction increases with increasing speed (up to a burned fraction of about 93.5% at 3000 rpm), which can make it difficult to operate the engine at part load and medium engine speeds.
- Properly tuned, the ammonia engine shows really interesting efficiencies (up to about 41% at the maximum allowable BP). At a given engine speed, the delivered power is comparable to that of the same engine fueled with gasoline. Of course, limits on the maximum speed greatly reduce the specific power compared to that of conventional engines.

It is worth noting that the analysis was carried out using a predictive model that does not allow evaluation the engine operating stability. This could further reduce the operating ranges found.

In the end, ammonia could be a very interesting fuel due to the absence of carbon dioxide emissions and low storage and transport costs. However, this analysis clearly shows that pure ammonia could excessively limit the operating range of a light-duty spark-ignition engine. On the other hand, the results suggest that this fuel could be suitable to power engines characterized by low engine speeds and high geometrical compression ratios.

**Author Contributions:** Conceptualization, D.L. and E.G.; methodology, D.L., E.G. and G.D.; formal analysis, D.L., E.G. and G.D.; investigation, D.L. and G.D.; data curation, D.L. and G.D.; writing—original draft preparation, D.L., E.G. and G.D.; writing—review and editing, D.L., E.G., G.F. and G.D.; supervision, E.G. and G.F. All authors have read and agreed to the published version of the manuscript.

**Funding:** This research received no external funding.

**Data Availability Statement:** Not applicable.

**Conflicts of Interest:** The authors declare no conflict of interest.

## Nomenclature

| | |
|---|---|
| 1D | One-dimensional |
| 3D | Three-dimensional |
| aTDC | After top dead center |
| BDC | Bottom dead center |
| BP | Boost pressure |
| BSFC | Brake specific fuel consumption |
| bTDC | Before top dead center |
| CA0-2 | 0–2% mass fraction fuel burn duration |
| CA0-90 | 0–90% mass fraction fuel burn duration |
| CA10-90 | 10–90% mass fraction fuel burn duration |
| CAD | Crank angle degree |
| CI | Compression ignition |
| $COV_{IMEP}$ | Coefficient of variation of IMEP |
| CR | Compression ratio |
| EVO | Exhaust valve opening |
| GDI | Gasoline direct injection |
| KI | Knock intensity |
| ICE | Internal combustion engine |

| | |
|---|---|
| IME | Indicated mean effective pressure |
| $NO_x$ | Nitrogen oxides |
| PFI | Port fuel injection |
| SA | Spark advance |
| SI | Spark ignition |
| TDC | Top dead center |
| WOT | Wide open throttle |
| $\phi$ | Equivalence ratio |

## References


1. Valera-Medina, A.; Xiao, H.; Owen-Jones, M.; David, W.; Bowen, P. Ammonia for power. *Prog. Energy Combust. Sci.* **2018**, *69*, 63–102. [CrossRef]
2. Giddey, S.; Badwal, S.P.S.; Munnings, C.; Dolan, M. Ammonia as a Renewable Energy Transportation Media. *ACS Sustain. Chem. Eng.* **2017**, *5*, 10231–10239. [CrossRef]
3. Cardoso, J.S.; Silva, V.; Rocha, R.C.; Hall, M.J.; Costa, M.; Eusébio, D. Ammonia as an energy vector: Current and future prospects for low-carbon fuel applications in internal combustion engines. *J. Clean. Prod.* **2021**, *296*, 126562. [CrossRef]
4. Thurston, R.H. A History of the Growth of the Steam-Engine. *Nature* **1879**, *19*, 381–382. [CrossRef]
5. Kobayashi, H.; Hayakawa, A.; Somarathne, K.K.A.; Okafor, E.C. Science and technology of ammonia combustion. *Proc. Combust. Inst.* **2019**, *37*, 109–133. [CrossRef]
6. Mørch, C.; Bjerre, A.; Gøttrup, M.; Sorenson, S.; Schramm, J. Ammonia/hydrogen mixtures in an SI-engine: Engine performance and analysis of a proposed fuel system. *Fuel* **2011**, *90*, 854–864. [CrossRef]
7. Onorati, A.; Payri, R.; Vaglieco, B.; Agarwal, A.; Bae, C.; Bruneaux, G.; Canakci, M.; Gavaises, M.; Günthner, M.; Hasse, C.; et al. The role of hydrogen for future internal combustion engines. *Int. J. Engine Res.* **2022**, *23*, 529–540. [CrossRef]
8. Mounaïm-Rousselle, C.; Bréquigny, P.; Medina, A.V.; Boulet, E.; Emberson, D.; Løvås, T. Ammonia as Fuel for Transportation to Mitigate Zero Carbon Impact. In *Engines and Fuels for Future Transport*; Springer: Berlin/Heidelberg, Germany, 2021; pp. 257–279. [CrossRef]
9. Hansson, J.; Brynolf, S.; Fridell, E.; Lehtveer, M. The Potential Role of Ammonia as Marine Fuel—Based on Energy Systems Modeling and Multi-Criteria Decision Analysis. *Sustainability* **2020**, *12*, 3265. [CrossRef]
10. Rodríguez, C.G.; Lamas, M.I.; Rodríguez, J.D.D.; Abbas, A. Possibilities of Ammonia as Both Fuel and $NO_x$ Reductant in Marine Engines: A Numerical Study. *J. Mar. Sci. Eng.* **2022**, *10*, 43. [CrossRef]
11. Chiong, M.-C.; Chong, C.T.; Ng, J.-H.; Mashruk, S.; Chong, W.W.F.; Samiran, N.A.; Mong, G.R.; Valera-Medina, A. Advancements of combustion technologies in the ammonia-fuelled engines. *Energy Convers. Manag.* **2021**, *244*, 114460. [CrossRef]
12. Xu, X.; Liu, E.; Zhu, N.; Liu, F.; Qian, F. Review of the Current Status of Ammonia-Blended Hydrogen Fuel Engine Development. *Energies* **2022**, *15*, 1023. [CrossRef]
13. Grannell, S.M.; Assanis, D.N.; Bohac, S.V.; Gillespie, D.E. The Operating Features of a Stoichiometric, Ammonia and Gasoline Dual Fueled Spark Ignition Engine. In Proceedings of the ASME 2006 International Mechanical Engineering Congress and Exposition, Energy Conversion and Resources, Chicago, IL, USA, 5–10 November 2006; pp. 15–27. [CrossRef]
14. Reiter, A.J.; Kong, S.-C. Combustion and emissions characteristics of compression-ignition engine using dual ammonia-diesel fuel. *Fuel* **2011**, *90*, 87–97. [CrossRef]
15. Dimitriou, P.; Javaid, R. A review of ammonia as a compression ignition engine fuel. *Int. J. Hydrogen Energy* **2020**, *45*, 7098–7118. [CrossRef]
16. Sun, X.; Liu, H.; Duan, X.; Guo, H.; Li, Y.; Qiao, J.; Liu, Q.; Liu, J. Effect of hydrogen enrichment on the flame propagation, emissions formation and energy balance of the natural gas spark ignition engine. *Fuel* **2021**, *307*, 121843. [CrossRef]
17. Duan, X.; Liu, Y.; Liu, J.; Lai, M.-C.; Jansons, M.; Guo, G.; Zhang, S.; Tang, Q. Experimental and numerical investigation of the effects of low-pressure, high-pressure and internal EGR configurations on the performance, combustion and emission characteristics in a hydrogen-enriched heavy-duty lean-burn natural gas SI engine. *Energy Convers. Manag.* **2019**, *195*, 1319–1333. [CrossRef]
18. Lhuillier, C.; Brequigny, P.; Contino, F.; Rousselle, C. Combustion Characteristics of Ammonia in a Modern Spark-Ignition Engine. In Proceedings of the Conference on Sustainable Mobility, Catane, Italy, 14–15 October 2019.
19. Lhuillier, C.; Brequigny, P.; Contino, F.; Rousselle, C. Performance and Emissions of an Ammonia-Fueled SI Engine with Hydrogen Enrichment. In Proceedings of the 14th International Conference on Engines & Vehicles, Capri, Italy, 15–19 September 2019.
20. Mounaïm-Rousselle, C.; Bréquigny, P.; Dumand, C.; Houillé, S. Operating Limits for Ammonia Fuel Spark-Ignition Engine. *Energies* **2021**, *14*, 4141. [CrossRef]
21. Guo, B.; Ichiyanagi, M.; Kajiki, K.; Aratake, N.; Zheng, Q.; Kodaka, M.; Suzuki, T. Combustion Analysis of Ammonia Fueled High Compression Ratio SI Engine with Glow Plug and Sub-Chamber. *Int. J. Automot. Eng.* **2022**, *13*, 1–8. [CrossRef]
22. Abd El Fattah, S.F.; Ezzat, M.F.; Mourad, M.A.; Elkelawy, M.; Youssef, I.M. Youssef Experimental Investigation of the Performance and Exhaust Emissions of a Spark-Ignition Engine Operating with Different Proportional Blends of Gasoline and Water Ammonia Solution. *J. Eng. Res.* **2021**, *5*, 38–45.

23. Şahin, Z.; Akcanca, I.Z.; Durgun, O. Experimental investigation of the effects of ammonia solution (NH3OH) on engine performance and exhaust emissions of a small diesel engine. *Fuel* **2018**, *214*, 330–341. [CrossRef]
24. Lanni, D.; Galloni, E.; Fontana, G. Numerical analysis of the effects of port water injection in a downsized SI engine at partial and full load operation. *Appl. Therm. Eng.* **2022**, *205*, 118060. [CrossRef]
25. Galloni, E.; Lanni, D.; Fontana, G.; D'Antuono, G.; Stabile, S. Performance Estimation of a Downsized SI Engine Running with Hydrogen. *Energies* **2022**, *15*, 4744. [CrossRef]
26. Metghalchi, M.; Keck, J.C. Burning Velocities of Mixtures of air with Methanol, Isooctane, and Indolene at High Pressure and Temperature. *Combust. Flame* **1982**, *48*, 191–210. [CrossRef]
27. Metghalchi, M.; Keck, J.C. Laminar Burning Velocity of Propane-Air Mixtures at High Temperature and Pressure. *Combust. Flame* **1980**, *38*, 143–154. [CrossRef]
28. Bozza, F.; Fontana, G.; Galloni, E.; Torella, E. 3D-1D Analyses of the Turbulent Flow Field, Burning Speed and Knock Occurrence in a Turbocharged SI Engine. *SAE Trans. J. Engines* **2007**, *116*, 1495–1507.
29. Livengood, J.; Wu, P. Correlation of autoignition phenomena in internal combustion engines and rapid compression machines. *Symp. (International) Combust.* **1955**, *5*, 347–356. [CrossRef]
30. Richard, S.; Bougrine, S.; Font, G.; Lafossas, F.-A.; Le Berr, F. On the Reduction of a 3D CFD Combustion Model to Build a Physical 0D Model for Simulating Heat Release, Knock and Pollutants in SI Engines. *Oil Gas Sci. Technol.-Rev. IFP Energies Nouv.* **2009**, *64*, 223–242. [CrossRef]
31. Scala, F.; Galloni, E.; Fontana, G. Numerical analysis of a downsized spark-ignition engine fueled by butanol/gasoline blends at part-load operation. *Appl. Therm. Eng.* **2016**, *102*, 383–390. [CrossRef]
32. Goldmann, A.; Dinkelacker, F. Approximation of laminar flame characteristics on premixed ammonia/hydrogen/nitrogen/air mixtures at elevated temperatures and pressures. *Fuel* **2018**, *224*, 366–378. [CrossRef]
33. Heywood, J.B. *Internal Combustion Engine Fundamentals*; McGraw-Hill: New York, NY, USA, 2018.
34. Chen, J.; Jiang, X.; Qin, X.; Huang, Z. Effect of hydrogen blending on the high temperature auto-ignition of ammonia at elevated pressure. *Fuel* **2020**, *287*, 119563. [CrossRef]
35. Nakamura, H.; Hasegawa, S.; Tezuka, T. Kinetic modeling of ammonia/air weak flames in a micro flow reactor with a controlled temperature profile. *Combust. Flame* **2017**, *185*, 16–27. [CrossRef]

energies

MDPI

Article

# Co-Combustion of Hydrogen with Diesel and Biodiesel (RME) in a Dual-Fuel Compression-Ignition Engine

**Wojciech Tutak *, Arkadiusz Jamrozik and Karol Grab-Rogaliński**

Department of Thermal Machinery, Czestochowa University of Technology, 42-201 Czestochowa, Poland
* Correspondence: wojciech.tutak@pcz.pl

**Abstract:** The utilization of hydrogen for reciprocating internal combustion engines remains a subject that necessitates thorough research and careful analysis. This paper presents a study on the co-combustion of hydrogen with diesel fuel and biodiesel (RME) in a compression-ignition piston engine operating at maximum load, with a hydrogen content of up to 34%. The research employed engine indication and exhaust emissions measurement to assess the engine's performance. Engine indication allowed for the determination of key combustion stages, including ignition delay, combustion time, and the angle of 50% heat release. Furthermore, important operational parameters such as indicated pressure, thermal efficiency, and specific energy consumption were determined. The evaluation of dual-fuel engine stability was conducted by analyzing variations in the coefficient of variation in indicated mean effective pressure. The increase in the proportion of hydrogen co-combusted with diesel fuel and biodiesel had a negligible impact on ignition delay and led to a reduction in combustion time. This effect was more pronounced when using biodiesel (RME). In terms of energy efficiency, a 12% hydrogen content resulted in the highest efficiency for the dual-fuel engine. However, greater efficiency gains were observed when the engine was powered by RME. It should be noted that the hydrogen-powered engine using RME exhibited slightly less stable operation, as measured by the $COV_{IMEP}$ value. Regarding emissions, hydrogen as a fuel in compression ignition engines demonstrated favorable outcomes for CO, $CO_2$, and soot emissions, while NO and HC emissions increased.

**Keywords:** dual-fuel; diesel; hydrogen; RME; combustion; emission

**Citation:** Tutak, W.; Jamrozik, A.; Grab-Rogaliński, K. Co-Combustion of Hydrogen with Diesel and Biodiesel (RME) in a Dual-Fuel Compression-Ignition Engine. *Energies* **2023**, *16*, 4892. https://doi.org/10.3390/en16134892

Academic Editors: Enzo Galloni and Davide Lanni

Received: 4 June 2023
Revised: 18 June 2023
Accepted: 21 June 2023
Published: 23 June 2023

**Copyright:** © 2023 by the authors. Licensee MDPI, Basel, Switzerland. This article is an open access article distributed under the terms and conditions of the Creative Commons Attribution (CC BY) license (https://creativecommons.org/licenses/by/4.0/).

## 1. Introduction

Due to their high efficiency, durability, and reliability, compression-ignition internal combustion engines are widely utilized to power various machines and generators [1]. However, these engines are also recognized as significant sources of environmental pollution, including emissions of NOx and soot [2]. In order to address health concerns and ensure air quality, stringent regulations have been implemented to limit exhaust emissions from these engines [3]. Over the years, extensive efforts have been made to mitigate the impact of human activities on the natural environment and combat global warming. One area of focus has been the exploration of alternative fuels as substitutes for petroleum-derived fuels, aiming to reduce exhaust emissions, achieve energy independence from fossil fuels, and diversify energy sources. Research is actively conducted on clean combustion systems and low-emission fuels. Hydrogen, as a carbon-free fuel that does not produce solid particles, hydrocarbons, or carbon oxides, has emerged as an appealing option for powering reciprocating engines [4]. However, hydrogen alone is not suitable for self-sustained combustion in compression-ignition engines due to its high self-ignition temperature and low cetane number. The combustion of hydrogen can result in significant increases in pressure and temperature, potentially leading to knocking combustion in the engine's combustion chamber and increased NOx emissions [5]. Therefore, in compression-ignition engines, hydrogen is co-combusted with diesel or biodiesel in a dual-fuel configuration. Typically, hydrogen is introduced into the intake system, and its mixture with air is ignited by a dose of liquid fuel injected directly into the engine's combustion chamber [6]. The ignition of hydrogen requires a smaller ignition dose, and due to its rapid combustion rate, the mixture in the combustion chamber is quickly consumed [7]. Many research studies have utilized hydrogen to shorten the combustion time and reduce the carbon-to-hydrogen (C/H) ratio, thereby significantly influencing engine exhaust emissions [8]. The high diffusivity of hydrogen positively impacts mixture homogeneity and the combustion process. However, in a dual-fuel engine where hydrogen is the primary fuel introduced into the intake manifold, the volumetric efficiency of the engine decreases as a portion of the air volume is replaced by hydrogen. This can be a limitation for high engine loads [9]. Biodiesel, on the other hand, is considered a potential replacement for diesel fuel in compression ignition engines. With its oxygen content in the molecular structure, biodiesel is regarded as a more environmentally friendly fuel. It can be produced through the chemical processing of both edible and inedible vegetable oils and animal fats [10]. However, biodiesel also has drawbacks such as a lower calorific value (LHV) and higher viscosity, which can impact the engine's fuel system [11].

Numerous studies have been conducted in the literature regarding the co-combustion of liquid fuels with hydrogen in compression ignition engines. Kanth and Debbarma [12] examined the co-combustion of hydrogen with diesel fuel and a mixture of diesel fuel and biodiesel. They observed a 2.5% increase in engine efficiency when fueled with diesel fuel and hydrogen, and a 1.6% increase when co-combusting with a mixture of diesel fuel and biodiesel (10% $v/v$). Additionally, there was a reduction in CO emissions (~30%) and UHC, but an increase in NOx emissions. Gomes Antunes et al. [13] investigated a dual-fuel engine powered by diesel and hydrogen, with both fuels injected directly into the combustion chamber. They achieved a 14% increase in engine power, but heating of the intake air was necessary for stable operation. Significant pressure increases during combustion were observed, accompanied by a significant improvement in engine efficiency. Liu et al. [14] also examined a dual-fuel compression-ignition engine utilizing direct hydrogen injection. They determined that the maximum energy share of hydrogen is 50%, and the co-combustion process with diesel fuel can be controlled within this range. It is crucial to inject hydrogen into the cylinder early enough to form a flammable mixture with air. Tests conducted with hydrogen injection ranging from 180 to 20 deg bTDC indicated that injecting hydrogen from 40 to 20 deg bTDC resulted in decreased engine efficiency, IMEP (Indicated Mean Effective Pressure), increased soot and CO emissions, likely due to insufficient time for mixture formation. Muralidhara et al. [15] presented the influence of hydrogen co-combustion with biodiesel on compression ignition engine parameters. Hydrogen was introduced into the intake manifold, and they observed a decrease in engine efficiency with an increase in hydrogen content. Akar et al. [16] employed hydrogen to enhance the combustion process during the disposal of biodiesel from waste oil transesterification. The addition of hydrogen positively affected engine performance and reduced emissions of toxic exhaust components, excluding NOx. Köse and Acaroğlu [17] investigated the effect of varying proportions of hydrogen co-combusted with diesel fuel, biodiesel, and a mixture of diesel fuel and biodiesel at low hydrogen concentrations. They found that the diesel-biodiesel mixture with 2.5% hydrogen resulted in the highest power output, while the greatest efficiency was achieved by the engine fueled solely by biodiesel with 2.5% hydrogen. Tutak et al. [18] studied the impact of enriching CNG with hydrogen on engine parameters and exhaust emissions. They established that a dual-fuel engine operated stably when powered by 90% CNG and 10% diesel fuel, with hydrogen replacing a portion of the CNG. The inclusion of hydrogen enhanced the combustion process, accelerating it and increasing engine efficiency. Up to nearly 20% of the energy share of hydrogen, there were no adverse effects on engine stability, and CO emissions from the exhaust gas were practically eliminated.

A synthetic summary of the results obtained by various authors is included in Table 1. Table 1 displays the effect of hydrogen on performance, emission and combustion parameters of hydrogen with diesel and biodiesel.

**Table 1.** Performance, emission and combustion characteristics of hydrogen with diesel or biodiesel.

| Ref. | Engine Type | Fuelling Type | Operating Parameters | Emission |
|---|---|---|---|---|
| [6] | 1-cylinder, 4-stroke, water cooled engine, constant speed 1500 rpm | diesel fuelling, dose of hydrogen into engine manifolds, hydrogen energy share: 0%, 5%, 10%, 20% | load (25, 50 and 75%), increase of $COV_{IMEP}$ at 25% load, stabilization and increase of $COV_{IMEP}$ at 50 and 75% load | decrease of unregulated emissions (formaldehyde and acetaldehyde), (propylene, ethylene and aromatic hydrocarbons) |
| [9] | 1-cylinder (normally aspirated, air-cooled). engine speed at 1500 rpm | hydrogen into the intake manifold, hydrogen energy share (HES): 0% (100% diesel), 6%, 12%, 18% and 24% | 5.2% decrease at 24% HES in the BSEC, BTE increase (by 7.85% with 24% HES) | NOx increase (3.42%), $CO_2$ (3.61%), CO (2.84%), and smoke (4.85%) decrease |
| [12] | 1-cylinder, 4-stroke, naturally aspired, constant speed 1500 rpm | diesel and biodiesel, hydrogen at a fixed flow rate of 7 lpm through the intake manifold, hydrogen-enriched diesel (D + H2), hydrogen-enriched 10, and 20% rice bran biodiesel blend (RB10 + H2) | engine load: 30%, 50%, 70%, 80%, 90%, 100%, 1.6% increase in the brake thermal efficiency, 6.35% decrease in fuel consumption | 4–38% lower CO emissions, 6–14% lower UHC emission, 6–13% higher NOx emission |
| [13] | single cylinder, constant engine speed, four-stroke, direct injection, naturally aspirated, air-cooled | conventional diesel, direct injection hydrogen-fuelled mode, varying the excess air ratio | higher power to weight ratio, 14% higher peak power, 43% fuel efficiency (28% for diesel engine) | 20% reduction in NOx emission |
| [14] | automotive-size, 1-cylinder, common-rail | hydrogen-diesel dual-fuel engine with dual direct injection, hydrogen energy fractions (0–50%) | intermediate fixed load of about 7.8 bar IMEP, indicated efficiency of 47% at 50% hydrogen substitution ratio | NOx emission below11 g/kWh |
| [15] | 1-cylinder, water cooled engine, constant speed 1500 rpm | B100 + hydrogen (variable ratio) hydrogen into the intake manifold, hydrogen-enriched biodiesels, flow rates of $H_2$ from 0.1 to 0.25 kg/h | 80% and 100% load reduction of brake thermal efficiency (BTE) | reduction of smoke, CO and HC emissions, increase NOx |
| [16] | naturally aspirated, water cooled, four stroke, 1-cylinder | B100 blended with diesel: 10 and 20%, pure hydrogen to intake manifold | B100 fraction deteriorated performance, Hydrogen improved performance | B100 fraction deteriorated emission, decrease CO and $CO_2$, increase of NOx |
| [17] | 4-cylinder, turbocharged, variable rpm | small dose of hydrogen into intake manifold, D100, B100, (B7 + H2.5), (D + H2.5); D100/B100 blends direct inj. | max power and torque for (B7 + H2.5), highest ITE (B100 + H2.5) | reduction HC; increase NOx, minimum CO for B100 and (B100 + H2.5) |
| [19] | 4-cylinder, water cooled, engine speeds ranging from 1200 to 2800 rpm | hydrogen into the intake manifold hydrogen-enriched pomegranate seed oil biodiesel (POB), hydrogen flow: 5 L/min | full load conditions, positive effects on power output and BSFC | improvement of CO emissions, deterioration of NOx emissions |

**Table 1.** *Cont.*

| Ref. | Engine Type | Fuelling Type | Operating Parameters | Emission |
|---|---|---|---|---|
| [20] | 4-cylinder, heavy-duty, (EGR) system and a boosting system | H2 energy share ratio of up to 98% | low and medium operating loads | for low load: reduction of CO and NOx emission of over 90%, soot decrease by 85%, for medium load increase NOx emission |
| [21] | four-stroke, water cooled, 4-cylinder, turbocharger, engine speeds 1000 to 2500 rpm | hydrogen into the inlet manifold (2.5%, 5% and 7.5% as volume) | full load, increase in engine torque, power, thermal efficiency, nitrogen oxides (NOx) and exhaust gasses temperatures | increase in NOx and EGT, decrease in HC, CO |
| [22] | 1-cylinder, four-stroke, water cooled, 4-cylinder, 1500 rpm | hydrogen into the inlet manifold (constant fraction 25%) biodiesel | full range of load, small increase in BTE | decrease in CO, $CO_2$, HC and NOx |
| [23] | 1-cylinder, four-stroke, water cooled | diesel and biodiesel, hydrogen flow into intake manifold 7 lpm | small increase in BTE | reduction in HC, CO and smoke, increase in NOx |
| [24] | 1-cylinder, water cooled, 1500 rpm | diesel and palm biodiesel blend (P20), hydrogen injected into the intake manifold at different flow rates of 7 lpm and 10 lpm, loading condition of 30%, 60%, 80%, 90%, and 100% | increase in BTE and decrease in BSEC, reduction in combustion duration (CD), best improvement achieved for 90% load. | increase in NOx, decrease in CO, $CO_2$ and HC, At 90% load, 10 lpm of H2-decreased HC od 51% and CO of 37% |

The literature on co-combustion of hydrogen with liquid fuels in dual-fuel engines highlights the positive impact of hydrogen on engine performance and exhaust emissions. Various types of liquid fuels, including diesel and biodiesels, are utilized. As shown in Table 1, the obtained results on different types of engines, under various conditions and operating points, often yield divergent outcomes. Each research setup and experimental procedure are to some extent distinct, and the results are often incomparable in terms of values. Both single-cylinder and multi-cylinder engines, naturally aspirated and turbocharged, are utilized in optimizing the combustion process or under constant settings. The study presents the results of a comparative analysis of the combustion process in an engine fueled by diesel fuel and hydrogen, as well as RME and hydrogen under very similar conditions on the same engine. The obtained results for both analyzed cases are fully comparable, reflecting the influence of fuel type on engine performance parameters and emissions over a wide range of hydrogen proportions. An estimation of variations in the occurrence of characteristic stages of combustion is provided. In the cited works in the Introduction, as well as in many others, angles are determined for ID (Ignition Delay), CD (Combustion Duration), and CA50 (Crank Angle at 50% heat release) for average combustion profiles from the dataset. This study includes a statistical analysis for these combustion stages.

This paper aims to provide a comparative analysis of the combustion process in a dual-fuel compression-ignition engine operating at maximum load, using hydrogen with diesel fuel and hydrogen with biodiesel (specifically, Rapeseed Methyl Esters-RME). The results presented demonstrate the variations in engine operation based on the specific liquid fuel co-combusted with hydrogen. For both examined scenarios, the same proportion of hydrogen was employed relative to the liquid fuel (4%, 7%, 12%, 23%, and 34%). The impact of hydrogen content on heat release, characteristic combustion stages, and exhaust emissions was evaluated.

## 2. Materials and Methods

The study on the co-combustion of hydrogen with diesel fuel and biodiesel (RME) was conducted on a test bench, primarily consisting of a single-cylinder, naturally aspirated, four-stroke, air-cooled diesel engine. The engine was coupled with a DC generator, which ensured a stable rotational speed and facilitated continuous measurement and control of the engine's load. The engine was specifically designed to operate at a constant speed under a consistent maximum load. To enable the co-combustion of the two fuels, an additional gas supply system was installed in the engine, injecting the gas into the intake manifold, with pressure control, thereby transforming it into a dual-fuel engine. The technical specifications of the test engine are presented in Table 2.

**Table 2.** Main technical specifications of the test engine.

| Item | Title 2 |
|---|---|
| Engine model | Andoria 1CA90 |
| Type of engine | Stationary, four stroke |
| Type of ignition | Compression ignition |
| Rated speed (rpm) | 1500 |
| Engine rated power @1500 rpm (kW) | 7 |
| Technique of cooling | Air cooled |
| Intake system | Naturally aspirates |
| Bore/stroke (mm) | 90/90 |
| Displacement (cc) | 573 |
| No. of cylinders | 1 |
| Injection system | Direct injection |
| Compression ratio | 17 |
| Start of injection (deg bTDC) | 20 |

In addition to the engine, the test setup included various components: a system for measuring and recording in-cylinder pressure using a Kistler 6061 pressure sensor, a Kistler 5011 charge amplifier, and a Measurement Computing USB-1608HS data acquisition card. The rotational speed was stabilized, measured, and recorded using a dedicated system. The power supply systems incorporated both liquid fuels (diesel and biodiesel) and gaseous fuel (hydrogen). Liquid fuel consumption was measured using a dedicated measurement system, while hydrogen and air consumption were monitored using Common CGR-01 gas rotor flow meters. An exhaust gas analyzer set, consisting of a Bosch five-gas exhaust gas analyzer Bosch BEA 350 and an opacimeter AVL 415SE, was employed. Figure 1 illustrates the test setup diagram, depicting the test engine and the accompanying measuring equipment.

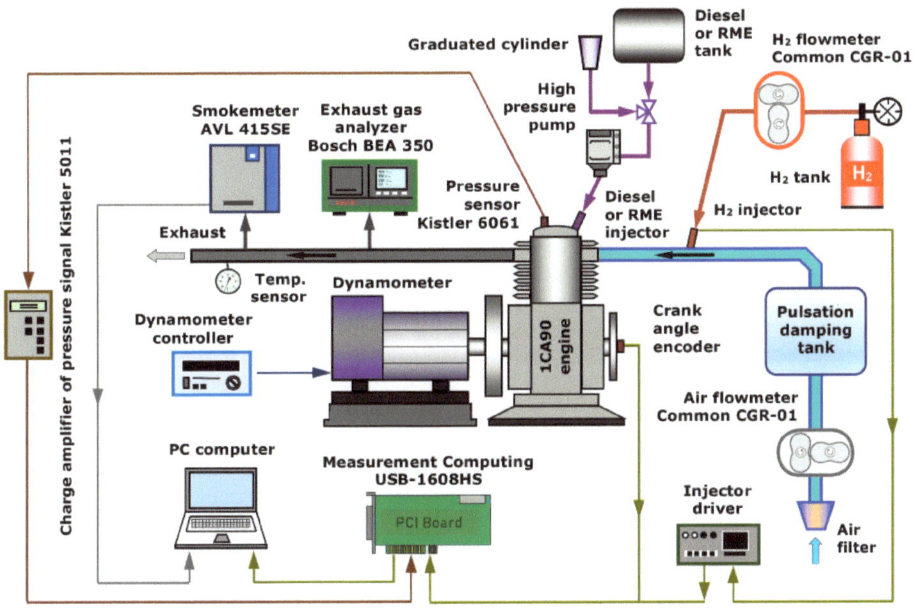


**Figure 1.** Diagram of the test stand with measuring equipment.

The study used the following measurement apparatus:

- pressure sensor Kistler 6061, range 0 . . . 250 bar, linearity < ±0.5% FS,
- charge amplifier Kistler 511, range ±10 . . . ±999,000 pC for 10 V FS, error < ±3%, linearity < ±0.05% FS,
- data acquisition module, Measurement Computing USB-1608HS-16 bits resolution, sampling frequency 20 kHz with software,
- air rotor flowmeter Common CGR-01 G40 DN50, measuring range 0.65 . . . 65 $m^3/h$, accuracy class 1,
- CNG rotor flowmeter Common CGR-01 G10 DN50, measuring range 0.25 . . . 25 $m^3/h$, accuracy class 1,
- Bosch BEA 350 analyzer.

Table 3 presents the characteristics of the exhaust gas analyzer, its measurement ranges and accuracy.

**Table 3.** Parameters of the Bosch BEA 350 analyzer.

| Apparatus | Measuring Range | Resolution | Accuracy from Measured Value | Absolute Accuracy |
|---|---|---|---|---|
| CO | 0.000–10.000% vol.<br>0.000–5.000% vol. | 0.001% vol.<br>0.001% vol. | ...<br>±5% | ...<br>±0.06% vol. |
| HC | 0–9999 ppm vol.<br>0–2000 ppm vol. | 1 ppm vol.<br>1 ppm vol. | ...<br>±5% | ...<br>±12 ppm vol. |
| $CO_2$ | 0.00–18.00% vol.<br>0.00–16.00% vol. | 0.01% vol.<br>0.01% vol. | ...<br>±5% | ...<br>±0.5% vol. |
| $O_2$ | 0.00–22.00% vol.<br>0.00–21.00% vol. | 0.01% vol.<br>0.01% vol. | ...<br>±4% | ...<br>±0.1% vol. |
| NO | 0–5000 ppm vol.<br>0–4000 ppm vol. | 1 ppm vol.<br>1 ppm vol. | ±4%<br>±8% | ±25 ppm vol.<br>±50 ppm vol. |
| λ | 0.500–9.999<br>0.700–1.300 | 0.001<br>0.001 | ...<br>±4% | ...<br>... |

### 2.1. Methodology

The main measurements conducted during the tests involved engine indication, which entailed recording changes in-cylinder pressure as a function of the crankshaft rotation angle. Additionally, fuel and air consumption were measured, and exhaust emissions, including NOx, HC, CO, $CO_2$, and soot concentrations, were monitored. The research focused on the co-combustion of hydrogen with diesel fuel and biodiesel (RME). For both cases, the energy share of hydrogen in the total fuel energy supplied to the engine was set at 4%, 7%, 12%, 23%, and 34%. In order to analyze the obtained results, combustion and emission tests were conducted for the engine operating solely on diesel fuel and RME as reference fuels. The tests were carried out under conditions of thermal stabilization with a constant maximum engine load. Engine indication allowed for the analysis of the combustion process, including the determination of key combustion stages such as ignition delay (ID), combustion duration (CD), and the angle of 50% heat release (CA50). Moreover, it facilitated the determination of essential engine operation parameters, including mean indicated pressure (IMEP), thermal efficiency (ITE), and specific energy consumption (SEC). The stability of the dual-fuel engine operation was assessed by analyzing changes in IMEP over 200 consecutive engine cycles.

### 2.2. Fuels Characteristics

The research involved the use of three fuels: diesel fuel, biodiesel (RME), and hydrogen. Table 4 presents the selected physical and chemical properties of these fuels. Diesel fuel is a typical distillation product of crude oil at atmospheric pressure. It consists of a mixture of naphthenic, paraffinic, and aromatic hydrocarbons, which are components of crude oil. One important parameter of diesel fuel, like any other fuel, is its calorific value, which is 42.7 MJ/kg. Another significant parameter for compression ignition engines is the cetane number, which indicates the fuel's self-ignition capability. Diesel oils commonly available at European petrol stations must meet the EN590 standard and have a cetane number of at least 51.

RME (Rapeseed Methyl Esters) is a biofuel derived from the methyl esters of rapeseed oil fatty acids. It possesses physicochemical properties similar to conventional oil-derived diesel fuel, allowing it to be used as a substitute fuel for compression ignition engines. Biodiesel (RME) is an environmentally friendly fuel that is renewable and biodegradable. Its production from rapeseed oil ensures a $CO_2$ balance, as part of the emitted carbon dioxide is absorbed during the growth of the oilseed rape used in production. RME has a low sulfur content and excellent lubricating properties, which positively impact engine operation and durability. It is often used as a bio-component in blends with conventional diesel fuel. RME meets the European quality standard EN590 and typically has a higher cetane number than diesel, reaching 56. With about 11% oxygen content, RME enables more efficient combustion and has a slightly lower calorific value than diesel (37.3 MJ/kg).

**Table 4.** Fuel specifications [25,26].

| | Diesel | RME | Hydrogen |
|---|---|---|---|
| Molecular formula | $C_{14}H_{30}$ | $CH_3(CH_2)nCOOCH_3$ | $H_2$ |
| Cetane number | 51 | 54 | 5–10 |
| Octane number | 0–15 | 25 | 130 |
| Molecular weight (kg/kmol) | 170–200 | 296 | 2.02 |
| Density at 15 °C (kg/m$^3$) | 840 | 820–845 | 0.084 |
| Lower heating value (MJ/kg) | 42.7 | 37.3 | 119.93 |
| Autoignition temperature (K) | 503 | 534 | 855 |
| Stoichiometric air-fuel ratio (kg/kg) | 14.53 | 12.3 | 34.4 |
| Kinematic viscosity at 20 °C (mm$^2$/s) | 4 | 7.4 | - |
| Flammability limits in air (at%) | 0.6–5.5 | - | 4–75 |
| Ignition energy in air (mJ) | 0.24 | - | 0.02 |
| Diffusion coefficient in air (cm$^2$/s) | 0.038 | - | 0.61 |
| Oxygen content (wt%) | 0 | 11 | 0 |
| Hydrogen content (wt%) | 15 | 12 | 100 |
| Carbon content (wt%) | 85 | 77 | 0 |

Hydrogen is recognized as the most abundant element in the universe, albeit usually not in its pure form but rather in various chemical compounds. It is a highly energetic fuel that can be produced without generating carbon emissions. The most common methods for hydrogen production are natural gas steam reforming, coal gasification, and electrolysis. Hydrogen possesses the highest calorific value (approximately 120 MJ/kg) and heat of combustion (based on mass) among all fuels. However, it's very low density presents a disadvantage when considering volume-related values. Hydrogen readily forms flammable and combustible mixtures with air due to its extensive flammability and explosive limits compared to other fuels. In theory, hydrogen boasts numerous advantages as an excellent energy carrier, with water vapor as the byproduct of its combustion. Nevertheless, the production of pure hydrogen is complex and energy-intensive, while its transportation and storage require extraordinary precautions. Due to its high octane number (130), hydrogen is a desirable fuel for spark-ignition engines. However, its low cetane number (5–10) prevents it from being used as a standalone fuel for compression ignition engines.

## 3. Results

Comparative research was conducted on the co-combustion of hydrogen with diesel fuel and hydrogen with biodiesel-RME in a compression-ignition internal combustion engine. The study focused on examining the effects of hydrogen content on engine efficiency, as well as evaluating the combustion process and exhaust emissions for both liquid fuels.

### 3.1. Combustion Characteristics

The recorded pressure profile in the cylinder serves as the primary and most reliable source of information for studying the combustion process in a reciprocating engine. The analysis of the indicator diagram, which depicts the heat release rate, is essential for evaluating the engine's performance. Figure 2 provides an overview of the pressure profiles in the cylinder and the heat release for different energy shares of hydrogen compared to the reference profiles obtained with diesel fuel (Figure 2a) and biodiesel (Figure 2b).

In both cases of co-combusting hydrogen with diesel fuel and biodiesel, the maximum pressure values exceeded those of the reference fuels. When hydrogen was co-combusted with diesel fuel, the engine achieved a pressure value of 71 bar for the maximum hydrogen share of 34% at 3 degrees after top dead center (aTDC). Similarly, for the co-combustion of hydrogen with biodiesel, the maximum pressure reached 73 bar at the same hydrogen share and timing. However, the combustion of biodiesel with hydrogen resulted in a more gradual pressure increase, as evidenced by the heat release rate curve.

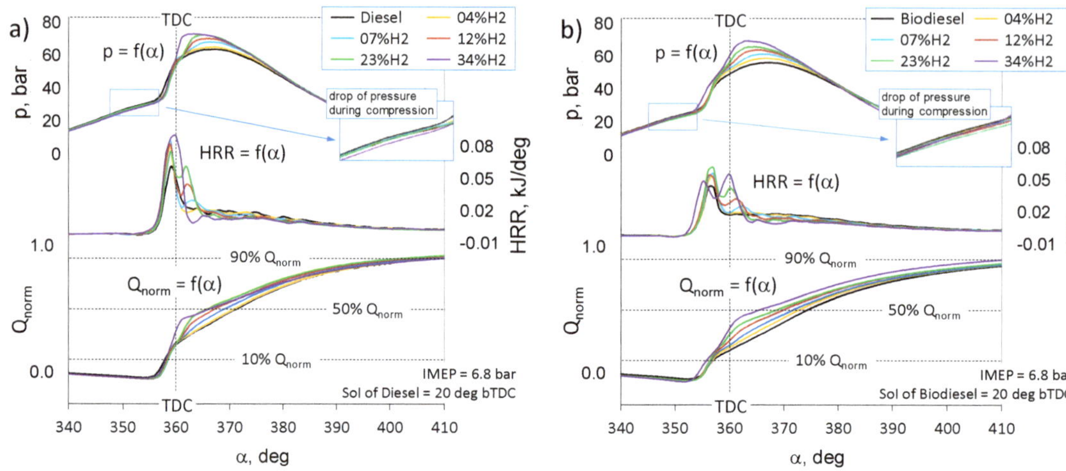


**Figure 2.** Comparison of the combustion pressure courses and the heat release rate and normalized heat release curves for a dual-fuel engine powered by hydrogen with diesel (**a**) and hydrogen with biodiesel (**b**).

For diesel/hydrogen combustion, hydrogen shares up to 23% led to two local maxima on the heat release rate curve, indicating a two-stage combustion process. At the maximum hydrogen share, the heat release rate curve resembled that of a spark-ignition engine, with the diffusion combustion phase diminishing. The maximum heat release rate value of 89.5 J/deg was recorded at a 34% hydrogen share. In the case of biodiesel/hydrogen combustion, the two-stage combustion occurred for each hydrogen percentage, and the second stage became more pronounced with higher hydrogen shares. The maximum heat release rate value of 61 J/deg was observed at a 23% hydrogen content. Increasing the hydrogen share resulted in a reduction of diffusion phases during combustion.

By integrating and normalizing the heat release rate (HRR) curves, the $Q_{norm}$ curves were derived, providing valuable insights into the combustion process and its stages. These curves were utilized to determine important parameters for engine optimization, including the ignition delay time (ID) at 10% $Q_{norm}$, combustion duration (CD) at 90% $Q_{norm}$, and the angle of 50% heat release (CA50) at 50% $Q_{norm}$.

From the $Q_{norm}$ curves, it is evident that, in both cases, the combustion process undergoes a noticeable change when the hydrogen share reaches 34%, resulting in faster combustion compared to lower hydrogen proportions. This effect is particularly pronounced in the initial phase of the combustion process. Figure 3 illustrates the impact of hydrogen content on the individual combustion phases for both analyzed scenarios.

Examining the influence of hydrogen share on ignition delay, it was observed that only for diesel fuel combustion, a 23% hydrogen share caused a 1-degree increase in ID, while for biodiesel co-combustion, a 34% hydrogen content led to a 2-degree decrease in ID compared to the values obtained with reference fuels.

Significant reductions in combustion duration were observed for both liquid fuels co-combusted with hydrogen. In the case of diesel fuel combustion with hydrogen, the CA90 combustion time decreased to 23% of the hydrogen share, and further increasing the hydrogen percentage to 34% led to a subsequent increase in combustion duration. Notably, for a 34% hydrogen share, a substantial acceleration in combustion rate was observed. However, prior to CA50 and at approximately 40% heat release, the combustion rate slowed down.

For the biodiesel engine, the combustion time was successively shortened as the hydrogen content increased across the entire range. At the maximum hydrogen share, the combustion time was reduced by 13 degrees compared to biodiesel combustion alone.

In both cases, as the hydrogen proportion increased, the CA50 angle shifted closer to the top dead center (TDC). However, it is important to note that during the tests, the injection angle for the liquid fuel remained constant at 20 degrees before top dead center (bTDC).

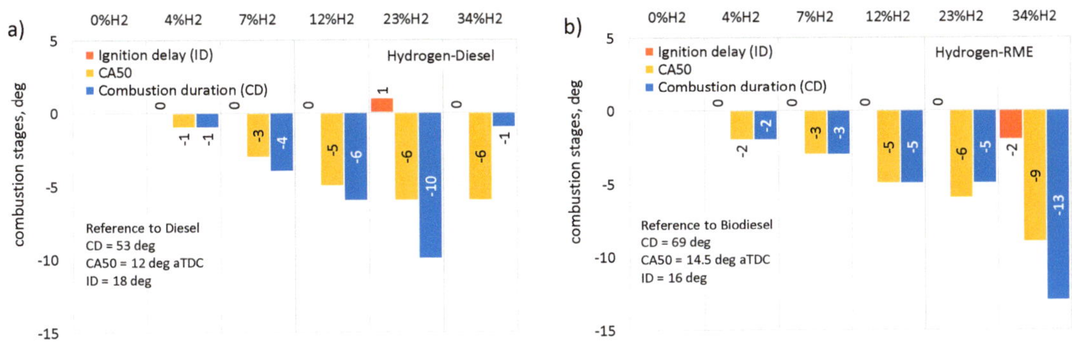


**Figure 3.** The stages of ignition delay (ID), CA50 and combustion duration (CD) for a dual-fuel engine powered by hydrogen with diesel (**a**) and hydrogen with biodiesel (**b**).

*3.2. Performance Characteristics*

The utilization of alternative fuels in reciprocating internal combustion engines aims to not only diversify energy sources but also improve their environmental aspects and often enhance engine efficiency. Hydrogen exhibits a high combustion rate, thereby reducing the duration of combustion in a dual-fuel engine. This reduction in combustion duration helps minimize heat loss across system boundaries, resulting in improved thermal efficiency of the engine. Figure 4a illustrates the impact of hydrogen content on engine efficiency for both analyzed cases. It was observed that engine efficiency increased across the entire range of hydrogen content in both cases. Comparing the reference fuels, the engine powered solely by diesel fuel achieved higher efficiency (ITE = 35.1%) than the engine powered by biodiesel (ITE = 32.2%). The optimal efficiency was achieved with a 12% energy share of hydrogen. Notably, the engine fueled by biodiesel experienced a greater efficiency increase for this hydrogen share, with a rise of 7.4%. Further increasing the hydrogen proportion did not yield such a substantial efficiency improvement, although it remained higher than that of the reference fuels. For a 12% hydrogen content in Hydrogen-RME fueling, a greater efficiency increase was achieved compared to diesel fuel, although the absolute value of ITE for diesel fuel was 41%, while for RME, it was 39.6%. For RME fueling, the CA50 angle was 9.5 degrees aTDC, while for diesel fuel, it was 7 degrees aTDC. The differences in these values are also reflected in the ITE values.

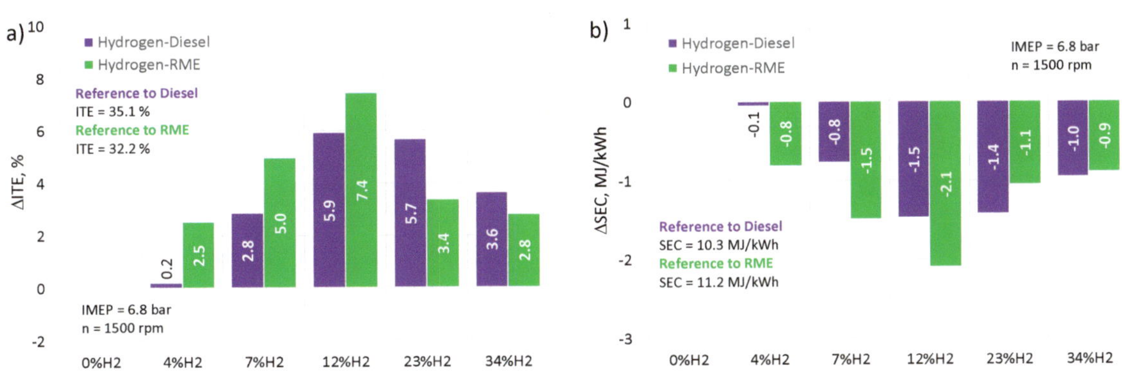


**Figure 4.** Engine indicated efficiency (ITE) (**a**) and specific energy consumption (SEC) (**b**) for a dual-fuel engine powered by hydrogen with diesel and hydrogen with biodiesel.

Figure 4b illustrates the impact of hydrogen content on the specific energy consumption (SEC) a parameter closely related to ITE. The most significant energy savings, while maintaining the same IMEP value, were achieved with a 12% hydrogen share. In the case of biodiesel, savings of 2.1 MJ/kWh compared to the reference fuel were realized.

An important parameter for evaluating the engine's performance is the consistency of successive engine cycles. The $COV_{IMEP}$ index, which measures the degree of variation in indicated mean effective pressure, is used for this assessment. According to literature [27], the value of this index should not exceed 5%. Figure 5 presents the influence of hydrogen content on the consistency of IMEP values, represented as differences from the values obtained for the reference fuels. For diesel fuel, the $COV_{IMEP}$ value was 1.61%, while for biodiesel, it was slightly higher at 2.6%.

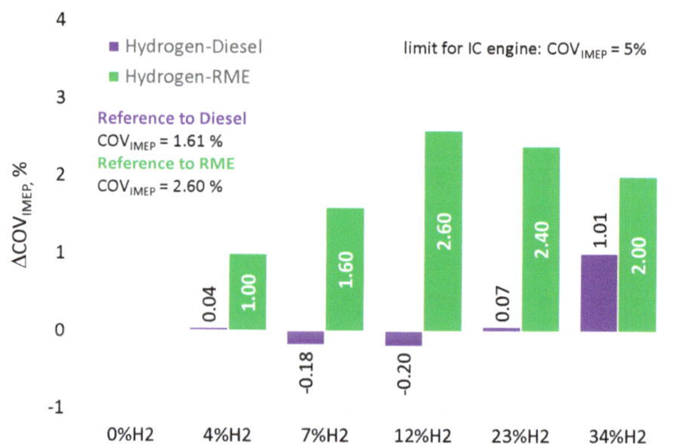


**Figure 5.** Uniqueness of IMEP ($COV_{IMEP}$) for a dual-fuel engine powered by hydrogen with diesel and hydrogen with biodiesel.

In the case of hydrogen co-combustion with diesel fuel, only a 34% hydrogen share resulted in a 1% increase in $COV_{IMEP}$, reaching a value of 2.61%, which is still acceptable. The engine demonstrated high repeatability in subsequent operating cycles. However, a more significant impact on the consistency of engine operation was observed for hydrogen co-combustion with biodiesel. The largest increase in $COV_{IMEP}$ was recorded at a 12% hydrogen share, with a 2.6% increase, resulting in a total $COV_{IMEP}$ value of 5.2%, slightly higher than the permissible value.

In the case of an engine fueled by biodiesel, an increase in the hydrogen content resulted in an increase in $COV_{IMEP}$, which could be attributed to the change in the combustion process dynamics of biodiesel due to the presence of hydrogen in the fuel-air mixture filling the engine combustion chamber. This led to a shift in the combustion phases compared to the values obtained for the reference fuel. As shown in Figure 6b, when fueled with Hydrogen-RME, there was a significant variation in the crank angle at the end of combustion (CA90), reaching as high as 41 degrees for a 12% hydrogen blend, while for the engine fueled with Hydrogen-Diesel, it was only 12 degrees. The distinctiveness of the RME combustion process with hydrogen compared to diesel fuel is also evident in the heat release rate curves. Due to a constant start of injection angle, the CA50 (Crank Angle at 50% heat release) angle for Hydrogen-Diesel was closer to the optimal value than for Hydrogen-RME fueling, which also contributed to the increase in $COV_{IMEP}$.

The combustion stages presented in Figure 6 were determined based on average profiles obtained from 200 consecutive engine cycles. Each cycle in a reciprocating engine exhibits some variability, which is reflected in other engine parameters derived from the indication results. Figure 6 displays the dispersion and range of variability for the characteristic combustion stages. The analysis was conducted based on the heat release curves of 200 consecutive engine cycles. To illustrate the matter, two sets of normalized heat release curves ($Q_{norm}$) are shown for the engine fueled with the reference fuel and a 23% hydrogen share. The analysis reveals that the start of combustion is highly repeatable for both fuels co-combusted with hydrogen. The greatest variation in ignition delay (ID) occurred at a 34% hydrogen share when co-combusted with biodiesel, but it did not exceed 2 degrees of crank angle (CA). Regarding the dispersion of the conventional end of combustion (CD), co-combustion with diesel fuel resulted in a range of 8–13 degrees CA for hydrogen shares up to 23%. Only at a 34% hydrogen share, this dispersion significantly increased to 42 degrees CA. For biodiesel/hydrogen co-combustion, an increase in the uniqueness of the combustion end was observed, reaching a maximum of 53 degrees CA for hydrogen shares up to 23%. However, it slightly decreased at a 34% hydrogen content.

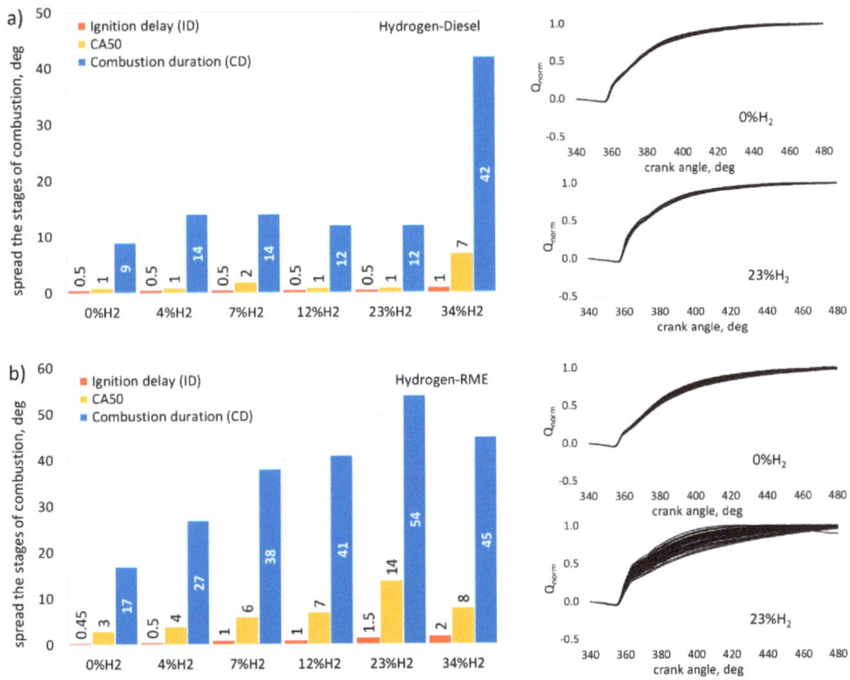


**Figure 6.** Spread of characteristic combustion stages for hydrogen co-combustion with diesel fuel (**a**) and hydrogen co-combustion with biodiesel (**b**).

The occurrence of the CA angle at 50% heat release for piston engines is typically determined after top dead center (TDC). In this regard, a similar trend to that observed for CD was also noticed. Based on this analysis, it can be concluded that diesel fuel exhibits better compatibility when co-combusted with hydrogen. There is a higher level of reproducibility and stability in the occurrence of the characteristic combustion stages. Hydrogen promotes the increase of combustion rate, shortening the total duration of this process, thus contributing to a significant reduction in heat loss from the combustion chamber.

For a 12% hydrogen content, CA50 was closest to the optimal position for the engine in both cases. This analysis should take into account not only the CA50 angle value itself (Figure 3) but also the variation of this crankshaft angle value's occurrence in consecutive engine cycles.

### 3.3. Emission Characteristics

The modern energy sector and automotive industry face new challenges that drive the development and improvement of technologies, materials, control systems, and the adoption of low-emission energy sources, including alternative fuels. Hydrogen has emerged as a promising fuel of the future, serving as an alternative to conventional hydrocarbon fossil fuels such as crude oil and natural gas. This gas is highly regarded as an excellent fuel for reciprocating engines due to its high combustion efficiency and negligible emission of harmful and toxic exhaust components. One way to utilize hydrogen in a reciprocating engine is through co-combustion with diesel or biodiesel in a dual-fuel compression-ignition engine. This study presents the exhaust gas emission results from a compression-ignition engine powered by hydrogen, diesel fuel, and a blend of hydrogen and biodiesel (RME), with the gaseous fuel contributing up to 34% of the energy share. The concentrations of NO, HC, CO, $CO_2$, and soot in the engine's exhaust were analyzed.

Nitrogen oxides. A reciprocating engine, operating by combusting hydrocarbon fuel, often in oxygen-rich conditions, leads to the formation of significant quantities of nitrogen oxides (NOx). Nitrogen oxides are a mixture of various nitrogen compounds with oxygen, produced both within the engine and in its exhaust system. The majority of NOx, approximately 95%, is in the form of nitrogen monoxide (NO), generated under specific conditions during the combustion of the air-fuel mixture within the engine cylinder [26,28]. Optimal conditions for NO formation are present in the high-temperature regions of the combustion chamber, where abundant oxygen and nitrogen are available. Nitrogen oxides are among the most hazardous components present in the exhaust gases of piston engines, including both compression-ignition and spark-ignition engines. Figure 7a illustrates the variations in NO concentration in the engine's exhaust when co-combusting hydrogen with diesel fuel and hydrogen with biodiesel (RME). These changes are compared to the emissions observed with the reference fuels, namely diesel fuel and RME. It is evident that the engine powered solely by biodiesel exhibited lower NO emissions (270 ppm) compared to the engine powered solely by diesel (476 ppm). The addition of hydrogen to both diesel and biodiesel fuels led to an increase in NO emissions from 4% to 34%. Although the concentration gains were higher when hydrogen was burned with biodiesel, the total emissions were higher when hydrogen was burned with diesel. The increased overall emissions associated with the presence of diesel fuel were primarily due to the heightened combustion intensity and higher heat release rate (HRR) during the combustion of the initially prepared mixture in the kinetic phase. Higher HRRs indicate elevated maximum cylinder temperatures, promoting NO formation. The NO concentration measurements were relatively accurate within a range of ±4% of the measured values [29,30].

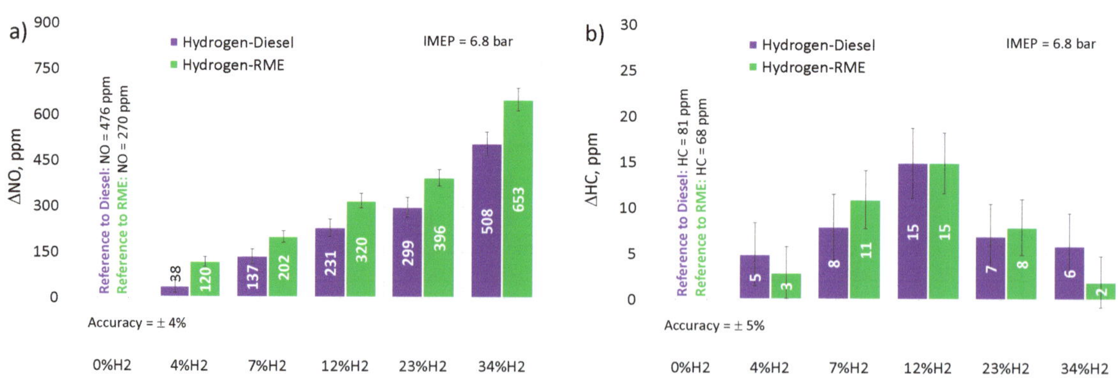


**Figure 7.** Nitric oxide (NO) (**a**) and hydrocarbons (HC) (**b**) emission for a dual-fuel engine powered by hydrogen with diesel and hydrogen with biodiesel.

Hydrocarbons. Unburned hydrocarbons (HC) are another group of toxic compounds released in the exhaust gases of piston engines. They result from incomplete combustion of the fuel in the engine cylinder and are present in both rich and highly lean mixtures. Excessive HC emissions are primarily caused by low combustion temperatures in lean mixtures, insufficient homogenization of the fuel-air mixture, and the presence of fuel-rich regions, which can be attributed to factors such as fuel injection characteristics. The shape and volume of the combustion chamber, including the size of crevices and cavities where fuel can accumulate with limited access to oxygen, also influence hydrocarbon emissions. Figure 7b illustrates the variations in HC concentration in the exhaust gas from a test engine fueled by hydrogen and diesel fuel, as well as hydrogen and biodiesel, compared to the emissions observed with diesel fuel and biodiesel alone. Upon analyzing the measurement results, it can be observed that the combustion of biodiesel resulted in lower HC emissions (68 ppm) compared to the combustion of diesel fuel (81 ppm). Similarly, the emission increments for hydrogen-biodiesel blends indicate lower total hydrocarbon emissions throughout the entire range of hydrogen content. The highest concentrations of hydrocarbons in the exhaust gas were recorded with a 12% hydrogen content for both tested mixtures. Beyond this concentration, HC emissions decreased due to improved homogeneity of the air-fuel mixture, which can be attributed to a significant increase in the amount of hydrogen supplied to the engine during the intake stroke. The relative accuracy of HC concentration measurements was within ±5% of the measured values [29,30].

Carbon monoxide. Carbon monoxide (CO) is a byproduct of incomplete combustion in reciprocating engines. It is formed in the engine cylinder under conditions of oxygen deficiency in fuel-rich regions and conditions of excess oxygen during the combustion of highly lean mixtures. In compression-ignition engines, where the fuel-air mixture is typically lean, insufficient mixing of fuel with air before ignition leads to inadequate oxidation. This results in the presence of fuel-rich and fuel-lean areas. In the former case, the lack of sufficient oxygen prevents complete oxidation of the carbon present in the fuel. In the latter case, the lower temperature and combustion rate in the lean fuel mixture hinder the complete oxidation of carbon. The carbon content in the hydrocarbon fuel supplied to the engine is a significant factor contributing to excessive CO emissions, which can be mitigated by using alternative fuels like hydrogen. Figure 8a illustrates the CO emissions from a test engine during co-combustion of hydrogen with diesel fuel and hydrogen with RME, compared to emissions measured when using diesel fuel and biodiesel alone. It can be observed that the combustion of biodiesel alone resulted in nearly twice as much CO emissions (1.5%) compared to diesel fuel combustion (0.76%), despite a decrease in the amount of elemental carbon in the cylinder due to the use of biodiesel (Figure 8a) and an increase in active oxygen. The decisive factor in this case was a significant reduction in the combustion intensity of biodiesel, as reflected in a decrease in the heat release rate, which hindered the complete oxidation of carbon. The addition of hydrogen to the reference fuels increased the combustion rate, reduced the amount of carbon in the fuel mixture, and had a positive impact on CO emissions. In both hydrogen-diesel and hydrogen-RME co-combustion, reductions in carbon monoxide concentration in the exhaust gas were observed for all analyzed hydrogen shares. The effect of hydrogen was more pronounced in the co-combustion with biodiesel. The relative accuracy of CO concentration measurements was within ±5% of the measured values [29,30].

Carbon dioxide. Carbon dioxide ($CO_2$) is a byproduct of complete combustion of hydrocarbon fuel and can indicate the proper progression of combustion in a piston engine cylinder. Although carbon dioxide is not classified as a toxic compound, it has a detrimental impact on the environment as one of the main greenhouse gases that contribute to climate change. High temperatures and an excess of oxygen are prerequisites for the formation of $CO_2$. Increased consumption of carbon-containing fuels leads to higher carbon dioxide emissions. Therefore, the utilization of alternative fuels such as hydrogen can help reduce $CO_2$ emissions. Figure 8b illustrates the variations in $CO_2$ concentration in the exhaust gas of a hydrogen-fueled diesel engine and a hydrogen-biodiesel (RME) engine. Similarly, these changes are compared to the reference fuels. It can be observed that the combustion of diesel fuel alone resulted in a comparable emission of carbon dioxide (7.80%) as the combustion of biodiesel alone (7.89%). The addition and increase of hydrogen content co-combusted with diesel fuel and biodiesel had a positive effect on $CO_2$ emissions, which was associated with a decrease in the amount of elemental carbon supplied by the fuel (Figure 9a). A more pronounced reduction in $CO_2$ was observed for RME, which has a lower carbon content compared to diesel fuel. The relative accuracy of $CO_2$ concentration measurement was within ±5% of the measured values [29,30].

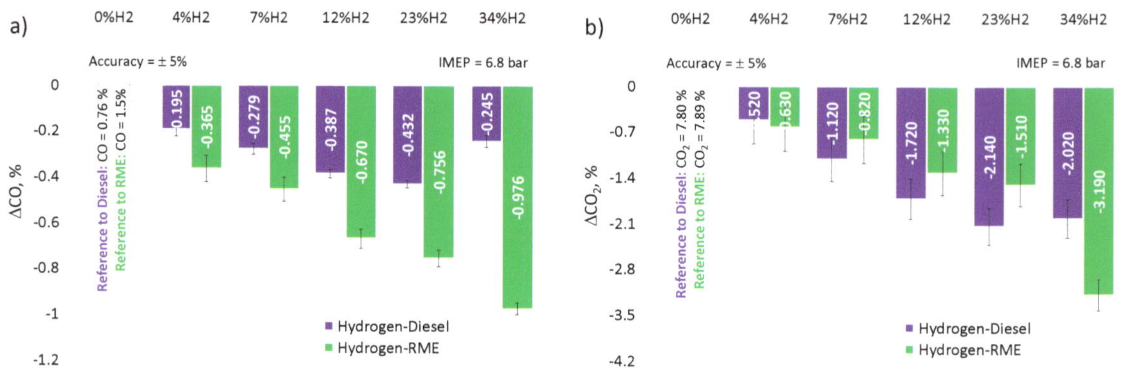


**Figure 8.** Carbon monoxide (CO) (**a**) and carbon dioxide ($CO_2$) (**b**) emission for a dual-fuel engine powered by hydrogen with diesel and hydrogen with biodiesel.

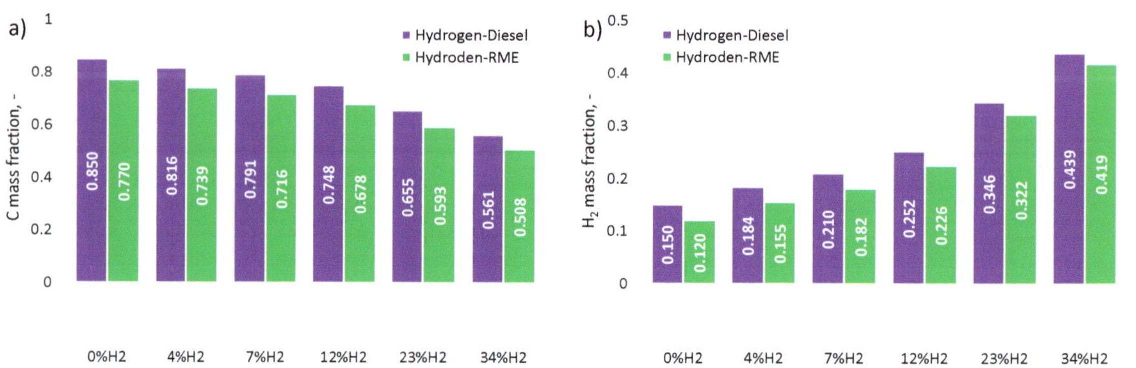


**Figure 9.** Mass fraction of carbon (**a**) and hydrogen (**b**) in fuel supplied to a dual-fuel engine powered by hydrogen with diesel and hydrogen with biodiesel.

Soot. In a compression-ignition engine, the formation of soot occurs due to the incomplete combustion of carbon-containing fuel, primarily during the diffusion phase, which is influenced by the fuel-air mixture's mixing. Inside the engine cylinder, soot is produced in the form of fine particles consisting of pure carbon. The exhaust system captures heavy hydrocarbons, nitrogen compounds, and sulfur, releasing them into the atmosphere as solid particles (PM). Optimal conditions for soot formation exist in areas of the combustion chamber with high temperatures and fuel-rich regions. Figure 10 illustrates the variations in soot concentration in the engine's exhaust gas when co-combusting hydrogen with diesel fuel and hydrogen from RME, compared to the concentrations obtained for the reference fuels. When considering the reference fuels alone, it is evident that higher soot emissions were associated with the combustion of pure biodiesel (1572 mg/m$^3$), while the corresponding value for diesel fuel alone was 863 mg/m$^3$. The higher emissions observed during the combustion of RME can be attributed to the lower heat release rate and the resulting reduced combustion intensity, which leads to fewer carbon particles being oxidized. The impact of hydrogen as a fuel in a compression-ignition engine on soot emissions was found to be positive, both in its co-combustion with diesel fuel and biodiesel. Increasing the amount of hydrogen supplied to the engine during the intake stroke improved the homogeneity of the air-fuel mixture in the cylinder and increased the rate of heat release. Additionally, as the amount of hydrogen increased (Figure 9b), the carbon content decreased (Figure 9a). The beneficial effect of using hydrogen was more pronounced in the case of its co-combustion with biodiesel, where the carbon content was lower compared to diesel-hydrogen combustion. An important factor enabling a greater reduction in soot concentration for hydrogen-RME was the additional oxygen supplied in the biofuel, which facilitated carbon oxidation. The relative accuracy of soot concentration measurement was $\pm 0.05$ mg + 3% of the measured value [29,30].

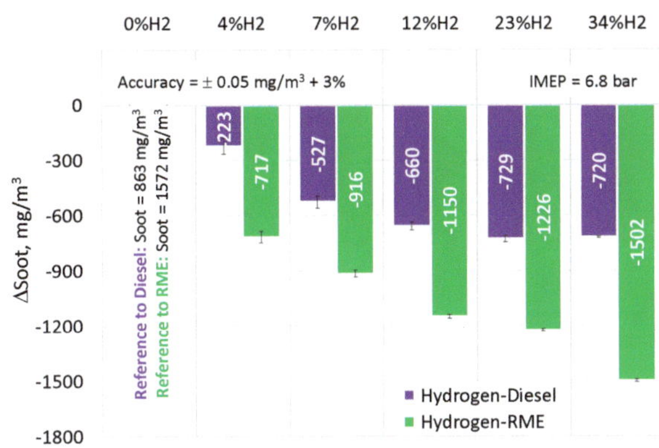


**Figure 10.** Soot emission for a dual-fuel engine powered by hydrogen with diesel and hydrogen with biodiesel.

## 4. Conclusions

Research on hydrogen as a fuel for reciprocating internal combustion engines holds significant importance in both the automotive industry and industrial power systems. This importance stems from the need to conserve Earth's natural resources and mitigate air pollution. The presented paper showcases the findings of a compression-ignition engine test involving the co-combustion of hydrogen with diesel fuel and RME biodiesel. Based on the analysis of the test results, the following conclusions can be drawn:

- Biodiesel (RME) co-combusted with hydrogen exhibits a milder combustion process, characterized by smaller pressure and heat release rate increments compared to diesel fuel, a shorter ignition delay and longer combustion duration in comparison to an engine fueled solely by diesel fuel. Both Hydrogen-Diesel and Hydrogen-RME combinations result in the CA50 angle shifting closer to top dead center (TDC) as the hydrogen share increases.
- The highest engine efficiency is achieved with a 12% energy share of hydrogen, but the engine fueled by biodiesel exhibits a more substantial efficiency increase for this hydrogen share.
- An engine powered by biodiesel slightly exhibits poorer work stability, as measured by the $COV_{IMEP}$ value. Hydrogen-RME combustion with a 12% hydrogen share leads to a 2.6% increase in variability.
- Hydrogen demonstrates a positive influence on soot emissions in a compression-ignition engine, both during its co-combustion with diesel fuel and biodiesel. The beneficial effect of using hydrogen is more notable in the Hydrogen-RME configuration.
- Hydrogen demonstrates a positive influence on soot emissions in a compression-ignition engine, both during its co-combustion with diesel fuel and biodiesel. The beneficial effect of using hydrogen is more notable in the Hydrogen-RME con-figuration.

Hydrogen can be efficiently utilized as a fuel in a compression-ignition engine, offering numerous benefits when co-fired with diesel fuel and biodiesel.

**Author Contributions:** Conceptualization, W.T., A.J. and K.G.-R.; Data curation, W.T. and A.J.; Formal analysis, W.T.; Investigation, W.T., A.J. and K.G.-R.; Methodology, W.T. and A.J.; Writing—original draft, W.T.; Writing—review & editing, W.T. and K.G.-R. All authors have read and agreed to the published version of the manuscript.

**Funding:** Research was financed by the Ministry of Science and Higher Education of Poland from the funds dedicated to scientific research No. BS/PB 1-100-316/2023/P.

**Data Availability Statement:** The data used in this study are reported in the paper figures and tables.

**Conflicts of Interest:** The authors declare no conflict of interest.

## References

1. Akal, D.; Öztuna, S.; Büyükakın, M.K. A review of hydrogen usage in internal combustion engines (gasoline-Lpg-diesel) from combustion performance aspect. *Int. J. Hydrog. Energy* **2020**, *45*, 35257–35268. [CrossRef]
2. Arbab, M.I.; Masjuki, H.H.; Varman, M.; Kalam, M.A.; Imtenan, S.; Sajjad, H. Fuel properties, engine performance and emission characteristic of common biodiesels as a renewable and sustainable source of fuel. *Renew. Sustain. Energy Rev.* **2013**, *22*, 133–147. [CrossRef]
3. Farhan, S.M.; Wang, P. Post-injection strategies for performance improvement and emissions reduction in DI diesel engines—A review. *Fuel Process. Technol.* **2022**, *228*, 107145. [CrossRef]
4. Shinde, B.J.; Karunamurthy, K. Recent progress in hydrogen fuelled internal combustion engine (H2ICE)—A comprehensive outlook. *Mater. Today Proc.* **2022**, *51*, 1568–1579. [CrossRef]
5. Şanlı, A.; Yılmaz, I.T. Cycle-to-cycle combustion analysis in hydrogen fumigated common-rail diesel engine. *Fuel* **2022**, *320*, 123887. [CrossRef]
6. Sharma, P.; Dhar, A. Effect of hydrogen fumigation on combustion stability and unregulated emissions in a diesel fuelled compression ignition engine. *Appl. Energy* **2019**, *253*, 113620. [CrossRef]
7. Zhou, S.; Luo, Z.; Wang, T.; He, M.; Li, R.; Su, B. Research progress on the self-ignition of high-pressure hydrogen discharge: A review. *Int. J. Hydrog. Energy* **2022**, *47*, 9460–9476. [CrossRef]
8. Zhang, K.; Luo, T.; Li, Y.; Zhang, T.; Li, X.; Zhang, Z.; Shang, S.; Zhou, Y.; Zhang, C.; Chen, X.; et al. Effect of ignition, initial pressure and temperature on the lower flammability limit of hydrogen/air mixture. *Int. J. Hydrog. Energy* **2022**, *47*, 15107–15119. [CrossRef]
9. Seelam, N.; Gugulothu, S.K.; Reddy, S.K.; Bhasker, B.; Panda, J.K. Exploration of engine characteristics in a CRDI diesel engine enriched with hydrogen in dual fuel mode using toroidal combustion chamber. *Int. J. Hydrog. Energy* **2022**, *47*, 13157–13167. [CrossRef]
10. Gerpen, J.V. Biodiesel processing and production. *Fuel Process. Technol.* **2005**, *86*, 1097–1107. [CrossRef]
11. Bukkarapu, K.R.; Krishnasamy, A. A critical review on available models to predict engine fuel properties of biodiesel. *Renew. Sustain. Energy Rev.* **2022**, *155*, 111925. [CrossRef]
12. Kanth, S.; Debbarma, S. Comparative performance analysis of diesel engine fuelled with hydrogen enriched edible and non-edible biodiesel. *Int. J. Hydrog. Energy* **2021**, *46*, 10478–10493. [CrossRef]
13. Gomes Antunes, J.M.; Mikalsen, R.; Roskilly, A.P. An experimental study of a direct injection compression ignition hydrogen engine. *Int. J. Hydrog. Energy* **2009**, *34*, 6516–6522. [CrossRef]
14. Liu, X.; Srna, A.; Yip, H.L.; Kook, S.; Chan, Q.N.; Hawkes, E.R. Performance and emissions of hydrogen-diesel dual direct injection (H2DDI) in a single-cylinder compression-ignition engine. *Int. J. Hydrog. Energy* **2021**, *46*, 1302–1314. [CrossRef]
15. Muralidhara, D.M.; Banapurmath, N.R.; Udayaravi, M.; Prabhakar Reddy, C.; Harari, P.A.; Karthik, T. Effect of hydrogen flow rates on the performance of two biodiesels fuelled dual fuel engine. *Mater. Today Proc.* **2022**, *49*, 2189–2196. [CrossRef]
16. Akar, M.A.; Kekilli, E.; Bas, O.; Yildizhan, S.; Serin, H.; Ozcanli, M. Hydrogen enriched waste oil biodiesel usage in compression ignition engine. *Int. J. Hydrog. Energy* **2018**, *43*, 18046–18052. [CrossRef]
17. Köse, H.; Acaroğlu, M. The effect of hydrogen addition to Cynara biodiesel on engine performance and emissions in diesel engine. *Energy Sources Part A Recovery Util. Environ. Eff.* **2020**, 1–20. [CrossRef]
18. Tutak, W.; Jamrozik, A.; Grab-Rogaliński, K. Effect of natural gas enrichment with hydrogen on combustion process and emission characteristic of a dual fuel diesel engine. *Int. J. Hydrog. Energy* **2020**, *45*, 9088–9097. [CrossRef]
19. Tuccar, G.; Uludamar, E. Emission and engine performance analysis of a diesel engine using hydrogen enriched pomegranate seed oil biodiesel. *Int. J. Hydrog. Energy* **2018**, *43*, 18014–18019. [CrossRef]
20. Dimitriou, P.; Kumar, M.; Tsujimura, T.; Suzuki, Y. Combustion and emission characteristics of a hydrogen-diesel dual-fuel engine. *Int. J. Hydrog. Energy* **2018**, *43*, 13605–13617. [CrossRef]
21. Köse, H.; Ciniviz, M. An experimental investigation of effect on diesel engine performance and exhaust emissions of addition at dual fuel mode of hydrogen. *Fuel Process. Technol.* **2013**, *114*, 26–34. [CrossRef]
22. Xia, C.; Brindhadevi, K.; Elfasakhany, A.; Alsehli, M.; Tola, S. Performance, combustion and emission analysis of castor oil biodiesel blends enriched with nanoadditives and hydrogen fuel using CI engine. *Fuel* **2021**, *306*, 121541. [CrossRef]
23. Kanth, S.; Ananad, T.; Debbarma, S.; Das, B. Effect of fuel opening injection pressure and injection timing of hydrogen enriched rice bran biodiesel fuelled in CI engine. *Int. J. Hydrog. Energy* **2021**, *46*, 28789–28800. [CrossRef]

24. Das, S.; Kanth, S.; Das, B.; Debbarma, S. Experimental evaluation of hydrogen enrichment in a dual-fueled CRDI diesel engine. *Int. J. Hydrog. Energy* **2022**, *47*, 11039–11051. [CrossRef]
25. Stępień, Z.A. Comprehensive overview of hydrogen-fueled internal combustion engines: Achievements and future challenges. *Energies* **2021**, *14*, 6504. [CrossRef]
26. Hosseini, S.E.; Butler, B. An overview of development and challenges in hydrogen powered vehicles. *Int. J. Green Energy* **2020**, *17*, 13–37. [CrossRef]
27. Heywood, J.B. *Internal Combustion Engine Fundamentals*, 2nd ed.; McGraw-Hill Education: New York, NY, USA, 2018.
28. Clapp, L.J.; Jenkin, M.E. Analysis of the relationship between ambient levels of $O_3$, $NO_2$ and NO as a function of NOx in the UK. *Atmos. Environ.* **2001**, *35*, 6391–6405. [CrossRef]
29. Jamrozik, A.; Tutak, W.; Grab-Rogaliński, K. An experimental study on the performance and emission of the diesel/CNG dual-fuel combustion mode in a stationary CI engine. *Energies* **2019**, *12*, 3857. [CrossRef]
30. Tutak, W.; Jamrozik, A.; Grab-Rogaliński, K. The effect of RME-1-butanol blends on combustion, performance and emission of a direct injection diesel engine. *Energies* **2021**, *14*, 2941. [CrossRef]

**Disclaimer/Publisher's Note:** The statements, opinions and data contained in all publications are solely those of the individual author(s) and contributor(s) and not of MDPI and/or the editor(s). MDPI and/or the editor(s) disclaim responsibility for any injury to people or property resulting from any ideas, methods, instructions or products referred to in the content.

energies

MDPI

Article

# Ammonia Combustion in a Spark-Ignition Engine Supported with Dimethyl Ether

**Wojciech Tutak *, Michał Pyrc, Michał Gruca and Arkadiusz Jamrozik**

Department of Thermal Machinery, Czestochowa University of Technology, 42-201 Czestochowa, Poland; michal.pyrc@pcz.pl (M.P.); michal.gruca@pcz.pl (M.G.); arkadiusz.jamrozik@pcz.pl (A.J.)
* Correspondence: wojciech.tutak@pcz.pl

**Abstract:** This paper presents the results of experimental tests with a spark-ignition engine powered using ammonia and DME (dimethyl ether). The tests were carried out on a CFR (cooperative fuel research) engine with a compression ratio of 10 and a rotational speed of 600 rpm. DME was used as a fuel to facilitate the initiation and then accelerate the combustion in the SI (spark-ignition) engine. It turned out that only about 10% of the energy share of DME ensures the correct combustion process. DME has a positive effect on the combustion stages, reduces the ignition delay time, and shortens the combustion duration. Thanks to this, for 18% of the energy share of $NH_3$, the highest engine efficiency (29.8%), the highest value of the average indicated pressure IMEP (712 kPa), and the minimum value of specific energy consumption (12.1 MJ/kWh) were obtained. Even the smallest DME content ensured the high repeatability of IMEP, below the permissible limit of 5%.

**Keywords:** ammonia; combustion; DME; combustion stages

## 1. Introduction

Piston combustion engines are still willingly used in drives of machines and devices, thanks to their compact structure and flexibility with the type of fuel powering them [1]. For environmental reasons, the users of combustion engines are forced to use non-petroleum fuels [2]. In the field of spark-ignition piston engines, renewable fuels, such as various types of alcohol fuels, have been used to power them for many years [3]. Gaseous fuels are also used here, e.g., hydrogen [4]. Such engines are also willingly used for the utilization of waste fuels such as post-fermentation gases (biogas) or process fuels (coke oven gas) [5]. Many researchers around the world are conducting intensive research on the use of carbon-free fuels to power combustion engines, the advantage of which is the lack of $CO_2$ emissions [6,7]. A carbon-free fuel capable of powering internal combustion piston engines is ammonia ($NH_3$). Ammonia may be a significant fuel for energy transformation in this area. Due to its relatively high energy density (13.72 MJ/$m^3$), this fuel can be an attractive carrier of chemical energy. Ammonia can be burned in a piston engine or used as a carrier for hydrogen obtained from its dissociation. Unfortunately, the process of ammonia dissociation into hydrogen is an expensive process; therefore, research into the fueling of piston engines with ammonia is justified. Ammonia, having no carbon in its molecular structure, is a chemical compound that does not generate greenhouse gases ($CO_2$) that enter the atmosphere. Ammonia is an attractive zero-emission fuel that can reduce the need for conventional hydrocarbon fuels and reduce the emissions of undesirable environmental pollutants such as carbon monoxide and dioxide, particulate matter, and unburned hydrocarbons. Ammonia, like any other fuel, also has its drawbacks: it is a gas harmful to humans and causes breathing problems. It is also a source of chemical compounds formed in exhaust gases, such as NO and $NO_2$, which are harmful to the natural environment. In order to reduce the negative impact of such an engine's exhaust gases on the environment, research should be carried out on the optimization of combustion in an engine powered only with ammonia or the combustion of this fuel supported with another fuel. Ammonia,

**Citation:** Tutak, W.; Pyrc, M.; Gruca, M.; Jamrozik, A. Ammonia Combustion in a Spark-Ignition Engine Supported with Dimethyl Ether. *Energies* **2023**, *16*, 7283. https://doi.org/10.3390/en16217283

Academic Editors: Enzo Galloni and Davide Lanni

Received: 8 October 2023
Revised: 22 October 2023
Accepted: 24 October 2023
Published: 26 October 2023

**Copyright:** © 2023 by the authors. Licensee MDPI, Basel, Switzerland. This article is an open access article distributed under the terms and conditions of the Creative Commons Attribution (CC BY) license (https://creativecommons.org/licenses/by/4.0/).

due to its properties, is a fuel with low reactivity, it is characterized by difficulty in ignition (ignition energy of approximately 8 mJ) and a low combustion speed of 0.015 m/s, and it is a difficult fuel with which to conduct a stable combustion process in an engine. For this reason, other fuels are added to improve the combustion process. A common case is the use of hydrogen as a fuel that improves the ignition process itself and improves the stability of engine operation [8–11].

Gu Xin et al. [11] used ammonia as an additional fuel, enabling the controlled combustion of hydrogen in a spark-ignition engine. They added ammonia to hydrogen at volume ratios of 5.2%, 7.96%, and 10.68%. They confirmed that the addition of ammonia extends the combustion period in such an engine and, thus, reduces the heat release rate. There was an increase in the average indicated pressure and engine efficiency. There was no clear impact of these volume fractions of ammonia on the increase in NOx emissions. Uddeen et al. drew attention to problems with ignition and a problem with the spread of the flame front in the ammonia combustion process in a piston engine [12]. They found that the use of multi-point ignition significantly improves combustion stability, shortens the duration of combustion in the engine cylinder, and increases engine power compared to an engine with a conventional ignition system. Flame nuclei formed at multiple ignition points resulted in higher NOx emissions in the exhaust system due to higher in-cylinder temperatures. Multipoint ignition tests have already been carried out for engines powered using conventional fuels [13]. In an ammonia-fueled engine, this solution, despite increasing costs, seems to be justified. The problem of ammonia combustion in a piston engine due to a low combustion speed was noticed by Ma et al. [14]. They indicated that the best way to compensate for this disadvantage of ammonia is to add fuel with high reactivity, e.g., hydrogen. The ignition delay period and combustion time are shortened, which has a positive effect on the engine's operating parameters. Chorowski et al. drew attention to the problems of using ammonia as a fuel for road transport [15]. The authors draw attention to the dangerous side of ammonia, which is its poisonous nature, which requires the use of systems controlling its presence, and it is necessary to control unburned ammonia that escapes from the exhaust system. Attention was also drawn to the lack of clear information on the impact of ammonia co-burned with other fuels on PM emissions. Ge et al., in their work, used a mixture of ammonia and hydrogen to power a spark-ignition piston engine [16]. Hydrogen and ammonia, considering their properties, are extremely different fuels. Hydrogen is characterized by a high combustion speed, which brings the engine to the limit of knocking combustion, which limits the engine parameters and increases the uniqueness of the engine's operation. Ammonia, however, causes ignition problems due to its very low combustion speed and low reactivity, but an appropriate combination of both fuels should yield satisfactory results. The authors used ammonia to modulate the combustion process by controlling the rate of flame spread. It was found that, after adding ammonia, engine knocking could be avoided, even under stoichiometric operating conditions. Mounaïm-Rousselle et al. presented the results of testing an engine powered with ammonia and hydrogen operating at partial loads [17]. For low engine loads, satisfactory operational stability was achieved for approximately 10% of the volume fraction of hydrogen in relation to $NH_3$. This share of $H_2$ allowed for NOx reduction of nearly 40% while maintaining stable engine operation. D'Antuono et al. conducted simulation tests of the operation of a spark-ignition engine powered with a mixture of ammonia and hydrogen (85% ammonia and 15% hydrogen by volume) at various operating points [18]. It was found that the addition of hydrogen contributed to widening the range of engine speed and load control. Ronan et al., in their work, presented the results of testing an ammonia spray stream in a constant-volume chamber in an atmosphere of different densities and temperatures; the results were compared with the parameters of ethanol and gasoline injection [19]. The parameters of the injected fuel stream were assessed: the range of the stream, the angle of the stream in the area of the injector outlet, and the half-length of the stream. A different geometry of ammonia injection compared to the reference fuels and a greater sensitivity to changes in air density and temperature in the tested space were noted. At the injector outlet, the ammonia stream was more compact than in the cases of other fuels by about 5 deg.

A synthetic summary of the results obtained by various authors is included in Table 1. Table 1 displays the effects of ammonia on performance, emission, and combustion parameters.

As can be seen from the several presented selected examples of the use of ammonia to power a spark-ignition piston engine, as well as the difficulties and challenges associated with it, most of these studies have referred to the use of hydrogen as a fuel facilitating the ignition process and shortening the combustion time. In our work, we used DME as a highly reactive fuel. It seems that DME may be an additional fuel that can successfully replace hydrogen in such an application. Meng et al. proposed DME as a renewable fuel that improves the ammonia combustion process [20]. In their model considerations, complex chemical mechanisms consisting of 1597 reactions were used, and the impact of the $NH_3$/DME share on the laminar combustion speed and combustion stages was assessed. An increase in DME addition caused an increase in NO emissions. It was found on the basis of theoretical research that DME can be a fuel that has a beneficial effect on the combustion control process in a spark-ignition engine.

**Table 1.** Performance, emission, and combustion characteristics of ammonia.

| Ref. | Engine Type | Fueling Type | Operating Parameters | Emission |
|---|---|---|---|---|
| [8] | One-cylinder, four-stroke, water-cooled engine, constant speed of 1500 rpm | $NH_3$ and $H_2$ into engine manifolds (PFI), hydrogen energy share: 0%, 5%, 10%, 20%, 40%, and 60% | Equivalence ratio of 0.6–1.2, the highest IMEP and ITE were achieved at low and moderate hydrogen ratios, and $H_2$ improved performance and stability | Minimal values were obtained for rich mixtures with high ammonia content, and maximal values were found for an equivalence ratio of 0.8–0.9 |
| [9] | One-cylinder, SI engine with variable ignition timings (18 deg bTDC to 32 deg bTDC) and WOT, CR changes (14–16) at 1400 rpm and 1800 rpm | $NH_3$ and $H_2$ (PFI) | Increasing the CR from 14 to 16 increased the brake power, brake thermal efficiency, NOx, cylinder pressure, and net heat release rate by 36.82%, 25.11%, 30.21%, 10.35%, and 9.53%, respectively | NOx increased |
| [10] | One-cylinder, four-stroke, naturally aspired, constant speed of 600 rpm | $NH_3$ and $H_2$ into engine manifold (PFI), $H_2$ energetic fraction: 0, 7, 12, 23, 46, 58, and 70 | 12% $H_2$ eliminated the instability of the ignition process, powering with ammonia alone resulted in a very high IMEP uniqueness, and the optimal $H_2$ fraction for CR10 was 7% | With the increase in the share of $H_2$, the NO emission increased |
| [11] | Four-cylinder, four-stroke gasoline direct injection (GDI) spark-ignition (SI) engine, 1300 rpm | $H_2$ and $NH_3$ fuels; ammonia via PFI; $H_2$ direct injection; the volume fraction of ammonia was 5.2%, 7.96%, and 10.68%; under part load and stoichiometric air–fuel ratio | The addition of ammonia increased the power of the engine and reduced the ITE | Increasing the ammonia volume fraction had little effect on the NOx emissions, and the NOx emissions gradually increased with the delay in the ignition timing |
| [12] | Optical spark-ignition engine (SI), multiple-spark-ignition (four spark plugs) | Pure ammonia fueling, three air–fuel equivalence ratios (λ) of 1.0, 1.2, and 1.4 | Single-spark-ignition resulted in lower in-cylinder pressure, a longer CD, and higher combustion instability due to the poor ammonia fuel combustion rate, and multiple spark plugs significantly improved these parameters | Multiple ignition sites resulted in higher NOx emissions, the maximum NOx level was obtained for λ: 1.2, and reducing the ammonia fuel and increasing the excess air to λ: 1.4 dramatically reduced NOx emission, |

**Table 1.** *Cont.*

| Ref. | Engine Type | Fueling Type | Operating Parameters | Emission |
|---|---|---|---|---|
| [16] | One-cylinder, 3000 rpm, CR10 | $H_2/NH_3$ mixture, Vol. of ammonia fractions: 0, 0.1, 0.2, 0.3, 0.4, and 0.5, | An increase in the ammonia fraction reduced the flame speed, and the maximum work output of blended ammonia/hydrogen can reach 90% of a pure gasoline SI engine under the same operating conditions | With more blends of ammonia, the NO can be partially converted to nitrogen, which reduces the engine-out NO emissions |
| [17] | Four-cylinder, four-stroke GDI-SI EP6 PSA engine modified to become an indirect-injection, single-cylinder engine, CR10.5 | $H_2/NH_3$ mixture, PFI, low load | The engine stably operated with less than 10% $H_2$ (of the total fuel volume) added to $NH_3$; for neat ammonia, the low load operation was difficult with a compression ratio of 10.5, requiring a higher minimum intake pressure | The maximum value of NOx was reached in a lean air/fuel mixture, the $NH_3$ at the exhaust increased with an equivalence ratio when the NOx decreased, and the addition of $H_2$ to the mixture with ammonia induced an $NH_3$ emission decrease |
| [21] | SI engine, CFR | Ammonia/hydrogen mixtures, PFI, varying excess air, different ammonia-to-hydrogen ratios | The fuel mixture with 10 vol.% hydrogen performed best with respect to efficiency and power; in comparison with a gasoline-fueled engine, efficiency and power increased due to the possibility of a higher compression ratio | To reduce the high NOx emissions, SCR exhaust was used |
| [22] | Four-stroke, two-cylinder SI engine of 505 $cm^3$ | $NH_3/H_2$ mixture | The addition of hydrogen to an air–ammonia mixture improved ignition and increased the combustion velocity; the hydrogen-to-ammonia energy ratio mainly depended on the load and less on the engine speed; and the efficiency of the engine fueled with $NH_3/H_2$ was lower than that with gasoline | - |

In the available literature, it is difficult to find works dealing with the use of DME as a fuel supporting the combustion of ammonia in a SI engine; hence, this work is innovative and expands the current scope of knowledge. This paper presents the results of experimental tests performed on a spark-ignition research engine (CFR) powered with ammonia, ammonia, and DME. Ether was intended to improve the ignition initiation process and shorten the combustion time. The impact of the DME share on engine operation indicators such as IMEP (indicated mean effective pressure), ITE (indicated thermal efficiency), and SEC (specific energy consumption), as well as the heat release process, combustion stages, and the stability of the engine operation, were assessed. The parameter determining the share of co-combusted DME with $NH_3$ was the achievement of optimal IMEP and ITE values. The tests were carried out for a compression ratio of 10 because, for lower compression ratios, it is not possible to operate an engine powered with only ammonia.

## 2. Materials and Methods

The tests were carried out on a single-cylinder spark-ignition research engine with a compression ratio of 10 and a constant rotational speed of 600 rpm. The assumption of the research was the application of such a fuel to power industrial engines operating at a constant rotational speed. The compression ratio of 10 was chosen based on previous studies on this engine powered with $NH_3$ and hydrogen fuels. The basis of the test stand was the UIT-85 research engine with a forced, open water-cooling system. The motor was braked via an asynchronous electric machine controlled with a frequency converter that gave energy to the braking resistors. The engine specifications are shown in Table 2.

**Table 2.** Technical data in the UIT-85 engine.

| Item | Data |
|---|---|
| Engine type | Four-stroke engine |
| Ignition | Spark-ignition |
| Number of cylinders | One |
| Cylinder position | Vertical |
| Timing type | OHV |
| Number of valves | Two |
| Cylinder diameter | 85 mm |
| Piston stroke | 115 mm |
| Connecting rod length | 266 mm |
| Displacement | 652.57 $cm^3$ |
| Compression ratio | $7 \div 13$ |

An encoder was mounted on the camshaft, serving as a digital transmitter of the signals of the instantaneous angular position of the engine shaft and the piston TDC (top dead center) marker, directed to the digital ignition controller and data acquisition card. This card also recorded the signal from the pressure sensor of the spark plug processed in the charge amplifier and amplified signals from thermocouples measuring the temperatures of exhaust gases, the fuel–air mixture, the surroundings, and the water jacket. These data were sent to a PC and analyzed and processed programmatically. Air was supplied to the engine intake manifold through a route consisting of a filter, a rotor flowmeter, and a pulsation dumping tank. Gaseous ammonia was taken from the cylinder with a pressure reducer through a flowmeter to the intake manifold, where it was mixed with gaseous DME that was also supplied from the cylinder with a pressure reducer through a second flowmeter. Both fuels were delivered to the intake manifold in gaseous form (PFI—port fuel injection). The research was conducted for the optimal CA50 angle, which required a change in the ignition timing. There was a pressure sensor in the intake manifold. A broadband lambda sensor with a measurement system was installed in the exhaust manifold, as well as an exhaust gas analyzer probe to determine NOx emissions. The diagram of the test stand is shown in Figure 1.

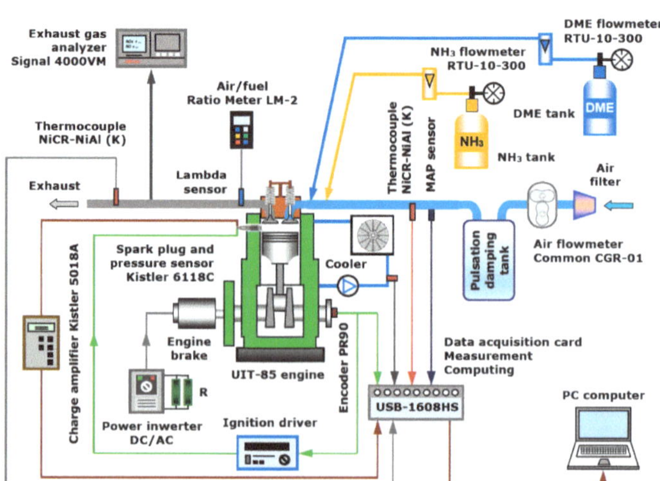


**Figure 1.** Diagram of the test stand with the UIT-85 engine.

The parameters of the measuring equipment and its accuracy are presented in Table 3.

**Table 3.** Main components of the measurement system.

| Instrument | Range | Accuracy |
|---|---|---|
| Exhaust gas analyzer: Signal 4000 VM | NOx: 0–1000 ppm<br>NOx: 0–4000 ppm<br>NOx: 0–10,000 ppm | <10 ppm<br><40 ppm<br><100 ppm |
| Air flowmeter: CGR-01 | 0.25–25.00 $m^3/h$ | 0.1% |
| $NH_3$ flowmeter: RTU-10-300 | 3–50 $dm^3/min$ | 1.5% |
| DME flowmeter: RTU-10-300 | 2–30 $dm^3/min$ | 1.5% |
| Charge amplifier: Kistler 5018A | 2 ÷ 10 pC<br>10 ÷ 100 pC<br>100 ÷ 2,200,000 pC | <±2%<br><±0.6%<br><±0.3% |
| Pressure sensor: Kistler 6118C | 0 ÷ 200 bar | <±1% |
| Data acq. system: USB-1608HS | ±10 V | ±0.07% |
| Thermocouple: NiCR–NiAl (K) | −40 ... 1140 °C | 1.5 °C or 0.004 t °C |
| Encoder PR90 | 360/rev | 0.1 deg |

The aim of the research was to determine the impact of the addition of DME to an ammonia-gas-powered SI engine on its operation and the emission of toxic exhaust gas components (NOx). The main research included checking the impact of different percentages of DME in a constant total energy dose of fuel composed of ammonia and DME. The percentage of DME energy in the $e_{DME}$ fuel was determined according to the following relationship:

$$e_{DME} = \frac{E_{DME}}{E_{DME} + E_{NH3}} 100\% \quad (1)$$

where $E_{DME}$ and $E_{NH3}$ represent the DME and ammonia energy in the total fuel dose per cycle.

Combustion tests in a spark-ignition engine powered with a mixture of DME and ammonia gas were carried out at a constant rotational speed of 600 rpm, with the throttle fully open and with a constant total energy fuel dose. The research included:

- The indication of an engine with a compression ratio of 10 for the energy percentages of DME in the fuel: 0%, 8%, 10%, 13%, 18%, 24%, and 29%. For each of these fuel compositions, the pressure in the engine cylinder was recorded during 150 consecutive cycles of its operation, with simultaneous measurements of the intake air temperature, engine temperature, exhaust gas temperature, ammonia, DME, intake air flows, and NOx concentration in exhaust gases;
- Analysis of the course of changes in cylinder pressure in subsequent cycles and the rate of heat release as a function of the fuel composition;
- Analysis of changes in IMEP and ITE as a function of fuel composition;
- Analysis of changes in the coefficient of variation of IMEP ($COV_{IMEP}$) as a function of fuel composition;
- Analysis of combustion phases divided into the initial combustion phase from ignition to CA10 and the main combustion phase from CA10 to CA90 as a function of fuel composition;
- Analysis of changes in nitrogen oxides (NOx) emissions and exhaust gas temperature as a function of fuel composition.

The ignition advance angle for all DME–ammonia mixtures was 22 deg bTDC, while in the case of operation using ammonia only, due to its difficult ignition, it was 46 deg bTDC. The excess air coefficient $\lambda$ was maintained in the range of 1.0–1.05. The tests were carried out at the optimal heat release angle of 50% (CA50), equal to approximately 10 deg CA after TDC. The cylinder pressure recording system used in the research also determined the instantaneous engine speed based on measuring the duration of pulses generated via the crankshaft angle marker. The recorded pressure waveforms were analyzed in real time, so it was possible to observe:

- The maximum pressure and the angle of its occurrence;
- The maximum speed of pressure changes and the angle of its occurrence;
- The engine speed;
- The unit-indicated work;
- The indicated cylinder power;
- A 50% heat release angle (CA50).

During measurements, the following were also displayed on an ongoing basis:

- A cylinder pressure indicator chart—open and closed;
- Graphs of the derivatives of pressure, temperature, and heat release and a graph of mass fraction burned as a function of the crankshaft rotation angle.

Pressure in the engine cylinder was recorded for every 1 degree of crankshaft rotation. Meanwhile, the temperatures of the intake air, engine, and exhaust gases were recorded as average values from 720 samples collected in each cycle. A 16-bit module with synchronously operating AC converters was used for signal acquisition.

The indication system also presented graphs of changes in the speed, indicated work, and indicated power in subsequent cycles and measured and recorded the temperature values for each engine cycle: the air in the intake duct, exhaust gases, and coolant. The exhaust gas analyzer allowed for the measuring and recording of the concentration of nitrogen oxide (NOx) in the engine exhaust gases.

The measurements of ammonia, DME, and air streams fed into the engine intake duct allowed the determination of the value of the excess air coefficient $\lambda$ for the DME and ammonia mixture:

$$\lambda = \frac{m_a}{(A/F)_s \cdot (m_{DME} + m_{NH3})} \tag{2}$$

where $m_a$ denotes the air mass, $m_{DME}$ represents the DME mass, $m_{NH3}$ indicates the mass of ammonia, and $(A/F)_s$ represents the stoichiometric air-to-fuel ratio, which is equal to:

$$(A/F)_s = \frac{(A/F)_{s1} m_{DME} + (A/F)_{s2} m_{NH3s}}{(m_{DME} + m_{NH3s})} \tag{3}$$

where $(A/F)_{s1}$ represents the stoichiometric air-to-fuel ratio for DME, $(A/F)_{s1} = 9$, and $(A/F)_{s2}$ denotes the stoichiometric air-to-fuel ratio for ammonia: $(A/F)_{s2} = 6.04$.

The measurements for each fuel composition were repeated three times, always after thermal stability was achieved, and their results were averaged. A detailed scope of the studies is given in Table 4.

**Table 4.** Main parameters of investigation.

| No. | $NH_3$ | DME | IT | n | λ | $NH_3$ Energy per Cycle | DME Energy per Cycle | Energy per Cycle | $e_{DME}$ | $P_{max}$ | $dp/d\varphi_{max}$ | IMEP |
|---|---|---|---|---|---|---|---|---|---|---|---|---|
| | l/min | | bTDC | rpm | - | J/cycle | J/cycle | J/cycle | % | MPa | MPa/deg | kPa |
| 1 | 32.5 | 0 | 46 | 598.3 | 1.0418 | 1574.7 | 0.0 | 1574.7 | 0 | 2.37 | 0.04 | 584.3 |
| 2 | 30 | 1.4 | 22 | 600.2 | 0.9969 | 1449.0 | 132.6 | 1581.6 | 8 | 3.84 | 0.17 | 703.7 |
| 3 | 29 | 1.63 | 22 | 600.0 | 1.015 | 1401.2 | 154.5 | 1555.6 | 10 | 3.72 | 0.16 | 702.1 |
| 4 | 28.5 | 2.1 | 22 | 600.1 | 0.9962 | 1376.8 | 199.0 | 1575.8 | 13 | 3.90 | 0.17 | 707.1 |
| 5 | 26.5 | 3 | 22 | 599.5 | 1.0118 | 1281.4 | 284.5 | 1566.0 | 18 | 4.02 | 0.18 | 714.0 |
| 6 | 25 | 4 | 22 | 599.5 | 0.9941 | 1208.9 | 379.4 | 1588.3 | 24 | 3.97 | 0.19 | 709.8 |
| 7 | 23 | 4.9 | 22 | 599.1 | 1.018 | 1112.9 | 465.1 | 1578.0 | 29 | 3.73 | 0.16 | 691.4 |

The parameter determining the share of co-combusted DME with $NH_3$ was the achievement of optimal IMEP and ITE values.

## 3. Fuel Characteristics

The fuels used in the tests were ammonia ($NH_3$) and dimethyl ether (DME) (Table 5). Ammonia ($NH_3$) is an inorganic chemical compound of nitrogen and hydrogen. It is a colorless gas with a pungent odor. It is an extremely toxic substance and may be very hazardous to health upon direct contact. It is a carbon-free compound, contains 17.6% hydrogen, and is characterized by a relatively high energy density (13.72 MJ/m$^3$). The very high octane number of ammonia (about 130) makes it possible to use this gas in a spark-ignition combustion engine with a higher compression ratio. Ammonia is characterized by good anti-knock properties, so as the main fuel ingredient, it should reduce the occurrence of this unfavorable phenomenon. These features make ammonia an attractive fuel in the transport and energy sectors. One of the main disadvantages of ammonia as a fuel is its lower reactivity than conventional coal-based fuel, which complicates the combustion process and may negatively affect the emission of toxic and harmful exhaust gas components. Its relatively low calorific value (LHV = 18.8 MJ/kg), high heat of vaporization (HoE = 1370 kJ/kg), low laminar combustion speed (LFS = 0.015 m/s), high auto-ignition temperature (651 °C), and narrow flammability range (15–28) hinder the use of ammonia in piston engines [21]. These properties create problems in effective combustion in both compression ignition and spark-ignition engines. The undoubted advantage of ammonia is its ease of storage. Ammonia can be used directly as a fuel, for example, in power turbines in thermal power plants or as a fuel in maritime transport. Ammonia can be easily transformed into hydrogen. Hydrogen, especially in the form of ammonia, is highly significant in the decarbonization process of the energy sector in many developed countries and is one of the methods to move away from fossil fuels. Ammonia can serve as a high-quality energy carrier, having a significant advantage over hydrogen. Hydrogen needs to be stored at a temperature of 20.15 K (−253 °C) as a liquid or at a pressure of about 70 MPa (700 bar) as a gas. Liquid ammonia can be stored at a reasonable temperature of 240.15 K (−33 °C) at standard pressure and 293.15 K (+20 °C) at a pressure of 0.9 MPa (9 bar). This makes storing and transporting this energy carrier much easier, cheaper, and simpler. Due to the large mass fraction of hydrogen in the $NH_3$ molecule, ammonia can be used as an efficient hydrogen energy carrier. Ammonia in nature is produced as a result of the microbiological decomposition of nitrogenous organic substances containing protein, while in industry, the primary raw materials used for its production are natural gas and air [22]. Currently, efforts are being made to produce ammonia from renewable sources (RES). In this case, energy from RES is first used to produce green hydrogen (water electrolysis), which is then converted into green ammonia. Hydrogen from water electrolysis can be used to produce ammonia, and the electrolysis itself can be performed using biomass energy, wind energy, solar energy, geothermal energy, or hydroelectric energy [23].

**Table 5.** Fuel specifications [24–27].

| Parameter | Ammonia | DME |
|---|---|---|
| Molecular formula | $NH_3$ | $CH_3OCH_3$ |
| Cetane number | - | 55–66 |
| Octane number (RON) | 130 | 60 |
| Molecular weight (g/mol) | 35.046 | 46.7 |
| Density at 15 °C and 1 bar (kg/m$^3$) | 0.73 | 0.668 |
| Lower heating value (MJ/kg) | 18.8 | 28.4 |
| Volumetric energy density (MJ/m$^3$) | 13.7 | 19 |
| Autoignition temperature (°C) | 651 | 235 |
| Stoichiometric air–fuel ratio (kg/kg) | 6.04 | 9 |
| Laminar flame speed (m/s) | 0.015 | 0.45 |
| Heat of evaporation (kJ/kg) | 1370 | 467 |
| Flammability limits in air (vol.%) | 15–28 | 3.4–27 |
| Ignition energy in air (mJ) | 8 | 0.29 |
| Hydrogen content (wt%) | 17.6 | 13 |
| Carbon content (wt%) | 0 | 52 |
| Nitrogen (wt%) | 82.4 | 0 |
| Oxygen (wt%) | 0 | 35 |

Dimethyl ether (DME) belongs to the organic chemical compounds from the group of ethers. It is a colorless gas with a characteristic odor and is moderately soluble in water. DME can be used as a standalone fuel in compression ignition engines or as a mixture in spark-ignition engines. DME was chosen as an assisting fuel to enhance the combustion process of $NH_3$ in a spark-ignition engine. The main reason for selecting this fuel lies in its properties, which largely contrast with $NH_3$, primarily the LFS value, wide flammability range in air, and ease of ignition. It plays a particularly good role as a fuel in diesel engines—its cetane number is over 55, while for diesel oil, it usually does not exceed 53. DME is the simplest ether obtained from both fossil and renewable raw materials. Dimethyl ether is usually produced via the dehydration of methanol. More environmentally friendly methods are increasingly being used, such as producing DME from syngas, a mixture of hydrogen and carbon monoxide. The synthesis of dimethyl ether is more efficient than the dehydration of methanol because it allows the production of this compound from waste or biomass, bypassing the production of methanol. Ecological DME production methods allow this gas to be classified as a renewable fuel. Dimethyl ether is produced from synthesis gas using two methods: indirect synthesis (through methanol) and direct synthesis. Dimethyl ether, through synthesis gas, can be produced from raw materials such as biogas, biomethanol, biomass, natural gas, hard coal, brown coal, oil shale, plastic waste, and municipal waste. Synthesis gas obtained from steam-reforming various raw materials and bioresources can be further utilized for the direct synthesis of dimethyl ether. If raw materials such as biomethanol, biomass, or biogas are used in DME production, the DME is formally recognized as a biofuel (BioDME). The application of BioDME in automotive transportation can reduce greenhouse gas (GHG) emissions by 92–95%. DME can be liquefied at room temperature and stored in liquid form. Dimethyl ether is characterized by a relatively low calorific value (LHV = 28.4 MJ/kg), a flash point comparable to that of diesel oil (235 °C), and a wide flammability range (3.4–27 vol.%). It burns without soot, has a low ignition temperature, and evaporates perfectly. One of the greatest advantages of DME is the lack of emission of dust pollutants, thanks to the absence of direct carbon bonding in the atomic structure, as well as the presence of oxygen in this bond (approx. 35%). Dimethyl ether (DME) may also be a potential fuel used in low-temperature fuel cells that operate below 100 °C.

## 4. Results and Discussion

The tests on the combustion of ammonia alone and the co-combustion of ammonia with DME were carried out for a full engine load, a constant rotational speed, and a compression ratio of 10. Due to the low value of the combustion speed of ammonia, it was examined how the addition of DME to it affects the combustion process. DME is characterized by a much higher combustion speed than ammonia and, therefore, should improve the ignition initiation process and shorten the combustion time. Ammonia is characterized by good anti-knock properties, so as the main fuel ingredient, it should reduce the occurrence of this unfavorable phenomenon. The analysis of the combustion process of both fuels was carried out based on the results of measurements of the instantaneous pressure in the engine cylinder and exhaust emissions. Heat release patterns were determined based on the pressure curves obtained. The standardized course of heat release was used to determine the characteristic stages of combustion, i.e., the ignition delay time (CA10) and combustion time (CA10-90). Ignition delay (CA10) is the period from the spark discharge at the spark plug to the point of 10% heat release. The combustion duration (CA10-90) is the period from the point of 10% heat release to the point of 90% heat release. The CA50 value defines the point of 50% heat release and is calculated from the TDC (top dead center). The tests were carried out at the optimal heat release angle of 50% (CA50) at the level of 10 ± 1 degrees CA after TDC. The ignition delay was determined in degrees CA as the angular interval from the moment of ignition to the moment of 10% heat release. The combustion time was the interval between the moment of releasing 10% of the heat and the moment of releasing 90% of the heat. Knowledge of these characteristic stages of combustion is crucial for optimal engine control.

Figure 2 presents a summary of the combustion pressure and pressure rise curves for an $NH_3$-fueled engine with various DME additives ranging from 0% to 29% of the energy share. For a compression ratio of 10, the engine powered by $NH_3$ alone worked, but its operation stability was unsatisfactory ($COV_{IMEP}$ = 30%). To improve the quality of engine operation and combustion stability, DME was added, and with its 8% share, a significant improvement in engine performance indicators was achieved.

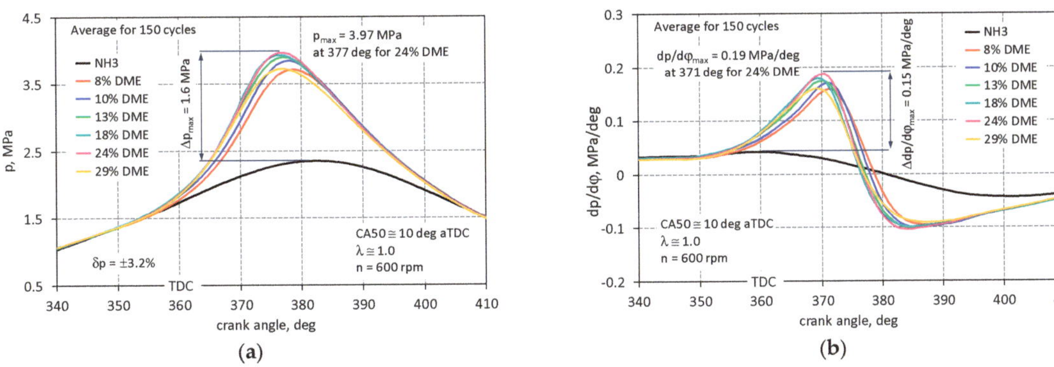


**Figure 2.** Pressure (**a**) and pressure rise (**b**) waveforms.

The addition of 8% DME (by energy share) resulted in an increase in the average (from 150 cycles) of the maximum pressure in the engine combustion chamber of 1.35 MPa compared to the engine powered by $NH_3$ alone. There was also a shift of $p_{max}$ towards TDC of 3.5 deg. Each research point was optimized according to the CA50 criterion. A further increase in the energy share of DME to 24% resulted in an increase in the average $p_{max}$ value, which for this DME share was 17 deg aTDC and amounted to 3.97 MPa. Increasing the DME share to 28% resulted in a significant decrease in the $p_{max}$ value to 3.72 MPa. Due to the $p_{max}$ criterion, it can be concluded that even an 18% DME share provides the optimal pressure value. The analysis of the pressure increase (Figure 2b) showed that the addition of DME provides the ($dp/d\varphi$) values desired in an SI engine for its proper operation. For all runs involving DME, peak values $(dp/d\varphi)_{max}$ occurred around 10 deg aTDC and were in the range of 0.16 to 0.19 MPa/deg. Only for an engine fueled only with $NH_3$, the average ($dp/d\varphi$) curve was too flat, which indicates very slow combustion, which does not ensure appropriate gas-dynamic parameters for exhaust gases in the engine cylinder.

Analysis of the course of heat release in a piston engine is the basis for assessing the combustion process in the engine. The heat release rate ($dq/d\varphi$) provides information about the dynamics of the combustion process, while the integrated waveform of the heat release rate allows for the determination of characteristic stages of combustion. Figure 3 shows the obtained heat release waveforms.

Figure 3a shows the waveforms of the heat release rate depending on the share of DME. For an engine fueled only with $NH_3$, the average waveform ($dq/d\varphi$) was very flat, i.e., the maximum heat release rate was very low and amounted to 22.8 J/deg, and 22 deg aTDC was achieved. Despite the optimization of the engine operation using the CA50 criterion, satisfactory results could not be achieved for fueling with $NH_3$ alone due to the very high uniqueness of the cycles. For the waveform averaged from a set of 150 cycles, the CA50 value was reached at 22 degrees aTDC, i.e., too late (Figure 3b). The addition of 8% DME resulted in a significant improvement in the heat release rate; for this share, a peak value of 58 J/deg was recorded, and the CA50 value was 12 deg, i.e., the engine operation was already close to optimal operation. For piston engines, it is assumed that CA50 within the limit of 10 deg is optimal [28]. The maximum value ($dq/d\varphi$) of 65.2 J/deg for an angle of 12 degrees after TDC was recorded for 24% of DME. The maximum values (dq/dφ) for the engine fueled with a DME addition were in the range of 10.4 J/deg. From the $Q_{norm}$ (normalized heat release) waveforms, it can be seen that the addition of DME clearly shifted the CA50 angle towards the optimal value, close to 10 deg aTDC; all obtained waveforms from 8 to 29% DME shares were focused in a narrow spread range (CA50) of 2.5 deg. The obtained $Q_{norm}$ waveforms were used to determine the location and length of characteristic combustion stages.

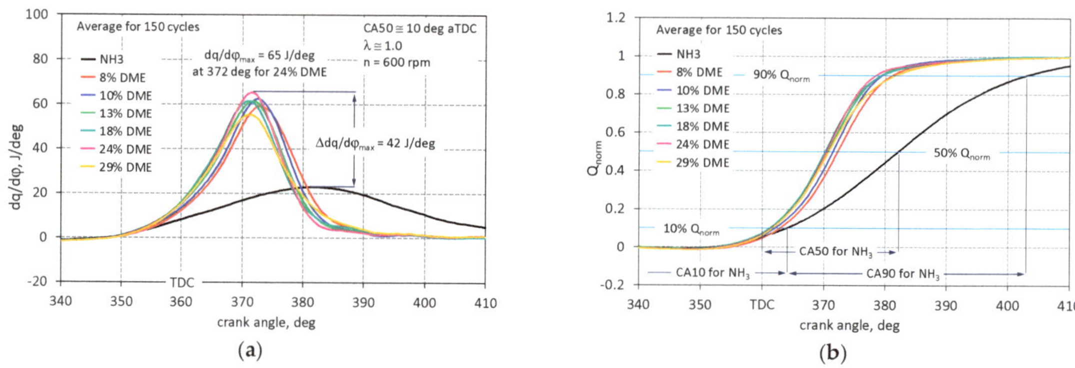


**Figure 3.** Waveforms of the heat release rate (**a**) and normalized heat release (**b**).

Knowing the values of individual combustion stages provides detailed information about the dynamics of this process. This is especially important for fuels that are difficult to burn, such as ammonia. Knowing the value of the ignition delay CA10 allows for appropriate corrections in the engine control system in terms of the ignition angle. The CA50 value is used to optimize the engine's thermal circulation, and knowledge of the CA10-90 angle provides indirect information about the engine's efficiency. Figure 4a presents a summary of individual combustion stages for the analyzed cases of the co-combustion of DME with $NH_3$. For the engine fueled only with ammonia, the CA10 value was 46.2 deg, and the 8% share of DME reduced the CA10 by 22 deg to 24.2 deg.

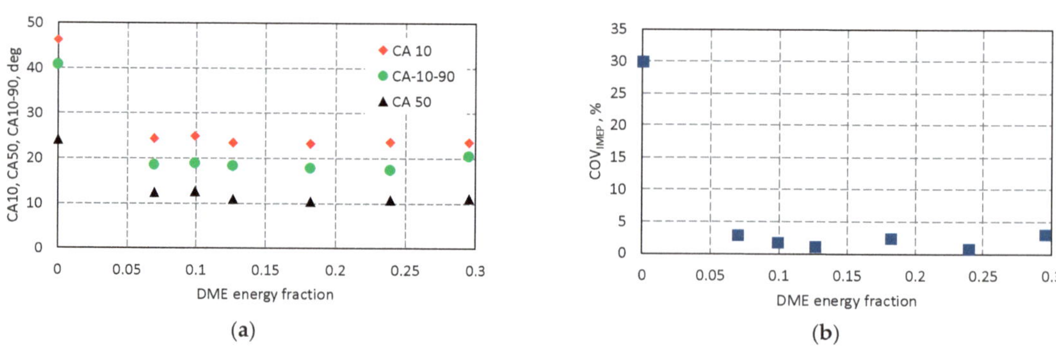


**Figure 4.** Characteristic stages of combustion CA10, CA10-90, and CA90 (**a**) and uniqueness of the average indicated pressure value (**b**).

Increasing the DME dose no longer caused such significant changes in CA10 values; for all DME shares, the CA10 value was within 1.5 deg. The lowest CA10 value of 23.1 deg was recorded for 18% DME.

When analyzing the combustion stages, it should be taken into account that the obtained values for ignition delay (CA10) and combustion time (CA10-90) were determined for the waveform averaged over 150 cycles. In the case of the high non-repeatability of the engine operation, especially when fueled only with ammonia, they do not reflect the actual fluctuations of these parameters in subsequent engine operation cycles. Therefore, the duration of CA10-90 combustion for an engine fueled only with $NH_3$ was as much as 41 deg. It should be noted that, for this operating point, there was very high non-repeatability of the engine operation ($COV_{IMEP}$ = 30%). This was mainly due to the low LFS of ammonia, which hinders the flame front propagation, which is especially crucial in the initial phase of combustion (the ignition delay period). The low flame front speed in the initial phase of combustion, just after spark-ignition initiation, contributes to an increase in the ignition delay time. Conversely, a significant ignition delay time in piston engines results in an increase in the variability of successive engine cycles. The addition of DME to $NH_3$ significantly shortened the average combustion duration, setting it at approximately 18 deg. The CA50 value, according to the optimization assumptions of the engine cycle, should be 10 deg aTDC. Shortening the duration of combustion in the engine results in an increase in its efficiency (ITE) and the value of the average indicated pressure and, at the same time, a decrease in specific fuel consumption [29,30]. The combustion time affects the use of the energy contained in the fuel, increasing the combustion efficiency, which will also affect the exhaust gas temperature and then the exhaust emissions [31]. The highest maximum temperature values during combustion of around 1750 K were obtained for DME contents ranging from 18–24%. Reducing the DME dosage resulted in a slowing down of the temperature increase during combustion and a decrease in its maximum value. With a DME dosage increased to 29%, the maximum temperature noticeably decreased.

When fueled with $NH_3$ alone, due to the high uniqueness of the engine operation, the CA50 criterion could not be met during the tests. The engine fueled with the addition of DME to $NH_3$ already provided the expected low operating uniqueness of less than 5%, i.e., below the permissible value for industrial engines [28]. Cycle-to-cycle variations often occur in spark-ignition engines, which are very sensitive to changes in the movement of the charge in the cylinder, the local swirls and turbulence of the charge, the uniqueness of the charge composition in subsequent cycles (lambda changes) in the area of the spark plug, and differences in the internal EGR [32]. For an engine powered only with ammonia, very high values of the $COV_{IMEP}$ index were obtained (Figure 4b). The reason for this is that, in many engine cycles, the $NH_3$–air mixture did not ignite, which is visible in Figures 5a and 6. When burning ammonia alone, frequent misfires occurred—on average, once every 15 cycles. Therefore, the engine operation was subject to high IMEP non-repeatability at a level of 30%. It was observed that, in a few cycles (approx. 5%), when ignition occurred with a minimum delay, which for pure ammonia was as much as 40–45 deg CA, ammonia combustion was very effective because the maximum pressure was 100% higher than its average value from all cycles, and the indicated work obtained was 30% higher than its average value. This phenomenon occurred in cycles immediately following cycles without ignition. They can be explained as the result of the accumulation of an increased dose of ammonia in the cylinder. The spread of maximum pressure values in individual cycles was very large, from 1.5 MPa to 5.5 MPa. Adding 10% DME to ammonia almost completely eliminated the instability of the ignition process of the flammable mixture (Figure 5b). Ignition occurred in all cycles, although the value of random ignition delay fluctuations was quite significant at 10 deg CA. But the IMEP uniqueness decreased to an acceptable level of 1.9%. The values of maximum pressure changes were still within a fairly wide range of 2.05–4.2 MPa. Increasing the energy dose of DME to 24% (Figure 5c) only resulted in further improvement of the combustion process. The range of ignition angle changes decreased to 10 deg CA, and the spread of maximum pressure values decreased to the range of 3.2–4.2 MPa. However, the uniqueness of IMEP decreased to 0.9%. With the increase in the DME share, an increasingly early location of the pressure maximum in relation to TDC was observed. This was mainly due to the low LFS of ammonia, which hinders the flame front propagation, which is especially crucial in the initial phase of combustion (the ignition delay period). The low flame front speed in the initial phase of combustion, just after spark-ignition initiation, contributes to an increase in the ignition delay time. Conversely, a significant ignition delay time in piston engines results in an increase in the variability of successive engine cycles.

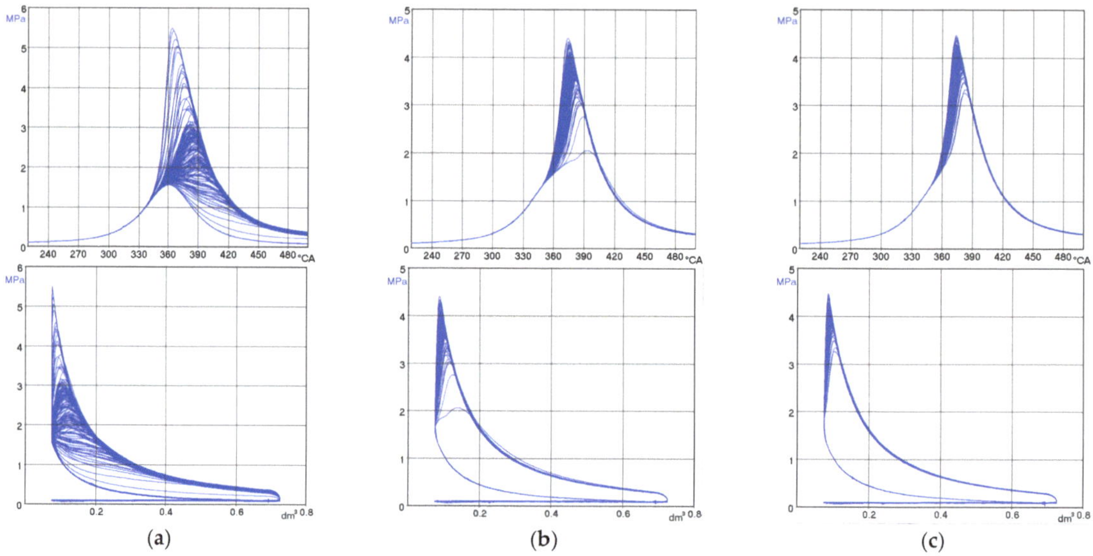

**Figure 5.** Indicator charts from 150 consecutive engine operation cycles for three DME energy shares: (**a**) 0%, (**b**) 10%, and (**c**) 24%.

Figure 6 shows the correlation between the value of the maximum pressure ($p_{max}$) in the engine cylinder and the angle of its occurrence for three cases, feeding the engine with ammonia only and two cases with 10% and 24% of the energy share of DME.

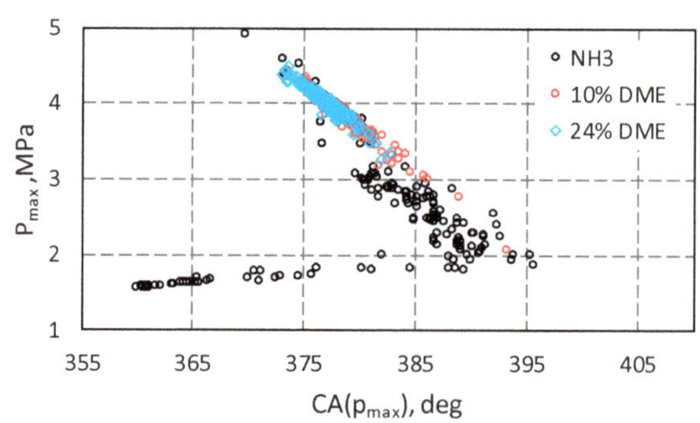


**Figure 6.** Variability of the maximum pressure in individual cycles, depending on the angle of its occurrence for three energy shares of DME (0%, 10%, and 24%).

For the engine fueled only with ammonia ($NH_3$), there were many cycles without combustion, which was reflected in the high $COV_{IMEP}$ value. For the combustion of ammonia only, cycles with very high pressure values were also observed, caused by the accumulation of fuel in the cylinder from previous cycles without combustion. The participation of DME provides control over the ammonia combustion process, which becomes more predictable and easier to control, e.g., to optimize CA50.

For an industrial combustion engine, its operational and economic values are very important. The engine should not only operate stably but also provide the expected performance (IMEP) and the highest possible efficiency (ITE), using as little fuel as possible to produce a unit of energy (BSEC). Figure 7 shows the influence of the DME share on the efficiency of an ammonia-fueled engine.

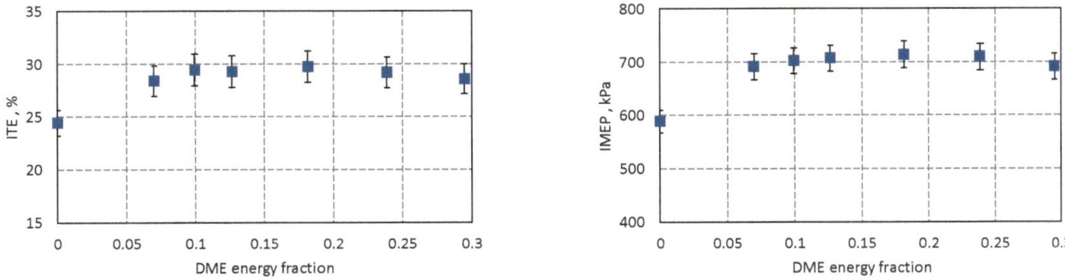


(a) (b)

**Figure 7.** Indicated thermal efficiency (**a**) and indicated mean effective pressure (**b**).

For an engine powered with $NH_3$ alone, an efficiency of 24.4% was recorded. The addition of DME significantly increased the efficiency of the engine because an 8% share of DME allowed the achievement of an efficiency of 28.4%, i.e., 16% higher than for an engine powered with ammonia alone. This is because DME significantly accelerates the combustion process and, thus, improves efficiency. For 8% DME content, a reduction in the ignition delay time (CA10) and the stabilization of engine operation ($COV_{IMEP} < 5\%$) were achieved. Below the limit of this DME, contribution cycles without combustion occurred, leading to a reduction in IMEP values. The highest efficiency of 29.8% was achieved for 18% of the energy share of DME. DME is characterized by a much higher LFS in comparison with ammonia, which significantly improves the flame propagation process in the combustion chamber. This likely allows the flame to quickly reach the boundaries of the working space, enhancing the burning of the charge and thereby increasing combustion efficiency. For this DME share, the highest IMEP value of 714 kPa and the minimum specific energy consumption of 12.1 MJ/kWh were recorded (Figure 8). Specific energy consumption is an indicator showing the economics of using the energy contained in the fuel.

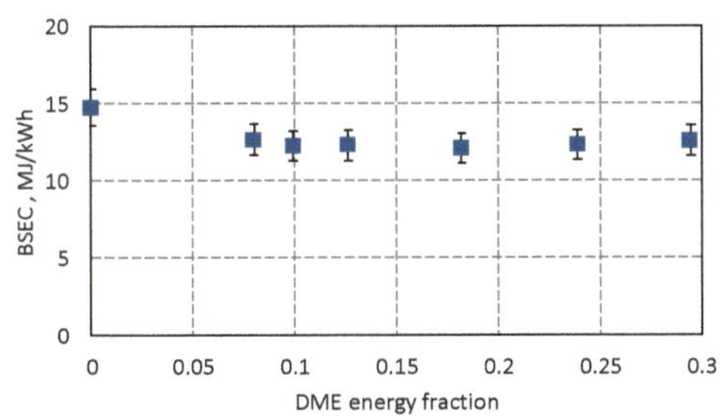


**Figure 8.** Brake-specific energy consumption (BSEC).

An important indicator for assessing the economic aspect of a piston engine is the specific fuel consumption (BSFC in g/kWh). This indicator is reliable only for an engine powered with a single fuel (constant heating value, LHV). In the case in which the engine is powered with two fuels with different heating values and their proportion changes, the BSFC indicator becomes less useful. For dual-fuel engines, the BSEC indicator should be used. It provides information about the energy consumption in the fuel blend to generate one kWh of energy.

As shown in Figure 8, for all cases of ammonia combustion with the addition of DME, the BSEC value was lower than for the engine fueled with $NH_3$ alone. This confirms the validity of using fuel with the addition of DME. For an 18% DME share, a BSEC decrease of nearly 19% was achieved. It should be added here that the basic fuel is $NH_3$, and DME is only intended to enable stable ammonia ignition and/or improve the quality of the combustion process. Varying the DME fraction in an ammonia-with-DME fuel blend for an SI engine affects the ignition timing, flame propagation in the combustion chamber, thermodynamic properties, emissions, and combustion stability. DME has a higher LHV, providing more thermal energy to combustion. Increasing the DME share will influence the heat release during the combustion process.

Due to the combustion of ammonia, i.e., a compound of hydrogen and nitrogen, nitrogen oxides (NOx) emissions were controlled. Figure 9 shows the results of the exhaust gas emission assessment and the exhaust gas temperature obtained during the co-combustion of ammonia with DME in a spark-ignition engine. The combustion temperature significantly impacts NOx emissions in the engine's combustion chamber. The NOx formation mechanism requires access to nitrogen, oxygen, and a sufficiently high temperature for chemical reactions to occur. Of course, finding the right balance between engine efficiency (including combustion efficiency) and NOx control is crucial to meet environmental standards while ensuring optimal engine performance. The highest NOx emissions were obtained using the co-combustion of ammonia with 10% DME.

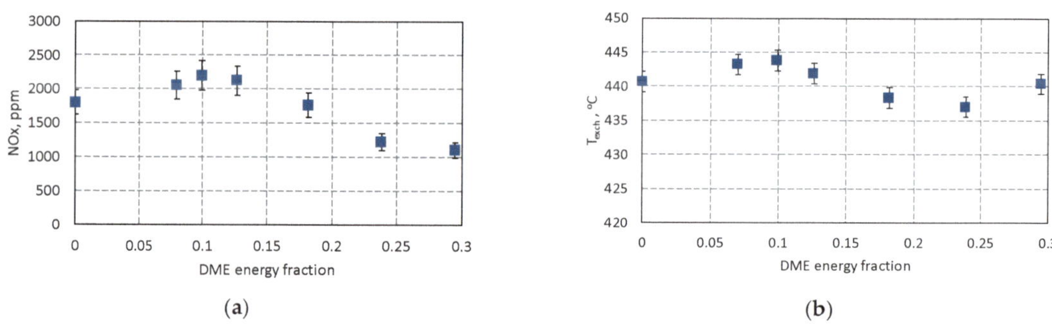


**Figure 9.** NOx emissions (**a**) and exhaust gas temperature (**b**).

In this case, the most intense combustion took place, characterized by the highest temperature (max $T_{exch}$) and almost the highest heat release rate ($dq/d\varphi$). A further increase in the DME share resulted in a decrease in the concentration of nitrogen oxides caused by a decrease in the combustion temperature. This decrease resulted from the depletion of the combustible mixture in the engine cylinder with additional oxygen supplied in DME. For an engine fueled only with $NH_3$, emissions of 1800 ppm were recorded, but this was under conditions of the very high non-repeatability of subsequent combustion cycles; there were cycles without combustion. The highest NOx emission was measured for a 10% share of DME and amounted to 2200 ppm, for this share of DME, the highest exhaust gas temperature was 443 °C. For the energy share of DME considered optimal, the NOx emission was 1764 ppm, which was less than when fueled with $NH_3$ alone. NOx emissions in an engine fueled with a mixture of $NH_3$ and DME can be significantly reduced, thanks to the use of selective catalytic reduction (SCR) technology, and a possible small excess of ammonia in the mixture may be beneficial for the better neutralization of NOx in the SCR reactor.

The oxygen content in the DME atomic structure enhances its combustion efficiency and allows it to burn more completely compared to hydrocarbon-based fuels. When DME burns, the oxygen atoms in its molecular structure participate in the combustion reaction, promoting more efficient and complete combustion. This contributes to a cleaner and more complete conversion of fuel into thermal energy.

## 5. Conclusions

This paper has presented the results of experimental studies on the combustion process in a spark-ignition engine powered with ammonia ($NH_3$) operating at a constant rotational speed and with a compression ratio of 10. As preliminary tests showed, for such a compression ratio, it is possible to achieve combustion of only ammonia in the engine. However, as the test results show, an engine powered with ammonia alone operates unstably, and in order to improve its operation, it is necessary to add a component with higher reactivity to the fuel, e.g., DME. Based on the results obtained using DME to improve $NH_3$ combustion in an SI engine, it was concluded that:

- The optimal value of the energy share of DME was 18%, for which the maximum value of IMEP = 714 kPa, the highest-engine-efficiency ITE = 29.8%, and the minimum-specific-energy-consumption BSEC = 12.1 MJ/kWh were recorded;
- DME had a positive effect on the increase in the repeatability of IMEP; a $COV_{IMEP}$ of less than 5% was achieved with just 8% DME;
- The DME share did not cause significant increases in NOx emissions; lower emissions were achieved for the optimal DME share than when fueled with $NH_3$ alone;
- The share of DME significantly reduced the ignition delay time and combustion duration;
- DME fuel effectively improves the combustion process in an SI engine, allowing for the acceptable operation of an industrial engine.

In conclusion, ammonia and DME offer potential environmental benefits, especially in terms of reduced carbon emissions and air pollutants. However, addressing challenges related to production emissions, safety, infrastructure, and sourcing sustainable raw materials is crucial to maximizing their positive environmental impact. Based on the conducted research, it can be concluded that DME can be an effective fuel supporting the ammonia combustion process in a spark-ignition engine.

**Author Contributions:** Conceptualization, W.T., M.P. and M.G.; data curation, M.G., M.P. and W.T.; formal analysis, W.T., M.G. and M.P.; investigation, M.P., M.G. and W.T.; methodology, W.T., M.P. and M.G.; writing—original draft, W.T., A.J. and M.G.; writing—review and editing, W.T., A.J. and M.G. All authors have read and agreed to the published version of the manuscript.

**Funding:** This research was financed by the Ministry of Science and Higher Education of Poland from the funds dedicated to scientific research No. BS/PB 1-100-301/2023/P.

**Data Availability Statement:** The data used in this study are reported in the paper figures and tables.

**Conflicts of Interest:** The authors declare no conflict of interest.

## Nomenclature

| | |
|---|---|
| $(A/F)_s$ | stoichiometric air-to-fuel ratio |
| $(A/F)_{s1}$ | stoichiometric air-to-fuel ratio for DME |
| $(A/F)_{s2}$ | stoichiometric air-to-fuel ratio for ammonia |
| BSEC | Brake-specific energy consumption |
| CA10 | ignition delay |
| CA50 | 50% heat release |
| CA90 | end of combustion |
| CFR | cooperative fuel research |
| CO | carbon monoxide |
| $CO_2$ | carbon dioxide |
| $COV_{IMEP}$ | coefficient of variation of the indicated mean effective pressure |
| DME | dimethyl ether |
| $dp/d\varphi$ | pressure rise |
| $dq/d\varphi$ | heat release rate |
| $E_{DME}$ | DME energy in the total fuel dose per cycle |
| $e_{DME}$ | percentage of DME energy in the fuel |

| | |
|---|---|
| EGR | exhaust gas recirculation |
| $E_{NH3}$ | ammonia energy in the total fuel dose per cycle |
| HoE | heat of evaporation |
| IMEP | indicated mean effective pressure |
| ITE | indicated thermal efficiency |
| LFS | laminar flame speed |
| LHV | lower heating value |
| $m_a$ | air mass |
| $m_{DME}$ | mass of DME |
| $m_{NH3}$ | mass of ammonia |
| $NH_3$ | ammonia |
| NO | nitrogen monoxide |
| $NO_2$ | nitrogen dioxide |
| $NO_x$ | nitrogen oxides |
| OHV | overhead valves |
| p | pressure |
| $p_{max}$ | maximum pressure |
| $Q_{norm}$ | normalized heat release |
| rpm | revolutions per minute |
| SCR | selective catalytic reduction |
| SI | spark-ignition |
| TDC | top dead center |
| $T_{exch}$ | exhaust temperature |
| THC | total hydrocarbons |
| $\lambda$ | excess air coefficient |

## References


1. Krishnan, M.G.; Rajkumar, S.; Thangaraja, J.; Devarajan, Y. Exploring the synergistic potential of higher alcohols and biodiesel in blended and dual fuel ombustion modes in diesel engines: A comprehensive review. *Sustain. Chem. Pharm.* **2023**, *35*, 101180. [CrossRef]
2. Awad, O.I.; Mamat, R.; Ali, O.M.; Sidik, N.A.C.; Yusaf, T.; Kadirgama, K.; Kettner, M. Alcohol and ether as alternative fuels in spark ignition engine: A review. *Renew. Sustain. Energy Rev.* **2018**, *82*, 2586–2605. [CrossRef]
3. Varma, J.R.; Katepalli, S.K.R.; Sreeja, M.; Hadagali, B. Comprehensive studies on alcohol using port fuel injection facilitated with spark plug engine. *Mater. Today Proc.* **2021**, *45*, 3219–3225. [CrossRef]
4. Qian, L.; Wan, J.; Qian, Y.; Sun, Y.; Zhuang, Y. Experimental investigation of water injection and spark timing effects on combustion and emissions of a hybrid hydrogen-gasoline engine. *Fuel* **2022**, *322*, 124051. [CrossRef]
5. Gupta, P.; Kurien, C.; Mittal, M. Biogas (a promising bioenergy source): A critical review on the potential of biogas as a sustainable energy source for gaseous fuelled spark ignition engines. *Int. J. Hydrogen Energy* **2023**, *48*, 7747–7769. [CrossRef]
6. Zhu, J.; Zhou, D.; Yang, W.; Qian, J.; Mao, Y.; Lu, X. Investigation on the potential of using carbon-free ammonia in large two-stroke marine engines by dual-fuel combustion strategy. *Energy* **2023**, *263*, 125748. [CrossRef]
7. Jamrozik, A.; Grab-Rogaliński, K.; Tutak, W. Hydrogen effects on combustion stability, performance and emission of diesel engine. *Int. J. Hydrogen Energy* **2020**, *45*, 19936–19947. [CrossRef]
8. Lhuilliera, C.; Brequignya, P.; Continod, F.; Mounaïm-Roussellea, C. Experimental study on ammonia/hydrogen/air combustion in spark ignition engine conditions. *Fuel* **2020**, *269*, 117448. [CrossRef]
9. Dinesh, M.H.; Kumar, G.N. Experimental investigation of variable compression ratio and ignition timing effects on performance, combustion, and NOx emission of an ammonia/hydrogen-fuelled SI engine. *Int. J. Hydrogen Energy* **2023**. [CrossRef]
10. Pyrc, M.; Gruca, M.; Tutak, W.; Jamrozik, A. Assessment of the co-combustion process of ammonia with hydrogen in a research VCR piston engine. *Int. J. Hydrogen Energy* **2023**, *48*, 2821–2834. [CrossRef]
11. Xin, G.; Ji, G.; Wang, S.; Meng, H.; Chang, K.; Yang, J. Effect of different volume fractions of ammonia on the combustion and emission characteristics of the hydrogen-fueled engine. *Int. J. Hydrogen Energy* **2022**, *47*, 16297–16308. [CrossRef]
12. Uddeen, K.; Tang, Q.; Shi, H.; Magnotti, G.; Turner, J. A novel multiple spark ignition strategy to achieve pure ammonia combustion in an optical spark-ignition engine. *Fuel* **2023**, *349*, 128741. [CrossRef]
13. Kociszewski, A.; Jamrozik, A.; Tutak, W.; Sosnowski, M. Computational analysis and experimental research into lean mixture combustion in multi-spark plug SI engine. *J. Kones* **2006**, *13*, 123–130.
14. Fei, M.; Guo, L.; Li, Z.; Zeng, X.; Zheng, Z.; Li, W.; Zhao, F.; Yu, W. A review of current advances in ammonia combustion from the fundamentals to applications in internal combustion engines. *Energies* **2023**, *16*, 6304.

15. Chorowski, M.; Lepszy, M.; Machaj, K.; Malecha, Z.; Porwisiak, D.; Porwisiak, P.; Rogala, Z.; Stanclik, M. Challenges of Application of Green Ammonia as Fuel in Onshore Transportation. *Energies* **2023**, *16*, 4898. [CrossRef]
16. Ge, H.; Bakir, A.H.; Zhao, P. Knock mitigation and power enhancement of hydrogen spark-ignition engine through ammonia blending. *Machines* **2023**, *11*, 651. [CrossRef]
17. Mounaïm-Rousselle, C.; Bréquigny, P.; Dumand, C.; Houillé, S. Operating limits for ammonia fuel spark-ignition engine. *Energies* **2021**, *14*, 4141. [CrossRef]
18. D'Antuono, G.; Lanni, D.; Galloni, E.; Fontana, G. Numerical modeling and simulation of a spark-ignition engine fueled with ammonia-hydrogen blends. *Energies* **2023**, *16*, 2543. [CrossRef]
19. Pelé, R.; Mounaïm-Rousselle, C.; Bréquigny, P.; Hespel, C.; Bellettre, J. First study on ammonia spray characteristics with a current GDI engine injector. *Fuels* **2021**, *2*, 253–271. [CrossRef]
20. Meng, X.; Zhang, M.; Zhao, C.; Tian, H.; Tian, J.; Long, W.; Bi, M. Study of combustion and NO chemical reaction mechanism in ammonia blended with DME. *Fuel* **2022**, *319*, 123832. [CrossRef]
21. Mørch, C.S.; Bjerre, A.; Gøttrup, M.P.; Sorenson, S.C.; Schramm, J. Ammonia/hydrogen mixtures in an SI-engine: Engine performance and analysis of a proposed fuel system. *Fuel* **2011**, *90*, 854–864. [CrossRef]
22. Frigo, S.; Gentili, R. Analysis of the behaviour of a 4-stroke SI engine fuelled with ammonia and hydrogen. *Int. J. Hydrogen Energy* **2013**, *38*, 1607–1615. [CrossRef]
23. Valera-Medina, A.; Gutesa, M.; Xiao, H.; Pugh, D.; Giles, A.; Goktepe, B.; Marsh, R.; Bowen, P. Premixed ammonia/hydrogen swirl combustion under rich fuel conditions for gas turbines operation. *Int. J. Hydrogen Energy* **2019**, *44*, 8615–8626. [CrossRef]
24. Chiong, M.-C.; Chong, C.T.; Ng, J.-H.; Mashruk, S.; Chong, W.W.F.; Samiran, N.A.; Mong, G.R.; Valera-Medina, A. Advancements of combustion technologies in the ammonia-fuelled engines. *Energy Convers. Manag.* **2021**, *244*, 114460. [CrossRef]
25. Kim, H.J.; Suh, H.K.; Lee, C.S. A study on an application of a hybrid break-up model for dimethyl ether atomization in a common-rail injection system. *Proc. Inst. Mech. Eng. Part D J. Automob. Eng.* **2009**, *223*, 1351–1359. [CrossRef]
26. Górski, K.; Olszewski, W.; Lotko, W. Alkohole i etery jako paliwa do silników o zapłonie samoczynnym. *Czas. Tech.* **2008**, *105*, 13–24.
27. Wu, J.; Huang, Z.; Qiao, X.; Lu, J.; Zhang, J.; Zhang, L. Study of combustion and emission characteristics of turbocharged diesel engine fuelled with dimethylether. *Front. Energy Power Eng* **2008**, *2*, 79–85. [CrossRef]
28. Heywood, J.B. *Internal Combustion Engine Fundamentals*, 2nd ed.; McGraw-Hill Education: New York, NY, USA, 2018.
29. Xuan, N.K.; Ocktaeck, L. Influence of combustion duration on the performance and emission characteristics of a spark-ignition engine fueled with pure methanol and ethanol. *ACS Omega* **2022**, *7*, 14505–14515.
30. Wildhaber, S.N. Impact of Combustion Phasing on Energy and Availability Distributions of an Internal Combustion Engine. Master's Thesis, Missouri University of Science and Technology, Rolla, MO, USA, 2011.
31. Qian, Y.; Wu, H.; Mi, S.; Zhao, W.; Zhou, D.; Lu, X. High-efficiency combustion of gasoline compression ignition (GCI) mode with medium-pressure injection of low-octane gasoline under wide engine load conditions. *Appl. Energy Combust. Sci.* **2023**, *15*, 100179. [CrossRef]
32. Shetty, S.; Shrinivasa Rao, B.R. In-cylinder pressure based combustion analysis of cycle-by-cycle variations in a dual spark plug SI engine using ethanol-gasoline blends as a fuel. *Mater. Today Proc.* **2022**, *52*, 780–786. [CrossRef]


**Disclaimer/Publisher's Note:** The statements, opinions and data contained in all publications are solely those of the individual author(s) and contributor(s) and not of MDPI and/or the editor(s). MDPI and/or the editor(s) disclaim responsibility for any injury to people or property resulting from any ideas, methods, instructions or products referred to in the content.

energies

*Article*

# A New Generation of Hydrogen-Fueled Hybrid Propulsion Systems for the Urban Mobility of the Future

**Ivan Arsie [1], Michele Battistoni [2], Pier Paolo Brancaleoni [3], Roberto Cipollone [4], Enrico Corti [3], Davide Di Battista [4], Federico Millo [5], Alessio Occhicone [1], Benedetta Peiretti Paradisi [5,*], Luciano Rolando [5,*] and Jacopo Zembi [2]**

[1] Department of Engineering, Università degli Studi di Napoli Parthenope, 80143 Naples, Italy; ivan.arsie@uniparthenope.it (I.A.); alessio.occhicone@uniparthenope.it (A.O.)
[2] Department of Engineering, Università degli Studi di Perugia, 06125 Perugia, Italy; michele.battistoni@unipg.it (M.B.); jacopo.zembi@unipg.it (J.Z.)
[3] Department of Industrial Engineering, Università degli Studi di Bologna, 40126 Bologna, Italy; pier.brancaleoni2@unibo.it (P.P.B.); enrico.corti2@unibo.it (E.C.)
[4] Department of Industrial and Information Engineering and Economics, Università dell'Aquila, 67100 L'Aquila, Italy; roberto.cipollone@univaq.it (R.C.); davide.dibattista@univaq.it (D.D.B.)
[5] Energy Department, Politecnico di Torino, 10129 Turin, Italy; federico.millo@polito.it
* Correspondence: benedetta.peiretti@polito.it (B.P.P.); luciano.rolando@polito.it (L.R.)

**Abstract:** The H2-ICE project aims at developing, through numerical simulation, a new generation of hybrid powertrains featuring a hydrogen-fueled Internal Combustion Engine (ICE) suitable for 12 m urban buses in order to provide a reliable and cost-effective solution for the abatement of both $CO_2$ and criteria pollutant emissions. The full exploitation of the potential of such a traction system requires a substantial enhancement of the state of the art since several issues have to be addressed. In particular, the choice of a more suitable fuel injection system and the control of the combustion process are extremely challenging. Firstly, a high-fidelity 3D-CFD model will be exploited to analyze the in-cylinder H2 fuel injection through supersonic flows. Then, after the optimization of the injection and combustion process, a 1D model of the whole engine system will be built and calibrated, allowing the identification of a "sweet spot" in the ultra-lean combustion region, characterized by extremely low NOx emissions and, at the same time, high combustion efficiencies. Moreover, to further enhance the engine efficiency well above 40%, different Waste Heat Recovery (WHR) systems will be carefully scrutinized, including both Organic Rankine Cycle (ORC)-based recovery units as well as electric turbo-compounding. A Selective Catalytic Reduction (SCR) aftertreatment system will be developed to further reduce NOx emissions to near-zero levels. Finally, a dedicated torque-based control strategy for the ICE coupled with the Energy Management Systems (EMSs) of the hybrid powertrain, both optimized by exploiting Vehicle-To-Everything (V2X) connection, allows targeting H2 consumption of 0.1 kg/km. Technologies developed in the H2-ICE project will enhance the know-how necessary to design and build engines and aftertreatment systems for the efficient exploitation of H2 as a fuel, as well as for their integration into hybrid powertrains.

**Keywords:** hydrogen; H2-ICE; decarbonization; public transport

**Citation:** Arsie, I.; Battistoni, M.; Brancaleoni, P.P.; Cipollone, R.; Corti, E.; Di Battista, D.; Millo, F.; Occhicone, A.; Peiretti Paradisi, B.; Rolando, L.; et al. A New Generation of Hydrogen-Fueled Hybrid Propulsion Systems for the Urban Mobility of the Future. *Energies* **2024**, *17*, 34. https://doi.org/10.3390/en17010034

Academic Editors: Davide Lanni and Enzo Galloni

Received: 16 November 2023
Revised: 9 December 2023
Accepted: 14 December 2023
Published: 20 December 2023

**Copyright:** © 2023 by the authors. Licensee MDPI, Basel, Switzerland. This article is an open access article distributed under the terms and conditions of the Creative Commons Attribution (CC BY) license (https://creativecommons.org/licenses/by/4.0/).

## 1. Introduction

In the context of decarbonization of the transportation sector, regulators are imposing increasingly stringent limitations in terms of $CO_2$ emissions, and the European Commission has declared its ambition to reach zero net Greenhouse Gas (GHG) emissions by 2050 [1]. As a matter of fact, 35% of the worldwide energy consumption is due to the transportation sector [2] and in 2020 road transport constituted 77% of all European Union (EU) transport GHG emissions [3]. In this sector, lorries, buses, and coaches are responsible for more than a quarter of GHG emissions in the EU (i.e., over 6% of total European GHG emissions) [4]. To mitigate this contribution, stronger $CO_2$ emission standards have been set by the EU for Heavy-Duty Vehicles (HDVs), including smaller trucks and urban or suburban buses. The proposal establishes a 90% decrease in $CO_2$ emissions per km from new HDVs by 2040 (compared to a mid-2019 to mid-2020 baseline) with intermediate targets for 2030 (45%) and 2035 (65%) [5].

In this framework, synergetic exploitation of decarbonized energy carriers and powertrain electrification could shift the public transportation fleets towards more sustainable mobility, compliant with current legislation targets. On the one hand, Battery Electric Vehicles (BEVs) may represent a promising alternative to improve the air quality of highly congested urban areas and to decarbonize the Light-Duty Vehicle (LDV) sector [6]. On the other hand, due to the limited specific energy of chemical batteries, BEVs are still not considered suitable for applications where the payload and the operative range are the main constraints (e.g., commercial vehicles or buses), requiring high initial costs and large batteries [7,8]. Among the possible alternatives for the HDV sector, the exploitation of hydrogen as an alternative fuel could play a key role in the EU energy transition [9], ensuring zero carbon-based emissions (e.g., CO, $CO_2$, and soot) and high energy efficiency. In addition, the production path of hydrogen could rely on a wide variety of available technologies and energy resources, many of which are renewables and/or have low carbon impact, strengthening the argument for hydrogen as a long-term sustainable energy vector to fuel the future [10]. In the context of the public urban transportation sector, hydrogen has already been exploited in Fuel Cell Electric Vehicles (FCEVs), demonstrating significant advantages with respect to BEVs, both in terms of weight reduction and range increase, with a shorter refueling time [11,12]. Moreover, using hydrogen as a fuel to power Internal Combustion Engines (ICEs) could add further benefits with respect to Fuel Cell (FC) technology, such as a great tolerance to fuel contamination, significant cost savings (up to one order of magnitude), and the possible exploitation of existing production lines, due to familiar technologies and parts to current Internal Combustion Engine Vehicles (ICEVs) [13]. Furthermore, one of the main concerns regarding hydrogen vehicles is the need for new infrastructures for hydrogen distribution and refueling, while in the case of urban buses, a quite limited range is typically requested, and tank refueling is not an issue, since captive fleets can centralize the fuel station.

Despite the great potential of H2 as a fuel for ICEs in public transport, in the last two decades, it has been implemented in only a few prototypes [14,15] and real bus fleets, such as in the two demonstration projects HyFLEET:CUTE [16] and SunLine Cleen Fleet [17]. In particular, the HyFLEET project involved the operation of 47 hydrogen-powered buses in regular public transport service in 10 cities on three continents, from 2006 to 2009, testing on 15 buses two different types of H2-ICEs (naturally aspirated or turbocharged). In addition, SunLine Cleen Fleet tested a prototype hybrid H2-ICE bus in California for three years (2006–2008). The outcomes of both projects have shown that hydrogen can be a low or even emission-free transport fuel, implementing today's technology. Nevertheless, some issues remain unresolved, such as the performance of the fuel injectors, the turbocharged operation, and the control of the engine management system. Moreover, a rising interest has been growing recently for the development of a new generation of ICEs designed to be operated at high boost levels with ultra-lean mixtures, thus simultaneously achieving extremely low NOx emissions (<0.05 g/kWh) along with high-efficiency values (>42%) at lambda > 2 [18–20]. A further enhancement could be obtained through the synergic combination with hybrid propulsion technologies, particularly suitable for urban bus applications, for which series hybrid architecture could significantly benefit from energy recovery through regenerative braking, as well as from a substantial downsizing of the ICE, with vehicle acceleration performance being decoupled from engine power output [21].

In this context, the present research project, the so-called H2-ICE project, aims to develop an innovative hybrid H2-fueled powertrain for urban bus application, giving insights into the current knowledge gap for H2 exploitation in ICEs. Specifically, the approach adopted for the H2-ICE project consists, at first, of defining the use case scenarios for the targeted application (i.e., 12 mt urban bus) and emission and fuel consumption targets to be met. This essential preliminary step allows the specification of the boundary conditions and the requirements delineating the development scope of the main vehicle subsystems. Then, in a second step, the H2 engine and its subsystems will be developed and optimized, starting from a state-of-the-art diesel engine. Afterwards, in a third step, the engine and its subsystems are tested and modeled to be integrated into a virtual high-fidelity simulation model and the energy management system will be developed exploiting Vehicle-To-Everything (V2X) connectivity. Then, the virtual demo vehicle will be evaluated to demonstrate its capabilities in real-world driving conditions. The present work is focused on the development phase of the engine and its subsystems and is organized as follows: Section 2 provides an overview of the project, with the case study description, a discussion of the challenging issues to be addressed, the overall fixed targets, and the hybrid powertrain technical specifications. Furthermore, Section 3 describes the analysis of the suitable combustion and fuel injection system, Section 4 the WHR system together with the aftertreatment technologies needed, and Section 5 the engine and powertrain management strategy. Finally, the actual findings and future steps of the project are presented in Section 6.

## 2. H2-ICE Project Overview

The H2-ICE project aims to enhance the know-how necessary to design and build engine and aftertreatment systems for the efficient exploitation of H2 as a fuel and for their integration into a hybrid powertrain for a series urban bus. In particular, the full exploitation of the potential of such a powertrain requires a substantial enhancement of the state of the art, since several challenging issues have to be addressed. At first, the combination of high efficiency, low emissions, high specific power output, and durability is not currently achievable in one H2 combustion concept. Starting from fuel injection, Port Fuel Injection (PFI) and Direct Injection (DI) are the two main considered and tested options [22,23]. On the one hand, PFI part load efficiency can be quite high with ultra-low NOx emissions, thanks to the possibility of ultra-lean operation allowed by the extremely high flame propagation and wide flammability limit characteristics of H2 combustion. However, the specific work output is typically limited to 14 bar Brake Mean Effective Pressure (BMEP) due to the low H2 density, even in highly boosted engines. On the other hand, DI engines have proved to be capable of achieving power densities 15% higher than PFI, while offering high efficiencies thanks to possible charge stratification at part load, with a lower risk of abnormal combustion [24]. Nevertheless, DI injectors still have to prove their durability as well as their operability with extra-high flow rates, typical of gaseous fuel such as H2. Moreover, the exploitation of fully optimized fuel injection strategies requires high injection pressures, typically above 100 bar, limiting H2 storage options [25]. Indeed, liquid H2 could be stored onboard in cryogenic tanks at a pressure of 16 bar, providing injection pressures onboard, or high-pressure tanks could be exploited to store gaseous hydrogen onboard with a working pressure generally ranging from 350 to 700 bar, but then the full tank capacity may not be utilized. As far as H2 combustion is concerned, the same properties that make H2 a suitable fuel for ICEs can be responsible for high risks of abnormal combustion phenomena: the wide flammability limits, the low ignition energy, and the high flame speeds can, as a matter of fact, result in combustion anomalies such as surface ignition, backfiring, preignition, and knock [26]. Although significant progress was made in improving computational tools for modeling and understanding mixture formation and combustion in H2 engines [27,28], the numerical simulation of mixture formation for DI H2 engines remains extremely challenging, because of the high-velocity gas jet, leading to a complex structure of shock waves and making charge stratification quite hard to achieve. Moreover, as far as combustion modeling is concerned, there is a shortage of data on H2 ignition and combustion properties in terms of engine conditions of pressure and temperature, and there has been little progress in extending the current dataset. Furthermore, both advanced WHR systems [29] and an innovative Selective Catalytic Reduction (SCR) aftertreatment system, based on hydrogen as the reductant element [30], need to be synergically developed to exploit the high exhaust gas enthalpy available and achieve near-zero tailpipe NOx. In addition, the use of H2 as a reducing fluid needs proper catalyst preparation, testing, and modeling to be then coupled to the engine. Finally, the elaboration of control strategies for the synergic operation of the developed WHR and H2-SCR aftertreatment systems and the exploitation of look-ahead information from V2X connectivity for the Energy Management System (EMS) of hybrid powertrains are just at the beginning of its exploitation [31]. The H2-ICE project thus aims not only to develop an innovative hybrid H2-fueled powertrain and related technological opportunities but also to contribute to filling the knowledge gap for H2 exploitation in ICEs, as summarized in Figure 1.

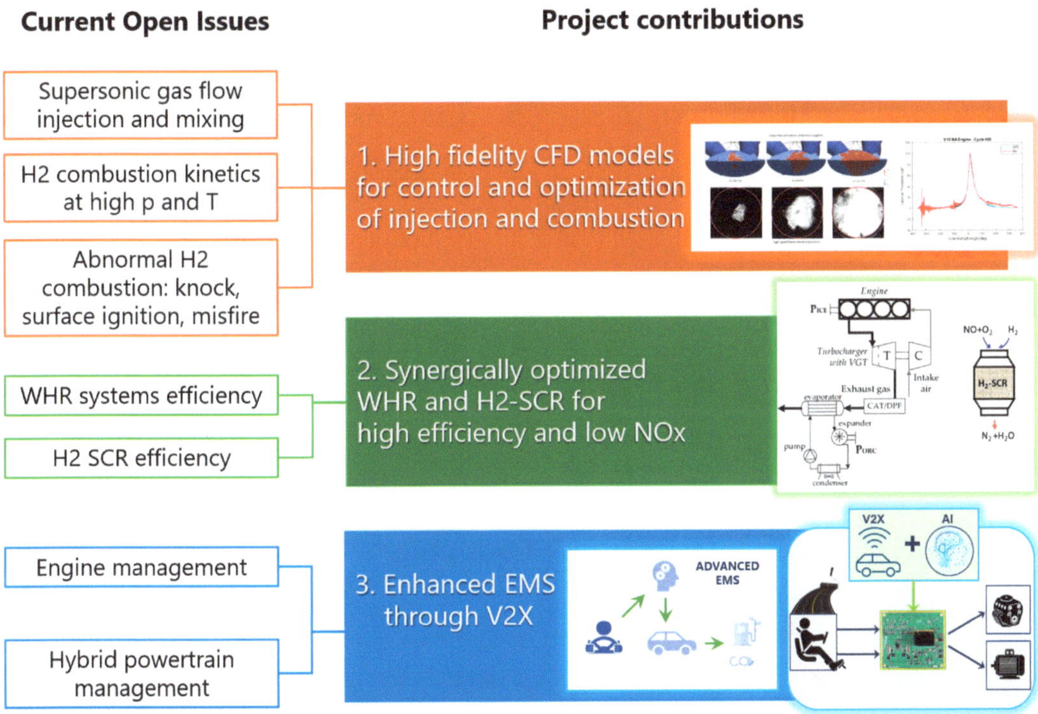


**Figure 1.** H2-ICE project state-of-the-art enhancements.

*Case Study*

First of all, the H2-ICE project defines the use case scenarios for the targeted application and emission and fuel consumption targets to be met, allowing the definition of the boundary conditions and the requirements delineating the development scope of the main vehicle subsystems. The vehicle developed is a 12 mt urban bus with a series hybrid architecture, featuring an H2-ICE derived from a state-of-the-art diesel engine and a tractive electric motor as power units. A scheme of the hybrid architectures is presented in Figure 2 and the vehicle technical specifications and target settings are then listed in Table 1.

In addition, Table 2 presents the overall targets of the project.

Both regulatory driving cycles and real-world mission profiles will be taken into account in the design of the hybrid traction system. In particular, three different mission profiles, representative of a typical urban bus operation, will be considered. The speed profiles are represented in Figure 3 and their specifications are listed in Table 3.

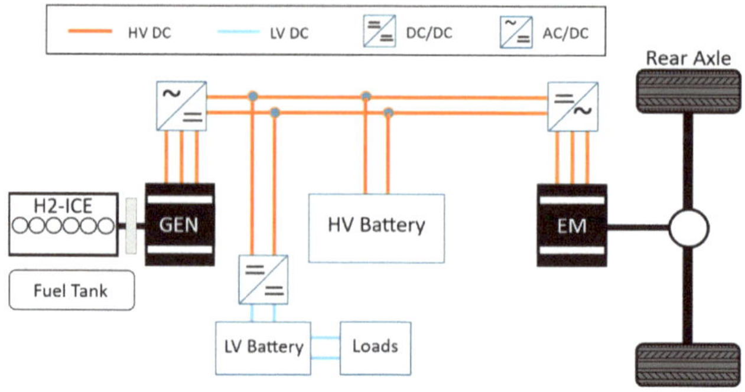


**Figure 2.** Scheme of the adopted series hybrid architecture.

**Table 1.** Hybrid bus specifications and target settings.

| | |
|---|---|
| **Vehicle length** | 12 m |
| **Curb weight** | 12 ton |
| **Fully loaded weight** | 18 ton |
| **Passenger capacity** | 90 |
| **Road load at 50 km/h** | 16 kW |
| **Road load at 80 km/h** | 43 kW |
| **Road load at 100 km/h** | 74 kW |
| **Rolling radius** | 0.54 m |

**Table 2.** H2-ICE targets for vehicle and engine efficiency and NOx emissions.

| | | |
|---|---|---|
| **Vehicle performance** | **Max speed** | 65 km/h |
| | **Min acceleration** | 1.1 m/s$^2$ |
| | **Max gradeability @ full load** | 14% |
| **Vehicle and engine efficiency** | **Specific power** | 40 kW/dm$^3$ |
| | **ICE efficiency** | 42% peak efficiency |
| | | 35% part load efficiency |
| | **Fuel consumption** | 0.1 kg/km |
| **Emissions** | **NOx tailpipe** | <0.05 g/kWh |
| **Auxiliaries** | **WHR system efficiency** | >4% |
| | **eTurbo efficiency** | >4% |

In particular, the Braunschweig is an urban driving cycle, employed in various research projects or equipment certification programs [32]. The MLTB cycle was developed by UK transport authorities in 1996 to verify the compliance of new vehicles with emissions and fuel economy standards [33]. Finally, the Gillingham Uphill is a real-world emission profile, obtained from GPS acquisitions, and features high elevation changes to also test challenging driving conditions [19].

The baseline reference engine is a diesel engine with 3.0 L of displacement, downsized to a rated power of 100 kW and modified in order to be exploited for H2 combustion, with a proper benchmark between the cost and the level of complexity [34]. Technical specifications of the different components have been set exploiting the outcomes of previous EU-funded projects such as ZeEUS [35] and based on an extensive literature review and market survey. Concerning the choice of the suitable Electric Motor (EM), at present almost the entire light-duty hybrid vehicle industry has shifted to Permanent Magnet (PM) machines in order to meet the increasing power density and efficiency requirements [36]. On the contrary, for medium- and heavy-duty vehicles induction machines remain an attractive option for their very low drag losses at part load. Nevertheless, the specific market of electrified buses is shifting towards PM electric machines with an average peak nominal power of 200 kW [35]. Finally, a PM synchronous machine has been chosen thanks to its high energy efficiency, delivering a peak power of 200 kW and continuous power of 140 kW. As far as the battery is concerned, at first, the battery sizing and energy content have been fixed, according to the route length and the final desired State of Charge (SoC) [37]. A total installed energy of 20 kWh is considered the most suitable for the present application, with a target maximum power of 200 kW, in consistency with the EM power. Regarding battery chemistry, a high-specific-power cell is desirable for a full hybrid series application [38]. In this field, the most exploited chemistries are the ones available with Li-ion batteries, granting high power densities and long life cycles, necessary for long traveled distances [39]. Among the different options, Lithium-Ferro-Phosphate (LFP) offers a longer life cycle, ensuring lower replacements, with a comparable carbon footprint of battery stack production with respect to other chemistries such as lithium Nickel Manganese Cobalt Oxide (NMC) batteries. In conclusion, the chosen cell is made of LFP, and the main characteristics are reported from [40], with 120 cells in series and 20 in parallel to achieve the target power and density. The main powertrain components' technical specifications are summarized in Table 4.

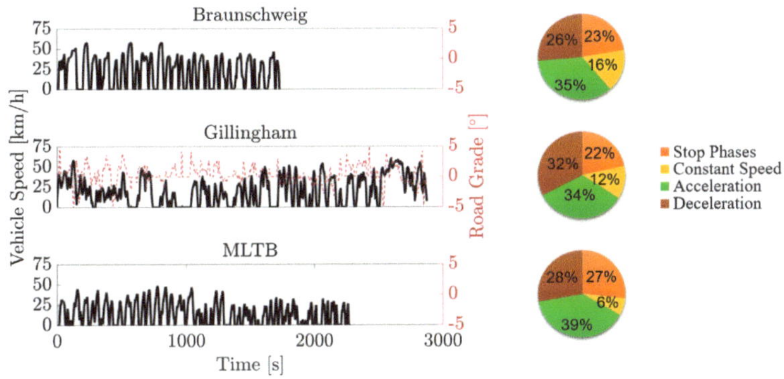


**Figure 3.** Vehicle speed of the considered driving cycles (black lines), and road grade when present (red line).

**Table 3.** Main Features of the considered mission profiles.

| | Braunschweig | Gillingham | MLTB |
|---|---|---|---|
| **Duration [s]** | 1740 | 2875 | 2281 |
| **Distance [km]** | 10.9 | 16.6 | 9.0 |
| **Average speed [km/h]** | 22.5 | 20.8 | 14.2 |
| **Maximum speed [km/h]** | 58.2 | 59.9 | 48.7 |
| **Average acceleration [m/s$^2$]** | 0.2 | 0.2 | 0.2 |
| **Maximum acceleration [m/s$^2$]** | 2.4 | 2.3 | 1.5 |
| **Specific energy demand [kwh/km]** | 0.90 | 1.05 | 0.94 |

**Table 4.** Hybrid powertrain technical specifications.

| | | |
|---|---|---|
| **ICE** | **Displacement** | 3.0 L |
| | **Bore × stroke** | 83 × 90 mm |
| | **Features** | Single stage Turbocharger w/VGT |
| | **Max. power** | 100 kW |
| **EM** | **Max. power** | 200 kW |
| | **Max. torque** | 1500 Nm |

Table 4. *Cont.*

| | | |
|---|---|---|
| Battery | **Capacity** | 19.8 kWh |
| | **Nominal voltage** | 396 V |
| | **Maximum current** | 2400 A |
| | **Maximum power** | 950 kW |
| | **Cell in series** | 120 |
| | **Cell in parallel** | 20 |

## 3. Combustion and Fuel Injection System

The present section describes the development and optimization of the fuel injection and combustion systems suitable for the H2-fueled internal combustion engine, through both numerical simulation and dedicated experiments in an optically accessible laboratory single-cylinder engine. In particular, the control of the fuel injection and the combustion process for a hydrogen engine is extremely challenging, due to the high risk of backfire in the case of port fuel injection and to the complexity of managing mixture formation through the supersonic flow of the gaseous fuel in the case of direct injection. Then, the development of high-fidelity CFD models to highlight the in-cylinder H2 fuel injection through supersonic flows and to describe accurately the hydrogen combustion, including anomalies such as knock, preignition, and misfire is fundamental to combine high efficiency, low emissions, high specific power output, and durability all in one concept.

### 3.1. Injection

Regarding the choice of the fuel injection system, different options have been analyzed, ranging from PFI to DI, as mentioned in Section 2. Indeed, to achieve high power output and efficiency, hydrogen DI is preferable, however, the low hydrogen density leads to supersonic chocked flows, with high penetration and reduced in-cylinder mixing times, despite hydrogen's extremely high laminar diffusivity. Injection pressure levels and nozzle design therefore need to be carefully scrutinized. In this context, high-fidelity Large Eddy Simulations (LESs) of the turbulent under-expanded H2 jets can play a crucial role in the fundamental understanding and prediction of the physical mixing process, allowing the design of an optimized injector nozzle and associated injection strategies. In addition to high-resolution CFD simulations, the Reynolds-Averaged Navier–Stokes (RANS) modeling approach could also be employed as a cost-effective engineering tool to predict and explore several nozzle designs and pressure ratios [41–44]. Boundary and operating conditions are derived from reduced-order models (cf. Sections 2 and 5).

As a first step, LES fundamental studies have been conducted on nitrogen jets for validation of the developed injection model, and predictions [41] have been compared with experimental data [44,45]. An example is shown in Figure 4, where the contours of the density gradient of the nitrogen jet are presented at different time steps, with 5.8 bar of Nozzle Pressure Ratio (NPR) and 1 bar of ambient pressure, and the predicted axial density is compared for validation against the two reference studies. Normalized centerline density was predicted relatively well, with a small phase difference after the Mach disk, also due to uncertainties in the actual nozzle conical shape [44]. The formation of the barrel-shocks and slip lines and the rolling up of the initial vortices are well visible.

By increasing the NPR up to 30, and considering actual hydrogen injections, the jet morphology changes substantially, as visible in Figure 5 (left). The higher NPR generates a larger barrel-shock, with only one shock cell. Immediately afterwards, the flow transitions to turbulent flow, which eventually reduces the tip penetration. The expected main characteristics of the jets are captured by the CFD model, such as the turbulent mixing and the emitted aerodynamic sound. Simulations have also been performed on pintle outward opening nozzles, with similar considerations. Less demanding RANS studies have then been conducted to explore a wide range of injection pressure and backpressure operating conditions, including the effect of nozzle shape, like inward vs. outward opening (Figure 5, right). In this modeling framework, even if capturing H2 far-field mixing remains challenging, RANS results provide an adequate description of the flow physics for engine CFD models allowing the screening of multiple design parameters at an affordable cost.

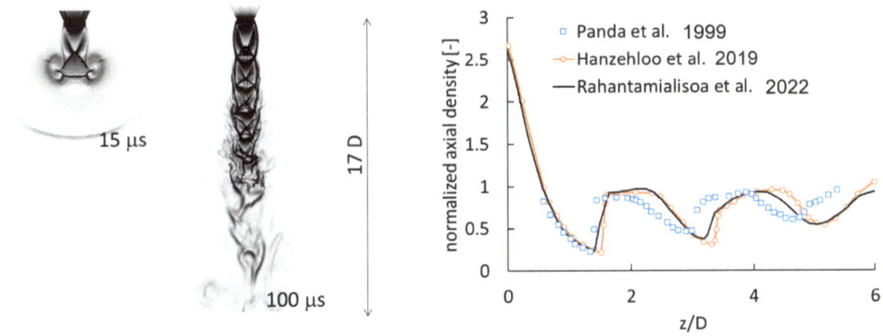


**Figure 4.** LES of nitrogen injection at 5.8 bar into 1 bar ambient backpressure, through a 1 mm diameter nozzle. Simulation results of the jet development in terms of density gradient over time (**left**). Normalized axial density vs. axial distance (**right**), for the study reported in [41] and two reference studies [44,45].

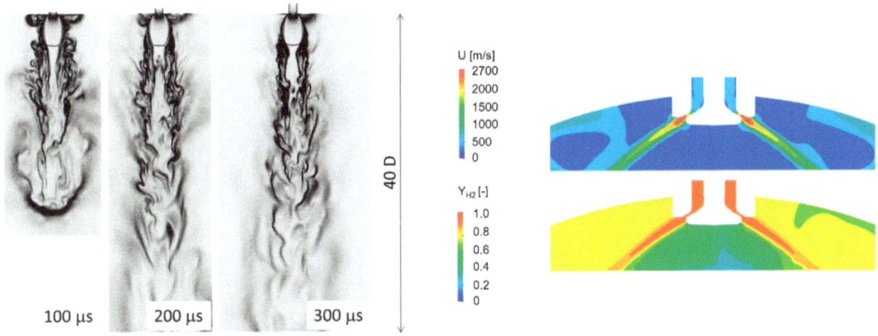


**Figure 5.** H2 injection simulations at 30 bar into 1 bar. LES simulation of a single-hole nozzle; predicted contours of density gradient over time (**left**) [41]. RANS simulation of a pintle injector; predicted velocity magnitude and fuel mass fraction (**right**).

### *3.2. Mixture Formation*

Still, limited experimental work is available in the literature dealing with H2 jet evolution, mixing, and combustion. To build a reliable CFD design workflow, quantitative measurements of H2 injection and combustion are needed for validation. Regarding hydrogen mixing in a motored engine, the most comprehensive dataset is available from Sandia's Optical Hydrogen Engine (SOpHy) [46]. Indeed, the CFD modeling capabilities have been tested against velocity and hydrogen mass-fraction fields measured in these experiments. The CFD software CONVERGE, version 3.0, is used for the analyses, with a second-order scheme for convective flux reconstruction and the implicit Euler scheme for temporal discretization. The considered engine features a central mounter single-hole injector, whose mesh resolution, geometry details, and boundary types were designed based on previous activities [47,48]. For the specified motored engine operation [49], various RANS turbulence models within the k-ε family, grid managements, wall treatments, and turbulent Schmidt numbers have been scrutinized.

The realizable k-ε model performed relatively better than others, and its corresponding results are shown in Figure 6 (left) in terms of mixture distribution. In the central column, the experimentally measured H2 distribution is presented at two different time steps for the optical accessible window, while in the right column, the simulation results are visible. Comparing the direction and spread of the hydrogen jet, especially at −130 deg., the numerical model can correctly predict the air/gas during the intake stroke, while a slight under-prediction of the mixing is visible at −30 deg. In Figure 6 (right), the results of the turbulent Schmidt number variation are presented in the velocity profile along the cylinder axis. Calibrating the Schmidt number, the hydrogen diffusion predictions are improved to some extent at the end of the compression stroke. Future project analyses regarding the fuel injection system will include the test of other turbulence models, such as low-Reynolds variants, or Reynolds stress model, still within the RANS framework for computationally efficiency.

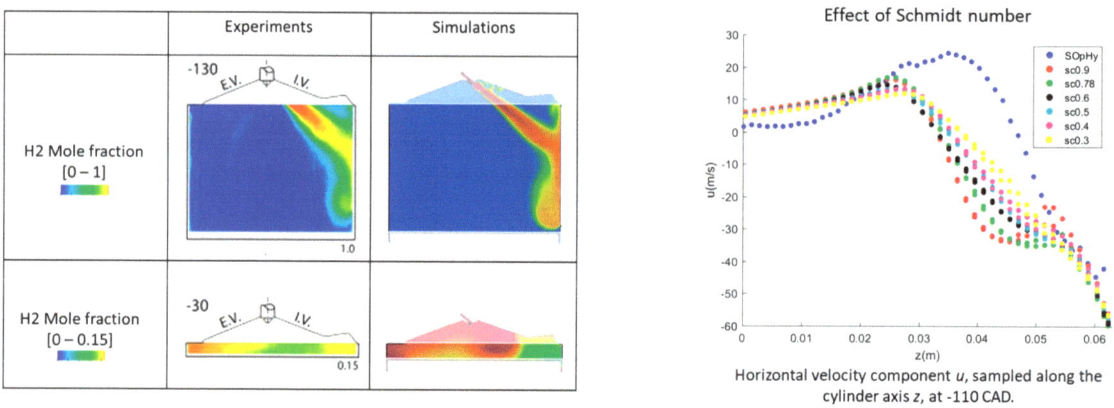


**Figure 6.** Hydrogen mixing in the motored engine SOpHy [46,49] experiments vs. simulation. Hydrogen mole fraction at two different timings (**left**), and velocity profile varying the Schmidt number (**right**).

### 3.3. Full Engine Model with Combustion

In the third step of the analysis, combustion conditions are considered, with numerical and experimental analyses on the single-cylinder naturally aspirated optical-access engine of the University of Perugia which allows combustion visualization through the transparent piston and side windows (see Table 5 for more engine details).

**Table 5.** Optical-access engine specifications.

| | |
|---|---|
| **Displacement** | 500 cc |
| **Bore × stroke** | 85 × 88 mm |
| **N. of valves** | 4 |
| **Chamber type** | Pent roof |

At the time of writing, tests are ongoing and experimental data are not yet available. All the results of the combustion measurements will serve as a validation data for both LES and RANS models to build a reliable and validated CFD model. The selected engine point is at 1000 rpm with 5 bar BMEP. In particular, various excess air ratios are considered (i.e., $\lambda = 2 \div 3.5$) with two main injection categories:

- premixed cases, meant to reproduce the behavior of a PFI engine assuming that hydrogen and air are perfectly mixed (limiting case);
- DI cases with side-mounted injector and central spark, at various Start of Injection (SoI) times.

Simulations have been carried out using the same software, CONVERGE, as detailed above. An assessment of the CFD model has been carried out via mesh sensitivity analysis. Furthermore, a validation of the laminar flame speed has been carried out against literature data, which led to the choice of the LLNL reaction mechanism developed by O'Conaire et al. [50]. Combustion is modeled using the well-mixed approach, and the energy deposition method is adopted to model ignition leveraging previous experiences and data [51]. Figure 7 presents a comprehensive summary of the combustion numerical analysis, with the mass fraction burned (left, center) at different values of lambda and SoI and the combustion efficiency (right). The port fuel injection cases (black lines) mimicked by a premixed assumption show the maximum combustion efficiency at all lambda values (Figure 7, right). Of course, volumetric efficiency will be lower, and a higher intake pressure will be needed to maintain the same load. With DI strategies, mixing tends to improve advancing SoI. At $\lambda = 3$, the fuel mass fraction burned was lower than the premixed case and SoI had no significant effect, while at $\lambda = 3.5$ SoI had a visible effect on the fuel mass fraction burned which increases as the SoI advances. The combustion efficiency at $\lambda = 3.5$ is very poor, and such a level of air–fuel ratio cannot be used, at least without further countermeasures.

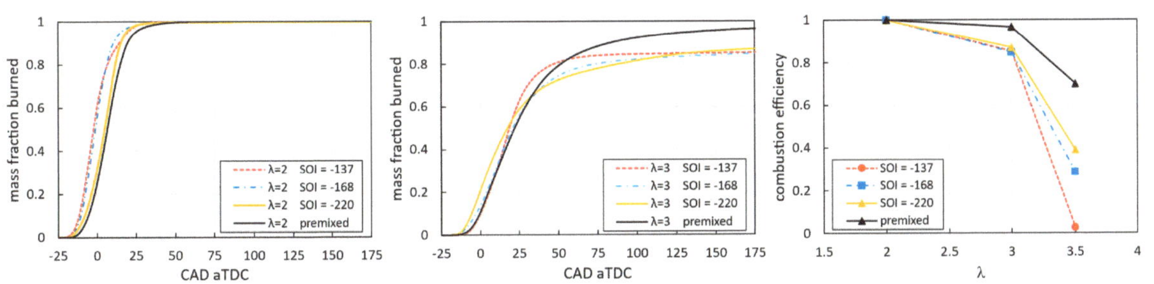


**Figure 7.** Effect of SoI timing on DI engine combustion at various air excess ratios. Mass fraction burned at $\lambda = 2$ (**left**) and $\lambda = 3$ (**center**). Combustion efficiency as a function of $\lambda$ (**right**).

The main task of this work will then be the metal engine design optimization carried out through thermo-fluid-dynamic simulations of the engine combustion system. RANS simulations can guide injection and combustion systems towards the achievement of targets specified in the previous section by tuning the injection system design with the piston and chamber geometry design. As a result of the 3D-CFD models, heat release rates can be derived as a function of the combustion system and operating conditions. These data can feed the 1D engine model developed in GT-Power which can be used for the overall engine performance assessment, such as generating engine maps of efficiency, performance, and emissions. Based on such calibration and validation steps, the 1D engine model will then be used to explore several additional parameters, such as injection strategies, ignition timings, lean stability, Exhaust Gas Recirculation (EGR) rate, and boost level, under various speed–load conditions. A crucial aspect will also be the prediction of abnormal combustions, due to preignition and end-gas autoignition, which will rely on calibrated combustion models against preliminary experimental tests on a prototype.

## 4. Waste Heat Recovery Technologies and Aftertreatment System

In parallel with the activities of engine design, the engine subsystems have been developed and optimized through both experimental and simulation activities. Different WHR technologies have been scrutinized, while for the aftertreatment system, an H2-SCR system has been chosen to achieve near-zero NOx emission levels, and the analysis is focused on the experimental characterization of the possible catalytic system to be implemented in this device.

## 4.1. WHR Technologies

The evaluation of the WHR has been preliminarily assessed on the reference engine: a 3.0 L turbocharged diesel one. The possibility to test a baseline-like engine mounted on the test bench at the ICE laboratory of the University of L'Aquila provides engine exhaust data, fundamental to evaluate the energy recovery potential. In particular, the temperature of the exhaust gases $T_{exh}$ is measured after the catalyst and particulate filter devices in order to preserve their operating conditions. Exhaust mass flow rate $m_{exh}$ is calculated by knowing the engine air intake and the fuel rate [52]. In the turbocharger diesel engine tested on the bench, both engine speed and torque have been varied in order to have a wide and comprehensive range of operating points. Engine power has been chosen as a reference parameter to hold engine speed and torque (Figure 8). Exhaust mass flow rate and temperature are almost linear with the engine power, since the control unit of the engine acts simultaneously on the boost pressure produced by the turbocharger and the fuel injection to achieve the desired power (torque and speed).

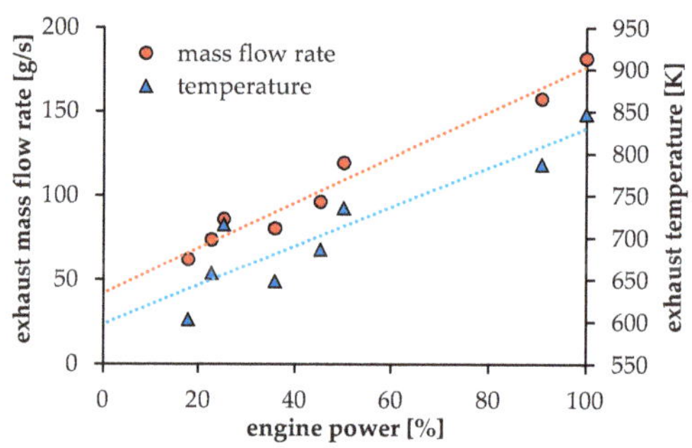


**Figure 8.** Reference data for engine exhaust mass flow rate and temperature as a function of the engine power, evaluated downstream of the aftertreatment devices. Red dots represent experimental values of engine exhaust mass flow rate, while blue triangles experimental data of exhaust temperature. Red and blue dashed lines are, respectively, linear fittings of data.

The turbocharger is managed to obtain a certain intake air flow rate, and fuel injection is more related to torque control. The resulting air/fuel ratio can be considered as a parameter for controlling the engine power. Moreover, the engine is operated in lean combustion conditions as highlighted in Figure 9, where the value of lambda is reported. This value ranges from 1.4 (in highest power condition) to 2.4 (at 20% of rated power). In Figure 8 is also reported the corresponding exhaust gas temperature.

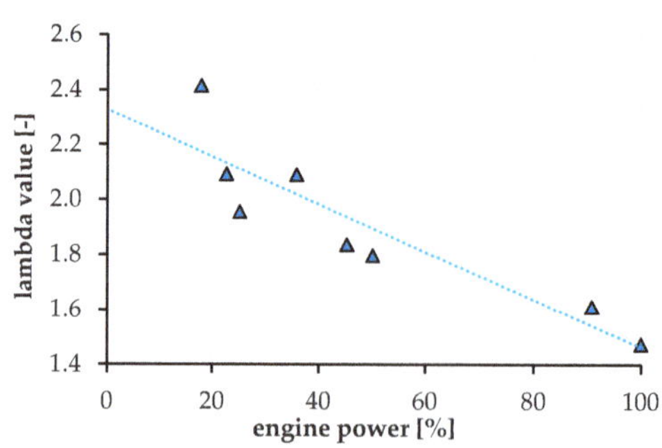


**Figure 9.** Values of lambda for different engine powers reproduced on test bench, triangles for observed values, dotted line to fit the data.

#### 4.1.1. ORC-Based Unit

The first option studied for WHR is the possibility to introduce an indirect waste heat recovery section, based on an Organic Rankine Cycle (ORC) unit. This is composed of an evaporator, where the energy of the exhaust gases can be exchanged with the organic fluid to vaporize and superheat it, an expander, which converts the thermo-dynamic energy of the fluid into mechanical and eventually electrical energy, a condenser, which exchanges heat towards the cold sink, and a pump to pressurize the fluid and close the thermodynamic cycle (Figure 10). The ORC-based unit proposed has a volumetric machine for the expander (of scroll type), suitable for its flexibility and capability to adapt in dynamic and off-design conditions [53,54], and it has been tested specifically to derive a numerical model of the whole ORC unit [55]. The model has been validated through experimental data [56] and is based on general assumptions: (a) the organic fluid is R245fa; (b) the cold sink is cooling water at 45 °C, to simulate the cooling conditions onboard; (c) the overall amount of fluid is equal to 7 kg; (d) pump and expander rotational speeds are controlled to match the thermal power available and to improve the overall performance considering the permeability of the ORC circuit.

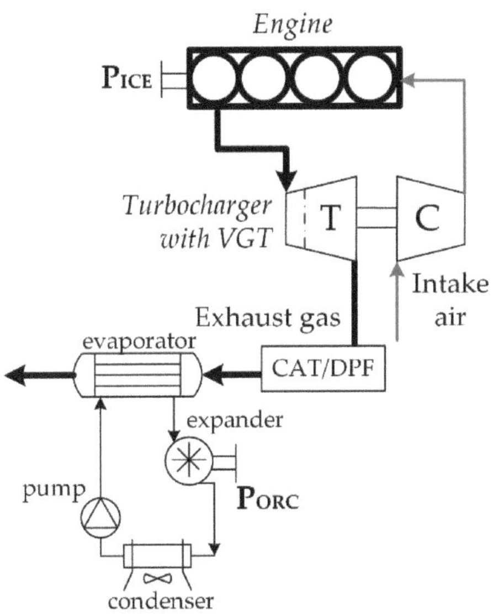


**Figure 10.** Scheme of the ORC-based unit proposed for exhaust heat recovery.

The control of the pump speed and the contextual one of the expander are needed to have the right mass flow rate of working fluid which matches the most suitable conditions at the evaporator. The evaporating pressure, moreover, is strictly related to the permeability of the circuit, mainly dependent on the WF flow rate and consequently from the entity of the thermal power recovered (i.e., the exhaust gas temperature entering the evaporator). Figure 11 shows the resulting values, which range from 10 to 25 bar, almost linearly with the exhaust temperature. It is just a point to notice that the maximum pressure achieved is not so far from the critical point of the R245fa fluid and also represents a technical limitation for safety and technology of the components and materials [57].

The evaporating pressure is also the parameter that sets the ORC efficiency since the condensing pressure is almost fixed at the value of 4 bar, related to the environmental conditions of the cold sink onboard. Figure 12 (left) shows the values of the ORC efficiency, evaluated according to the Equation (1), and ranging from 4% to 9.5%, at the maximum pressure:

$$\eta_{ORC} = \frac{P_{ORC}}{P_{exh}} = \frac{P_{expander} - P_{pump}}{m_{exh} c_{p,exh} (T_{exh} - T_{exh,out})} \tag{1}$$

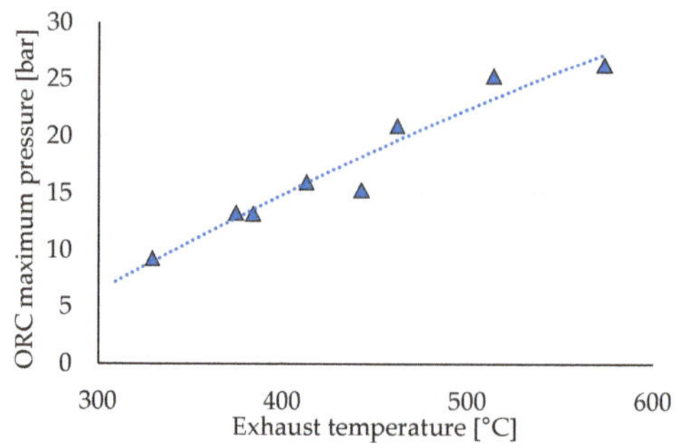


**Figure 11.** ORC evaporating pressure and exhaust temperature, triangles for observed values, dotted line to represent the fit the data.

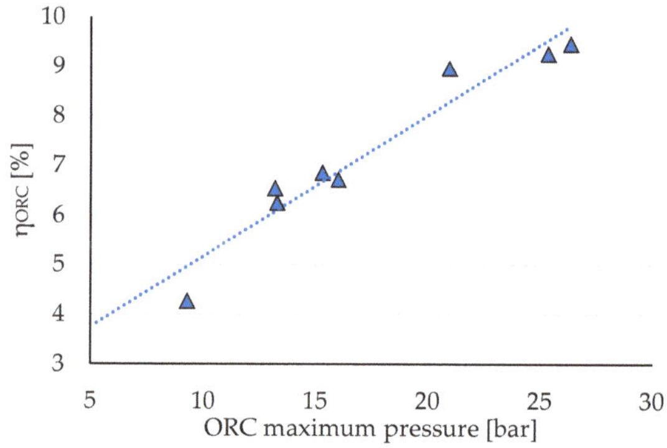


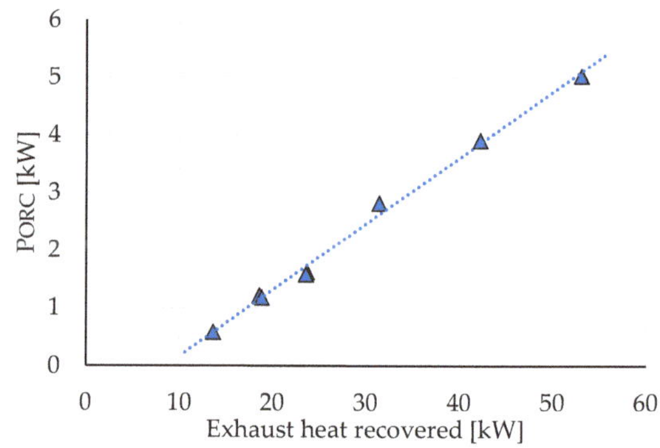


**Figure 12.** (**Left**) ORC efficiency vs. evaporating pressure; (**right**) ORC net power recovered, triangles for observed values, dotted line to fit the data.

Finally, the net ORC power $P_{ORC}$ is evaluated and plotted vs. the exhaust thermal power recovered from the engine in Figure 12, right. It is up to 5 kW in higher engine power conditions, where the recovered thermal power $P_{exh}$ is equal to 53 kW.

#### 4.1.2. Turbocompound

The second option investigated is a direct recovery, where the same exhaust gases are expanded in an additional turbine, placed downstream of the main turbine of the turbocharger (Figure 13). The possible recovery has been evaluated thanks to a 0D model, where the recovery turbine $T_{aux}$ is represented by a characteristic map (mass flow rate vs. pressure ratio vs. efficiency vs. rotational speed), derived from an existing one and adapted to the specific case study.

The first result is the pressure ratio of the turbocompound $\beta_{TC}$, which is related to the pressure admissible on the exhaust line since the outlet pressure is close to the environmental one. The growth in turbocompound pressure ratio increases the engine backpressure, although the presence of a Variable Geometry Turbocharger (VGT) mitigates this value on the exhaust manifold of the engine [58,59] and it is considered in the net power estimation ($P_{TC,\,net}$). Figure 14 shows the turbocompound pressure ratio, which is linearly dependent on the exhaust temperature $T_{exh}$ (i.e., turbocompound inlet temperature) and ranges from 1.2 to 1.6. For engine exhaust temperatures lower than 350 °C, the available thermal power is not sufficient to drive the turbocompound, and the resulting pressure ratio is too low.

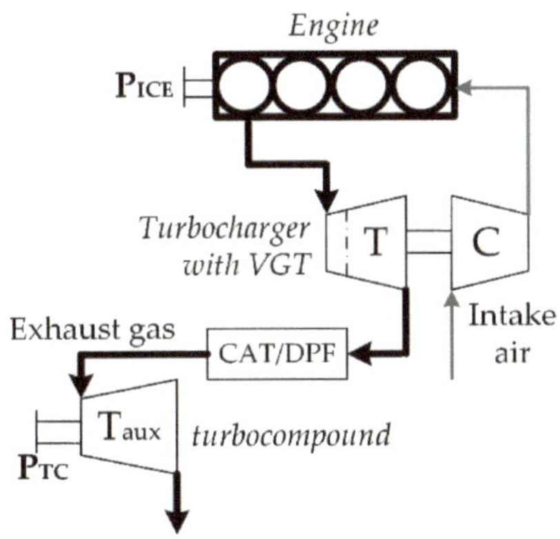


**Figure 13.** Layout of the turbocompound under the engine exhaust.

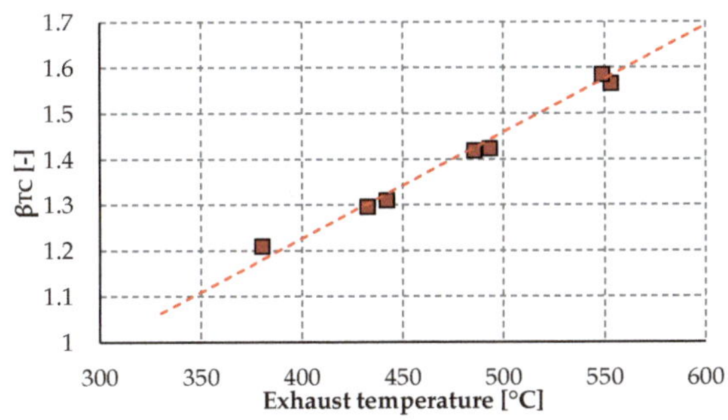


**Figure 14.** Turbocompound pressure ratio and exhaust temperature (i.e., turbocompound inlet temperature), squares for observed values, dotted line to fit the data.

The operation of the turbocompound is evaluated thanks to a characteristic map of the additional turbine, where the efficiency $\eta_{TC}$ is a function of the specific working point in terms of mass flow rate and pressure ratio. In Figure 15 (left), the efficiency is plotted as a function of the exhaust mass flow rate $m_{exh}$: the values range from 55% to 68%, and this provides a possible optimization of the design of the machine for the specific application.

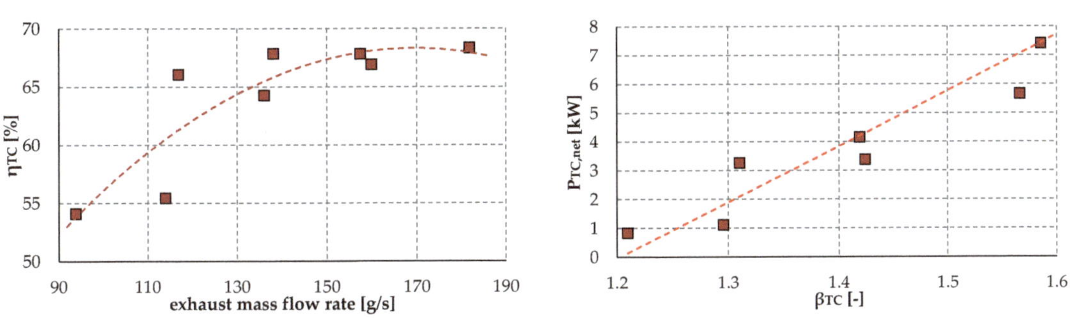


**Figure 15.** (**Left**) Turbocompound efficiency vs. exhaust mass flow rate crossing the auxiliary turbine. (**Right**) Final net power recovered by the turbocompound vs. pressure ratio of the turbine, squares for observed values, dotted line to fit the data.

Finally, the net turbocompound power has been reported in Figure 15 (right). The values are up to 8 kW and linearly dependent on the pressure ratio across the auxiliary turbine. As already stated, this dependence is limited by the backpressure effect on the engine: if the pressure ratio on the turbine is higher, the side effect on the engine can overcome the increase in turbocompound power and nullify the recovery.

#### 4.1.3. Engine Side Effects

The introduction of the WHR section on the exhaust gas line will produce an unavoidable engine backpressure increase. This could be more significant for the turbocompound specifically if it is placed downstream of the turbocharger (series turbocompound). However, if a parallel turbocompound is also considered, the rearrangement of the equilibrium of the turbocharger, to meet the request of the boost pressure and leave room for possible recovery, could produce a backpressure increase in the exhaust manifold. In this case, the possibility of introducing two sections of waste heat recovery, combining direct and indirect ways, can significantly increase the recovery and better exploit the available thermal energy from the exhausts [60].

Considering the ORC-based unit, the backpressure increase is strongly related to the permeability of the heat exchanger used as an evaporator on the gas side. In this case, the technology chosen for this component is crucial. Finned tube heat exchangers seem to have the best performances in this regard [61] and, if these pressure drops are limited (e.g., <100 mbar), the negative effect of the engine can be neglected. However, the introduction of the ORC unit is more invasive than the turbocompound and requires more space onboard. Moreover, it introduces a non-negligible extra weight, whose influence on the propulsive power should be estimated. But, considering the application of the H2-ICE project (i.e., 12 m bus), this extra weight may not be so significant if related to the overall weight of the vehicle and its payload.

### *4.2. SCR System*

Although the H2-fueled engine is intended to operate with an extremely lean mixture, thus producing low NOx emissions, the target of the hybrid bus application is to reach almost zero emissions. Therefore, $NO_X$ abatement by the aftertreatment system is still mandatory and is at present based on SCR [62]. Currently, the SCR system is based on a technology that consists of reductive treatment with ammonia ($NH_3$-SCR) to catalytically convert $NO_x$ into nitrogen ($N_2$) and water ($H_2O$) molecules. Due to the toxicity and hazardous nature of ammonia, it is produced in vehicles using an aqueous solution, called AdBlue, containing 32.5 wt.% urea $CO(NH_2)_2$ [63]. This additive is stored in a separate tank and injected before the SCR converter into the gas stream leaving the particulate filter, where the urea thermally decomposes to produce isocyanic acid (HNCO) and ammonia ($NH_3$) [64]. Although ammonia and urea are currently the preferred choice for the selective catalytic reduction of NOx, hydrogen has recently attracted strong interest as a reducing agent, representing a rather interesting alternative. Indeed, in recent years, several researchers have focused their studies on the development of a suitable catalytic system for the selective catalytic reduction of NOx using hydrogen ($H_2$-SCR [65–67]). The latter technology has several advantages: zero carbon footprint, as there is no formation of CO and $CO_2$ molecules during $NO_x$ removal; formation of water from hydrogen combustion ($O_2$ present in the gas stream); high $NO_x$ conversion activity at lower temperatures ($T < 200$ °C) with respect to $NH_3$-SCR counterparts ($T = 350$–$400$ °C) [68]; possibility of drawing from a single tank using hydrogen also as a fuel to partially replace diesel or petrol in the combustion mixture. Four main reactions occur in selective catalytic reduction with hydrogen in the presence of oxygen, but the desiderated one has two NO molecules reacting with four hydrogen molecules to form $N_2$ [62,65]. Studies reported so far in the literature have shown a low selectivity towards $N_2$ production with undesirable productions of partially reduced species such as $N_2O$ and $NH_3$.

As a first step, the preparation of suitable catalytic systems using palladium (Pd) as the active phase and metal oxides as supports is the object of the study. These catalytic systems could be deposited on monolithic cordierite support [69] for the realization of a system that can be implemented in the last stage of treatment in diesel engine flue gas purification. In particular, the catalytic support that is to be developed consists of the mixture of the two metal oxides $CeO_2$ and $ZrO_2$ and of a Pd-based active catalytic phase (hereinafter called PCZ). The catalytic systems were made by wet impregnation using 60 wt% $CeO_2$ and 40 wt% $ZrO_2$ properly mixed with deionized water at 40 °C for 2 h. Then, the catalytic support was dried in an oven at 120 °C for 12 h until completely dry. To the resulting powder, palladium nitrate hydrate ($Pd(NO_3)_2\,2H_2O$) was added so as to obtain a Pd loading of 1 wt.%, as from previous studies it appears to be the best in terms of $NO_x$ conversion and selectivity toward $N_2$. Wet impregnation was performed at 70 °C at a constant pH of 9.5 (dropwise addition of a 0.25 M $NH_4OH$ solution), where a color change from white to brownish due to the presence of palladium is observed. The obtained catalytic system is first dried in an oven at 120 °C for 12 h and then calcined in air at 500 °C for 4 h. Figure 16 shows the diffraction patterns of the raw $CeO_2$ (a) and $ZrO_2$ (b) powders, of the mixture $CeO_2$-$ZrO_2$ (c), and the complete catalytic system (active phase and catalytic support) Pd/$CeO_2$-$ZrO_2$ (d). The $CeO_2$-$ZrO_2$ catalytic support (c) shows peaks attributable to the crystalline phases of the starting metal oxides and the Pd/$CeO_2$-$ZrO_2$ (PCZ) catalytic system (d) shows a diffraction pattern with narrower peaks than the only support. The crystalline phases of Pd cannot be appreciated due to the small quantity of metal on the catalytic support.

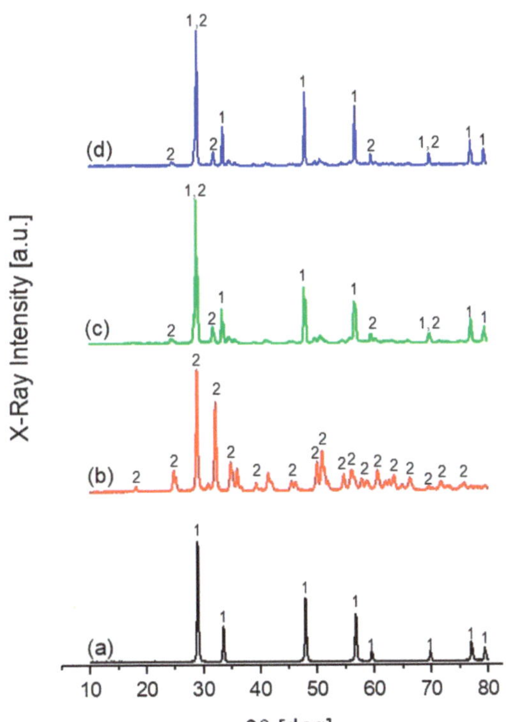


**Figure 16.** X-ray diffraction pattern of $CeO_2$ (**a**) and $ZrO_2$ (**b**) precursor powders, $CeO_2$-$ZrO_2$ catalytic support (**c**) and catalytic system Pd/$CeO_2$-$ZrO_2$ (**d**); 1 = Cerianite, $CeO_2$; 2 = Zirconium Oxide, $ZrO_2$.

Figure 17 shows micrographs at various magnifications of the PCZ catalytic system. From these micrographs, it is possible to see a very regular and compact surface in which there is good interaction between the two metal oxides that act as the catalytic support. Specifically, one can observe areas (circled in red) with a regular spherical morphology attributable to cerium oxide and areas (circled in green) with a smaller average crystalline domain size attributable to zirconium oxide. The structure of the produced support material thus presents the two support oxides evenly distributed in space with exposed sites of both precursors.

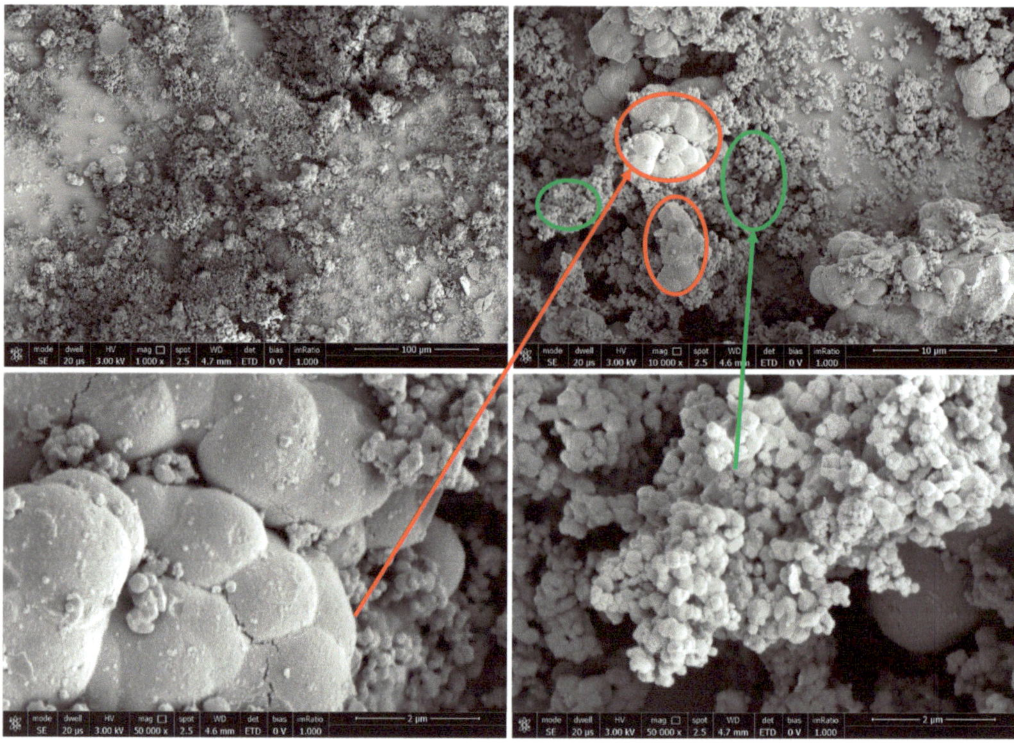

**Figure 17.** SEM micrographs at various magnifications of the Pd/$CeO_2$-$ZrO_2$ catalytic system, regular spherical morphology circled in red and smaller averga cristallin domain circled in green.

As a second step, small-scale prototypes of coated cordierite substrate will be tested with real currents emitted by a diesel engine at the test bed, studying the degree of conversion and selectivity of the catalyst under different engine operating conditions and with different flow rates of $H_2$. The experimental setup is schematized in Figure 18. The reagent mixture is obtained by flowing the exhaust gases downstream of the turbine. The hydrogen is injected into the prototype exhaust gas line upstream of the H2-SCR reactor.

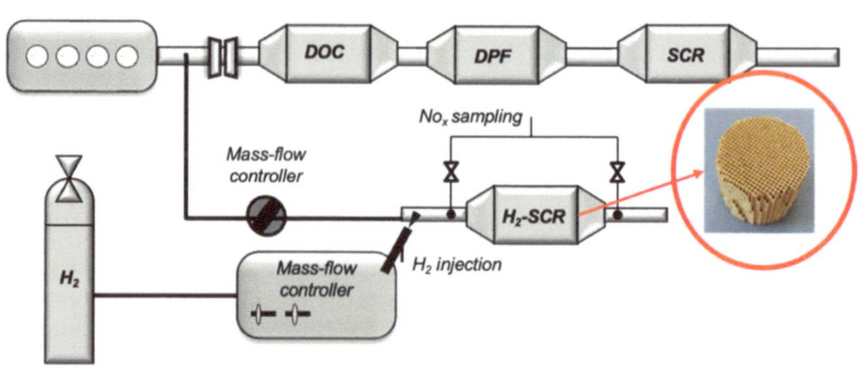


**Figure 18.** Experimental setup scheme for the small-scale substrate prototype.

## 5. Engine and Hybrid Powertrain Management

The present section describes the development of the engine and hybrid powertrain management strategies. On the one hand, the load control of H2-ICEs is crucial to ensure good performance and low emissions: a proper control strategy is required to avoid the risk of combustion anomalies (e.g., knock and preignition) while limiting NOx emissions and enhancing engine efficiency. Since the 1D GT Power engine model of the H2-ICE featured in the present work has been already discussed [70] such a model has been exploited to generate a set of reference data that accurately represent the behavior of the engine. As a starting point, a fast-running engine model based on Artificial Neural Networks (ANNs) has then been developed, based on GT Power results. The advantage of such a model is the low computational time, which makes it suitable for the development of the H2-ICE control strategy using a Model-in-the-Loop (MIL) approach. On the other hand, since in a hybrid powertrain both fuel economy and pollutant emission reduction strongly depend on the exploitation of each energy source (i.e., fuel and battery), a supervisory high-level control strategy is mandatory to determine the power which has to be delivered by each source. Two methodologies have been developed for the design of the energy management strategies, the former based on a heuristic rule-based approach, the latter based on the minimization of the equivalent fuel consumption (ECMS) by means of Pontryagin's Minimum Principle (PMP). Both approaches are intrinsically causal as they do not require the a priori knowledge of the mission profile, nevertheless this information is accounted for in the design process to improve the accuracy of the strategy in reaching a target SoC. It is worth noting that for urban buses, for which the route is defined a priori, this could be a non-critical issue.

### *5.1. Engine Model Description*

The engine model has been developed in Matlab/Simulink starting from available data of the H2-ICE. To cover as much as possible the whole operating envelope, the engine model has been validated in a wide operational range for each of the controllable variables as shown in Table 6.

**Table 6.** Engine Model Limits.

| Variable | Range |
|---|---|
| Speed [rpm] | [1000–4000] |
| BMEP [bar] | 2–max BMEP available for given RPM |
| CA50 [deg] | [−30–30] |
| Lambda [-] | [2–2.7] |
| EGR [%] | [0–10] |

The engine model is necessarily coupled to the control strategy: indeed, the inputs of the model are the outputs of the control strategy, thus all the modeled quantities have to be directly or indirectly (e.g., derived from an intermediate block) calculated based on the control actions: this makes the simulation a loop including the engine model and the engine controller. The model outputs concern combustion metrics, and intake and exhaust conditions, as shown in the I/O sketch reported in Figure 19.

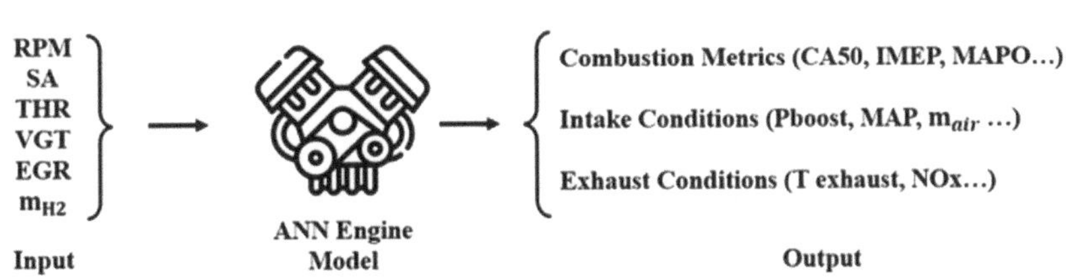


**Figure 19.** Input and output of the engine model.

The modeling approach allows representing the cycle-by-cycle behavior: it can be explained by taking as an example the generation of Indicated Mean Effective Pressure (IMEP) as shown in Figure 20:

1. The intake conditions, in terms of air mass and EGR percentage, are estimated from the intake block;
2. RPM, Spark Advance (SA), mass of fuel, and the intake conditions are the input parameters for the Crank Angle of 50% Heat Release (CA50) ANN;
3. CA50, intake conditions, mass of fuel, RPM, and VGT position are the input for the IMEP ANN.

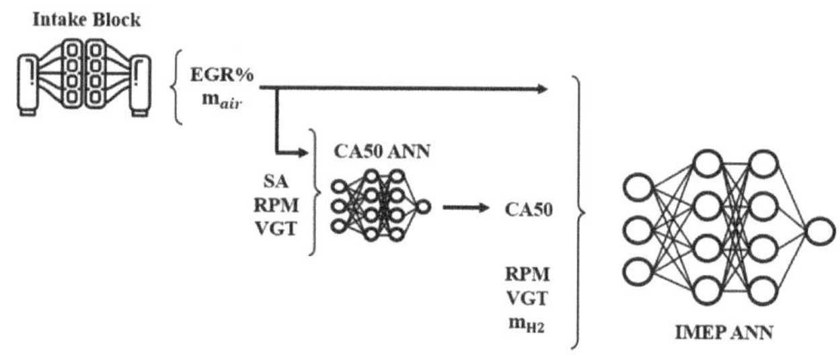


**Figure 20.** Input and output of the engine model, IMEP example.

Table 7 summarizes the accuracy of the engine model (compared to reference data) in steady-state conditions, in terms of Root Mean Square Error (RMSE) and percentual RMSE.

**Table 7.** Engine Model Accuracy.

| Variable | RMSE | RMSE % |
|---|---|---|
| CA50 [deg] | 0.34 | 0.56 |
| IMEP [bar] | 0.033 | 0.33 |
| BSFC [g/kWh] | 2.26 | 0.75 |
| Pcyl, max [bar] | 0.51 | 0.43 |
| Tcyl, max [K] | 6.57 | 0.41 |
| NOx (below 2000 ppm) [ppm] | 11 | 0.56 |
| P exhaust [bar] | 0 | 0.3 |
| T exhaust [K] | 2.3 | 0.57 |

As shown in Table 7, the percentage RMSE is always lower than 1% for each predicted variable. Once the model has been validated in steady-state conditions, the capability of representing cycle-by-cycle variations is introduced. Specifically, two additional ANNs have been introduced to properly represent the statistical distribution of CA50, which propagates to other combustion metrics. For knock intensity, a further statistical dispersion is used to represent Maximum Amplitude Pressure Oscillation (MAPO) variations at fixed CA50 levels. Finally, the model has been completed with intake manifold and turbocharger dynamics. Figure 21 shows a comparison, in terms of BMEP, Brake Specific Fuel Consumption (BSFC), and NOx, between the GT Power and the Simulink model outputs in a transient simulation where RPM, VGT, Throttle Valve (THR), SA, and mass of injected fuel change. As it is possible to see, the model is capable of replicating the engine behavior in terms of torque production, consumption, and emissions.

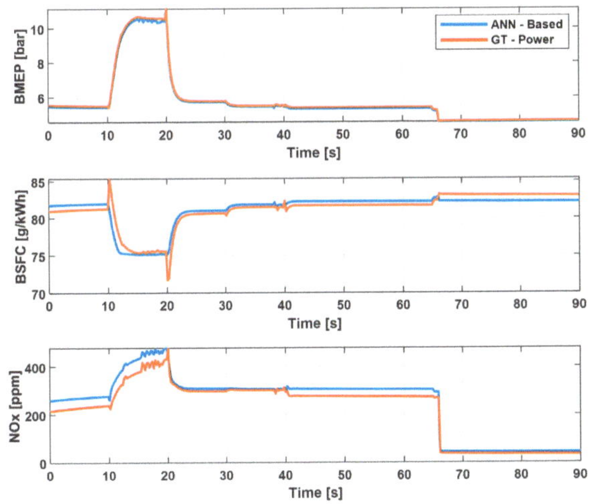


**Figure 21.** BMEP, BSFC, and NOx simulated in GT Power in a transient simulation and BMEP simulated from the presented engine model.

Control Strategy Description

Once the engine model has been validated, a torque control strategy can be implemented. The inputs of the latter are the requested BMEP at the current RPM, the manifold and boost pressure, and the feedback from a combustion analyzer (CA50, Pcyl, max, and MAPO), as shown in Figure 22. Based on these inputs, the control strategy sets the position of the actuators (e.g., SA, THR, VGT, etc.).

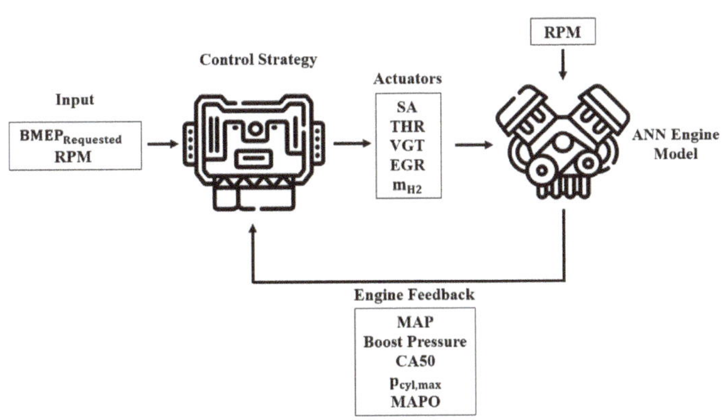


**Figure 22.** Input and output of the control strategy.

In order to avoid abnormal combustions (e.g., knock, preignition, etc.) and limit NOx emissions, the lambda value has to be above 2.2, thus the load control can be directly managed with the amount of injected fuel. However, the fuel request has to be saturated between the knock and misfire conditions to keep the engine in safe operating conditions. As regards combustion phasing, proper control of the CA50 position results is crucial in order to maximize efficiency without compromising reliability. For this purpose, a CA50-based control strategy has been implemented. The target CA50 is defined as a function of the current operating condition (RPM, load request, EGR, and lambda) and is saturated under knocking conditions. The SA allowing achievement of the target CA50 for the current conditions is then calculated using a dedicated ANN. Moreover, the open loop positions of the THR, VGT, and EGR settings are calculated from calibrated maps as a function of engine speed and load request. Finally, a closed loop control has been developed, on manifold pressure and boost pressure, acting respectively on THR and VGT. Figure 23 shows an example of the control strategy behavior. As can be noticed, the engine is also capable of following the BMEP request under transient conditions.

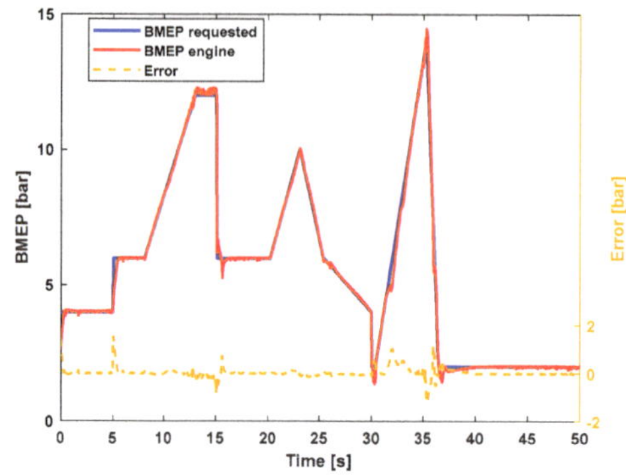


**Figure 23.** BMEP request, BMEP provided, and error.

Figure 24 shows the actuator (THR, VGT, SA, fuel) settings and the corresponding achieved CA50. It is worth noticing how after a sudden drop in the requested BMEP, the control strategy saturates the minimum requestable fuel and retards the CA50 to keep the BMEP on target.

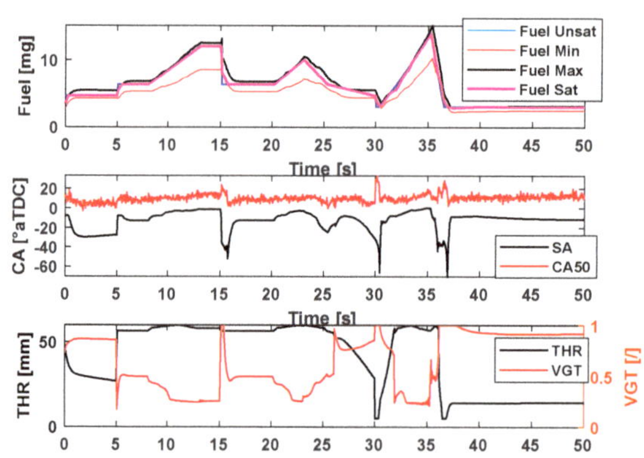


**Figure 24.** Fuel, SA, CA50, THR, and VGT over time.

## 5.2. Hybrid Powertrain Management

The Energy Management Strategy (EMS) of the hybrid powertrain is aimed at controlling the power provided by the range extender (i.e., ICE-EG) in order to guarantee sustaining the battery charge with the minimum energy consumption. A vehicle model has been developed for the EMS design and simulation, according to the data flow shown in Figure 25. Starting from fitting the experimental data of the various vehicle components and vehicle parameters, the optimal operating points of the range extender were identified as a function of speed and torque through an optimization process. This is made possible by the series architecture of the powertrain, which allows its fixed-point operation.

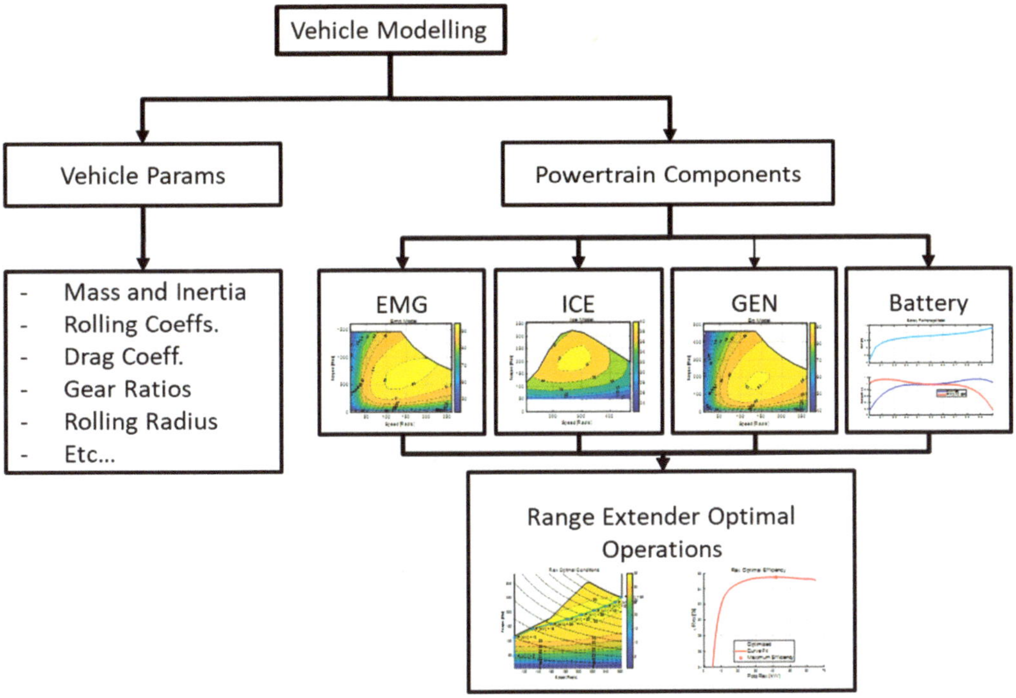


**Figure 25.** Vehicle Modeling Data Flow.

Two Energy Management Strategies (EMSs) have been developed. Specifically, a heuristic rule-based strategy and an optimal strategy based on the minimization of the equivalent fuel consumption by means of the Pontryagin Minimum Principle (PMP). The rule-based strategy has been implemented to achieve a causal non-optimal benchmark to be used as a reference for tuning the optimized PMP strategy.

5.2.1. Rule-Based Strategy

The EMS consists of a Rule-Based (RB) controller of the ICE-EG, in which the state variable is the battery SoC [16], as can be seen in Equation (2):

$$\left\{ \begin{array}{c} \xi_{target} = \xi_{start} + \frac{\xi_{end} - \xi_{start}}{distance} \times \int v dt = \xi_{start} + \xi_{rate} \times \int v dt \\ \xi_{Max} = \xi_{target} \times toleranceMax \\ \xi_{Min} = \xi_{target} \times toleranceMin \\ \xi_{HardMin} = \xi_{target} \times toleranceHardMin \end{array} \right\} \quad (2)$$

The initial State of Charge (SoC) level, denoted as $\xi_{Start}$, represents the starting point, while $\xi_{End}$ is the final SoC desired. The ICE-EG is activated when SoC decreases down to the lower threshold, $\xi_{End}$, while it is deactivated when the SoC reaches the upper threshold, $\xi_{Max}$. These thresholds vary with time and are calculated based on the target SoC ($\xi_{target}$) multiplied by predefined *tolerance* (*Max* and *Min*) factors. This ensures that the target SoC stays within a defined window throughout the simulation. The two thresholds and $\xi_{target}$ decrease along the driving transient, in accordance with the expected battery discharge rate ($\xi_{rate}$).

An additional activation threshold for the ICE-EG ($\xi_{HardMin}$) is also provided, in which, in the event of excessive battery discharge, more power is requested to cope with an unexpected increase in load. The flow chart of the rule-based EMS is shown in Figure 26.

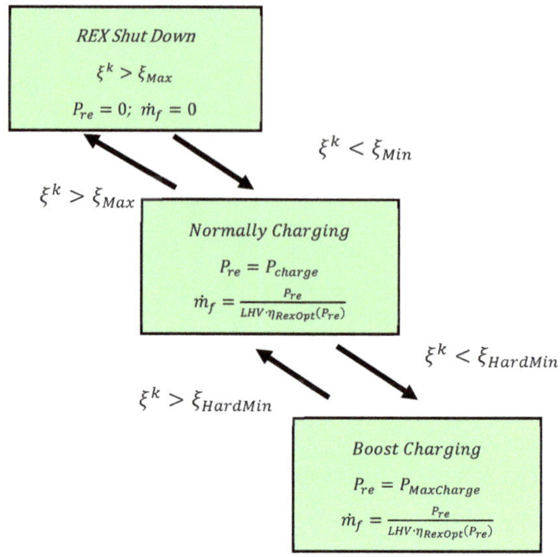


**Figure 26.** Rule-Based Flow Chart.

$P_{\rm re}$ denotes the selected delivered power by ICE-EG, which can fluctuate between two power levels: $P_{\rm chg}$ where maximum efficiency is achieved, and $P_{MaxCharge}$ representing the maximum delivered power by the ICE-EG when the SoC reaches the second lower threshold ($\xi_{HardMin}$). $\dot{m}_f$ is the fuel consumption. Figure 27 shows the time trajectory of the battery SoC with the different SoC thresholds for a generic driving cycle.

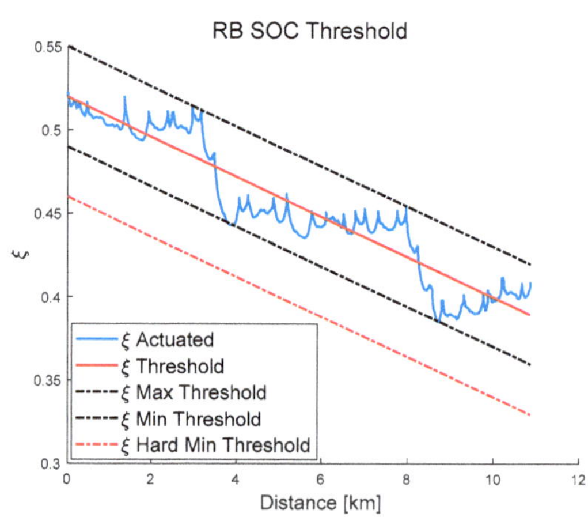


**Figure 27.** SoC Thresholds.

5.2.2. Equivalent Consumption Minimization Strategy

The Pontryagin Minimum Principle (PMP) is a crucial tool in the study of optimal control problems for dynamic systems. It addresses the challenge of finding the best trajectory that minimizes a given performance criterion, subject to the system's dynamics and constraints. Further theoretical details on the principle can be found in [71], while examples of its application to the HEV powertrain control can be found, e.g., in [72,73].

In the current application, the problem is formulated by considering the battery State of Charge (SoC) as the state variable and the battery power ($P_b$) as a control variable. The Hamiltonian can be defined as the sum of the fuel mass consumed by the engine and an equivalent fuel mass consumed by the electric motor weighted by an equivalence factor, represented by the co-state (λ), thus leading to an Equivalent Consumption Minimization Strategy (ECMS). The co-state drives the solution alternatively towards battery charging or battery depletion and the ECMS-based technique minimizes the Hamiltonian while adjusting the equivalent factor dynamically over time to guarantee the charge is sustained throughout the whole reference driving cycle. Figure 28 shows the results of the co-state identification and the deviation of the final SoC vs. the target value corresponding to the initial one.

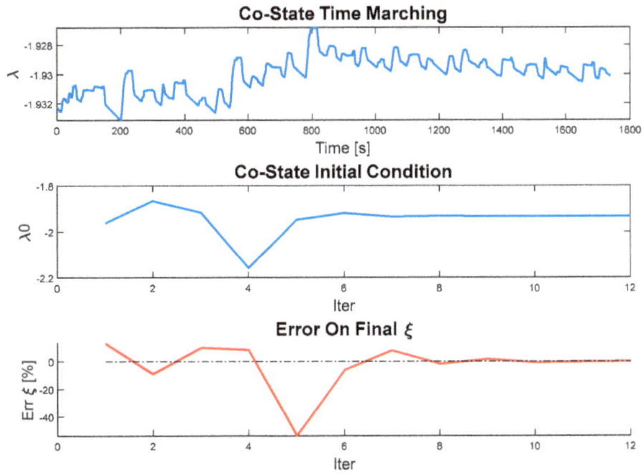


**Figure 28.** Co-State Identification results.

Table 8 compares the results in terms of fuel consumption and powertrain efficiency achieved by the two proposed EMSs (i.e., RB and PMP) for the three driving cycles considered. It can be noted that the use of the optimization strategy leads to an improvement (on average, 9%) compared to the use of the causal strategy. It is worth noting that the fuel consumption is still higher than the imposed target of 10 kg/100 km, nevertheless, a significant reduction is expected to be achieved by further improvements in engine combustion system and control, implementation of WHR technologies, and further refinement of the EMS. Figure 29 shows the time trajectories of the SoC and cumulative fuel consumption, evidencing that both RB and PMP guarantee suitable charge sustaining throughout the whole driving cycle though the former EMS exhibits a larger SoC deviation from the mean value, especially in Braunschweig and MLTB cycles.

**Table 8.** Simulations results: PMP vs. RB.

| **Simulation Results** | | | | | | |
|---|---|---|---|---|---|---|
| **Road Cycle** | **Braunschweig** | | **Gillingham Uphill** | | **MLTB** | |
| **EMS** | PMP | RB | PMP | RB | PMP | RB |
| **Fuel consumption (kg/100 km)** | 11.578 | 12.694 | 13.604 | 14.882 | 14.803 | 16.342 |
| **Δ (%)** | −8.789 | | −8.586 | | −9.415 | |
| $\overline{\eta}_{powertrain}$ **(%)** | 24.208 | 22.039 | 25.915 | 23.715 | 26.259 | 23.766 |

Where the powertrain efficiency, $\overline{\eta}_{powertrain}$, is defined as in Equation (3):

$$\overline{\eta}_{powertrain} = \frac{E_{wheels} + E_{Aux}}{E_{Rex} + E_{battery}} \quad (3)$$

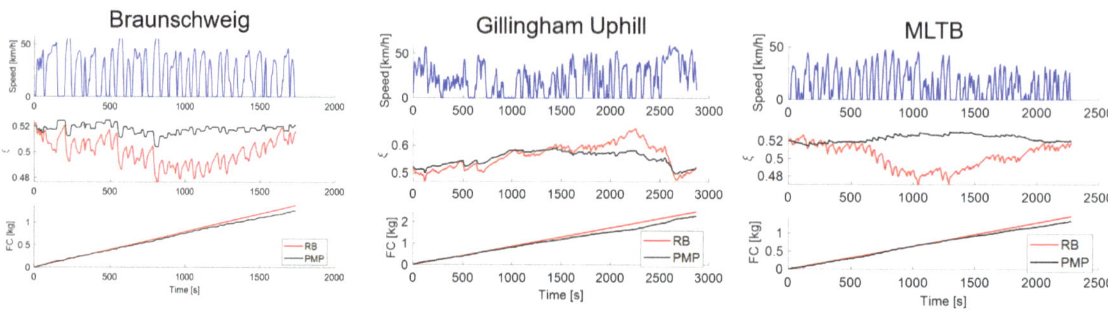


**Figure 29.** Fuel consumption and state of charge time trajectories achieved by RB and PMP for the three driving cycles. Braunschweig (**left**), Gillingham (**center**), MLTB (**right**).

## 6. Conclusions and Future Steps

In the present work, preliminary results achieved to date from the H2-ICE project were presented. A comprehensive methodology has been developed to support the design of engine concepts and vehicle subsystems for the efficient exploitation of H2 as a fuel, as well as for their integration in a hybrid powertrain for an urban bus application. High-fidelity CFD models have been developed and validated to assess the potential of different injection options from Port Fuel Injection (PFI) to Direct Injection (DI), with different levels of nozzle designs and pressure ratios, and to control and optimize the combustion, setting a maximum of $\lambda = 3$ to keep the combustion efficiency above 85%. Two different Waste Heat Recovery (WHR) systems have been analyzed and tested to enhance engine efficiency at higher levels: for a direct heat recovery system based on the Organic Rankine Cycle (ORC), a net recovered power of 5 kW has been obtained in higher-engine-power conditions, while for an indirect turbocompound it is up to 8 kW recovered power, limited by engine backpressure effects. Regarding the aftertreatment system, the preparation of suitable catalytic systems for the Selective Catalytic Reduction (SCR) of NOx using hydrogen as a reducing agent has been completed. Moreover, a dedicated control strategy has been developed exploiting a fast-running engine model based on Artificial Neural Networks (ANNs) together with the torque control strategy. Finally, two approaches to an Energy Management System (EMS) have been tested to target the desired fuel consumption (i.e., 0.1 kg/km) and to guarantee a suitable SCR operation: a Rule-Based (RB) approach and an optimal strategy based on the equivalent fuel consumption minimization (ECMS). A gain in the fuel consumption near 10% is obtained with the PMP strategy with respect to the RB one in the different considered driving cycles. Future steps of the project will be focused on:

- further developing the engine combustion system, also exploiting the outcomes of the upcoming experimental campaign on the mentioned single-cylinder optical-access engine, to further decrease the hydrogen consumption,
- assessing the capability of the two WHR systems tested for different operating conditions and considering the possibility of introducing two sections of WHR, combining direct and indirect ways to increase the recovery,
- giving additional insights on the behavior of the H2-SCR reactor with the experimental testing of small-scale prototypes in a wide range of operating conditions,
- optimizing the Energy Management System (EMS) to reach the desired target in terms of fuel consumption, together with the H2-ICE optimization, over the different mission profiles.

Then, the final outcome of the project would be not only the development of a traction system for a new generation of H2-ICE but also the enhancement of the know-how for sustainable H2 exploitation in the context of public transport mobility.

**Author Contributions:** Conceptualization, I.A., M.B., R.C., E.C. and F.M; methodology, P.P.B., D.D.B., B.P.P., A.O., F.M., L.R. and J.Z.; validation, P.P.B., D.D.B., A.O., L.R. and J.Z., formal analysis, I.A., M.B., R.C., E.C. and F.M.; investigation, P.P.B., D.D.B., A.O., L.R. and J.Z.; writing—original draft preparation, I.A., M.B., P.P.B., D.D.B., A.O., B.P.P. and J.Z.; writing—review and editing, I.A., B.P.P., M.B., R.C., E.C., F.M. and L.R.; supervision, I.A., M.B., R.C., E.C. and F.M.; project administration, I.A., M.B., R.C., E.C. and F.M.; funding acquisition, I.A., M.B., R.C., E.C. and F.M. All authors have read and agreed to the published version of the manuscript.

**Funding:** Research funded by Ministero dell'Istruzione, dell'Università e della Ricerca (MIUR) (2020R92Y3Z).

**Data Availability Statement:** The data presented in this study are partially available on request from the corresponding author.

**Conflicts of Interest:** The authors declare no conflict of interest.

## Abbreviations

| | |
|---|---|
| ANN | Artificial Neural Networks |
| BEV | Battery Electric Vehicle |
| BMEP | Break Mean Effective Pressure |
| BSFC | Brake Specific Fuel Consumption |
| CA50 | Crank Angle of 50% Heat Release |
| CFD | Computational Fluid Dynamics |
| DI | Direct Injection |
| DP | Dynamic Programming |
| EGR | Exhaust Gas Recirculation |
| EM | Electric Motor |
| EMS | Energy Management System |
| EU | European Union |
| FCEV | Fuel Cell Electric Vehicle |
| GHG | Greenhouse Gas |
| HDV | Heavy-Duty Vehicle |
| ICE | Internal Combustion Engine |
| ICE-EG | Internal Combustion Engine—Range Extender |
| IMEP | Indicated Mean Effective Pressure |
| IVO | Intake Valve Opening |
| LDV | Light-Duty Vehicle |
| LES | Large Eddy Simulation |
| LFP | Lithium-Ferro-Phosphate |
| MAPO | Maximum Amplitude of Pressure Oscillations |
| MIL | Model-in-the-Loop |
| NMC | Nickel Manganese Cobalt Oxide |
| NPR | Nozzle Pressure Ratio |
| ORC | Organic Rankine Cycle |
| Pcyl, max | Maximum In-Cylinder Pressure |
| PFI | Port Fuel Injection |
| PM | Permanent Magnet |
| PMP | Pontryagin Minimum Principle |
| RANS | Reynolds-Averaged Navier–Stokes |
| RB | Rule Based |
| RMSE | Root Mean Square Error |
| RPM | Revolutions Per Minute |
| SA | Spark Advance |
| SCR | Selective Catalytic Reduction |
| SoC | State of Charge |
| SoI | Start of Injection |
| SOpHy | Sandia's Optical Hydrogen Engine |
| Tcyl, max | Maximum In-Cylinder Temperature |
| THR | Throttle Valve |

| | |
|---|---|
| V2X | Vehicle To Everything |
| VGT | Variable Geometry Turbocharger |
| WF | Working Fluid |
| WHR | Waste Heat Recovery |

## References


1. UNECE. European Green Deal: Commission Proposes Transformation of EU Economy and Society to Meet Climate Ambitions. 2021. Available online: https://ec.europa.eu/commission/presscorner/detail/en/IP_21_3541 (accessed on 31 July 2023).
2. IEA. Largest End Uses of Energy by Sector in Selected IEA Countries. Available online: https://www.iea.org/data-and-statistics/charts/largest-end-uses-of-energy-by-sector-in-selected-iea-countries-2019 (accessed on 31 July 2023).
3. EEA. Greenhouse Gas Emissions from Transport in Europe. 2023. Available online: https://www.eea.europa.eu/ims/greenhouse-gas-emissions-from-transport (accessed on 31 July 2023).
4. ACEA. Fact Sheet: CO2 Standards for Heavy-Duty Vehicles. 2023. Available online: https://www.acea.auto/fact/fact-sheet-co2-standards-for-heavy-duty-vehicles/ (accessed on 31 July 2023).
5. European Commission. *Proposal for a Regulation of the European Parliament and of the Council Amending Regulation (EU) 2019/1242 as Regards Strengthening the CO2 Emission Performance Standards for New Heavyduty Vehicles and Integrating Reporting Obligations, and Repealing Regulation (EU) 2018/956*; European Commission: Brussels, Belgium, 2023.
6. Zhang, C.; Zhao, X.; Sacchi, R.; You, F. Trade-off between critical metal requirement and transportation decarbonization in automotive electrification. *Nat. Commun.* **2023**, *14*, 1616. [CrossRef] [PubMed]
7. Mareev, I.; Becker, J.; Sauer, D.U. Battery dimensioning and life cycle costs analysis for a heavy-duty truck considering the requirements of long-haul transportation. *Energies* **2018**, *11*, 55. [CrossRef]
8. Çabukoglu, E.; Georges, G.; Küng, L.; Pareschi, G.; Boulouchos, K. Battery electric propulsion: An option for heavy-duty vehicles? Results from a Swiss case-study. *Transp. Res. Part C Emerg. Technol.* **2018**, *88*, 107–123. [CrossRef]
9. Moriarty, P.; Honnery, D. Prospects for hydrogen as a transport fuel. *Int. J. Hydrogen Energy* **2019**, *44*, 16029–16037. [CrossRef]
10. Sharma, S.; Ghoshal, S.K. Hydrogen the future transportation fuel: From production to applications. *Renew. Sustain. Energy Rev.* **2015**, *43*, 1151–1158. [CrossRef]
11. Kim, H.; Hartmann, N.; Zeller, M.; Luise, R.; Soylu, T. Comparative tco analysis of battery electric and hydrogen fuel cell buses for public transport system in small to midsize cities. *Energies* **2021**, *14*, 4384. [CrossRef]
12. Vodovozov, V.; Raud, Z.; Petlenkov, E. Review of Energy Challenges and Horizons of Hydrogen City Buses. *Energies* **2022**, *15*, 6945. [CrossRef]
13. Heid, B.; Martens, C.; Orthofer, A. *How Hydrogen Combustion Engines Can Contribute to Zero Emissions*; McKinsey & Company: Chicago, IL, USA, 2021.
14. Verhelst, S. Recent progress in the use of hydrogen as a fuel for internal combustion engines. *Int. J. Hydrogen Energy* **2014**, *39*, 1071–1085. [CrossRef]
15. Verhelst, S.; Wallner, T. Hydrogen-fueled internal combustion engines. *Prog. Energy Combust. Sci.* **2009**, *35*, 490–527. [CrossRef]
16. HyFLEET:CUTE, Hydrogen Transport, Bus Technology & Fuel for Today and for a Sustainable Future. 2023. Available online: https://cordis.europa.eu/project/id/19991 (accessed on 31 July 2023).
17. SunLine Transit Agency Hydrogen-Powered Transit Buses: Third Evaluation Report. Available online: https://www.nrel.gov/docs/fy08osti/43741-1.pdf (accessed on 31 July 2023).
18. Matthias, N.S.; Wallner, T.; Scarcelli, R. A Hydrogen Direct Injection Engine Concept that Exceeds U.S. DOE Light-Duty Efficiency Targets. *SAE Int. J. Engines* **2012**, *5*, 838–849. [CrossRef]
19. Golisano, R.; Scalabrini, S.; Arpaia, A.; Pesce, F.; Vassallo, A.; Borgia, F.; Cubito, C.; Biasin, V.; Knichel, T.; Millo, F.; et al. Hydrogen Internal Combustion Engine & KERS: An Appealing Value-Proposition for Green Power Pack. In Proceedings of the 42nd International Vienna Motor Symposium, Online, 29–30 April 2021.
20. Koch, T.D.; Sousa, A.; Bertram, D. *$H_2$-Engine Operation with EGR Achieving High Power and High Efficiency Emission-Free Combustion*; SAE Technical Papers 2019-01-2178; SAE: Warrendale, PA, USA, 2019. [CrossRef]
21. Millo, F.; Rolando, L.; Fuso, R.; Zhao, J. Development of a new hybrid bus for urban public transportation. *Appl. Energy* **2015**, *157*, 583–594. [CrossRef]
22. Wang, L.; Yang, Z.; Huang, Y.; Liu, D.; Duan, J.; Guo, S.; Qin, Z. The effect of hydrogen injection parameters on the quality of hydrogen–air mixture formation for a PFI hydrogen internal combustion engine. *Int. J. Hydrogen Energy* **2017**, *42*, 23832–23845. [CrossRef]
23. Li, Y.; Gao, W.; Zhang, P.; Ye, Y.; Wei, Z. Effects study of injection strategies on hydrogen-air formation and performance of hydrogen direct injection internal combustion engine. *Int. J. Hydrogen Energy* **2019**, *44*, 26000–26011. [CrossRef]
24. Dhyani, V.; Subramanian, K.A. Fundamental characterization of backfire in a hydrogen fuelled spark ignition engine using CFD and experiments. *Int. J. Hydrogen Energy* **2019**, *44*, 32254–32270. [CrossRef]
25. Keck, M.; Bessey, D.; Buehler, F.; Manuel, D.; Faiß, E. Hydrogen Storage Technologies. In *International MTZ Conference on Heavy-Duty Engines*; Springer Fachmedien Wiesbaden: Wiesbaden, Germany, 2022; pp. 123–130.

26. Ringler, J.; Gerbig, F.; Eichlseder, H.; Wallner, T. Insights into the development of a hydrogen combustion process with internal mixture formation. In Proceedings of the Sixth International Symposium on Internal Combustion Engine Diagnostics, Baden Baden, Germany, 15–16 June 2004; pp. 290–304.
27. Scarcelli, R.; Kastengren, A.L.; Powell, C.F.; Wallner, T.; Matthias, N.S. High-pressure gaseous injection: A comprehensive analysis of gas dynamics and mixing effects. In Proceedings of the ASME 2012 Internal Combustion Engine Division Fall Technical Conference, ICEF 2012, Vancouver, BC, Canada, 23–26 September 2012; pp. 793–801. [CrossRef]
28. Scarcelli, R.; Wallner, T.; Matthias, N.; Salazar, V.; Kaiser, S. Mixture Formation in Direct Injection Hydrogen Engines: CFD and Optical Analysis of Single-and Multi-Hole Nozzles. *SAE Int. J. Engines* **2011**, *4*, 2361–2375. [CrossRef]
29. Di Battista, D.; Di Bartolomeo, M.; Villante, C.; Cipollone, R. On the limiting factors of the waste heat recovery via ORC-based power units for on-the-road transportation sector. *Energy Convers. Manag.* **2018**, *155*, 68–77. [CrossRef]
30. Väliheikki, A.; Petallidou, K.C.; Kalamaras, C.M.; Kolli, T.; Huuhtanen, M.; Maunula, T.; Keiski, R.L.; Efstathiou, A.M. Selective catalytic reduction of NOx by hydrogen ($H_2$-SCR) on WOx-promoted $Ce_zZr_{1-z}O_2$ solids. *Appl. Catal. B* **2014**, *156–157*, 72–83. [CrossRef]
31. Hegde, B.; Ahmed, Q.; Rizzoni, G. Velocity and energy trajectory prediction of electrified powertrain for look ahead control. *Appl. Energy* **2020**, *279*, 115903. [CrossRef]
32. Kivekäs, K.; Lajunen, A.; Vepsäläinen, J.; Tammi, K. City bus powertrain comparison: Driving cycle variation and passenger load sensitivity analysis. *Energies* **2018**, *11*, 1755. [CrossRef]
33. Transport for London: London Exhaust Emissions Study: A summary of the Drive Cycle Development, Test Programme and Comparison of Test Data Compared with Type Approval Data. Available online: https://content.tfl.gov.uk/london-exhaust-emissions-study-drive-cycle-development.pdf (accessed on 31 July 2023).
34. Vassallo, A.; Pesce, F.; Arpaia, A.; Millo, F.; Rolando, L.; Piano, A.; Bianco, A. Ultra-lean Combustion System Optimization for H2-fuelled ICEs via Synergistic Application of 1D- and 3D-CFD. In Proceedings of the SIA International Conference and Exhibition on Powertrain & Power Electronics, Paris, France, 9–10 June 2021; pp. 1–9.
35. ZeEUS eBus Report #2 An Updated Overview of Electric Buses in Europe. Available online: https://zeeus.eu/uploads/publications/documents/zeeus-ebus-report-2.pdf (accessed on 31 July 2023).
36. EL-Refaie, A.M. Electrical Machines for Traction and Propulsion Applications. In *Transportation Electrification: Breakthroughs in Electrified Vehicles, Aircraft, Rolling Stock, and Watercraft*; John Wiley & Sons: Hoboken, NJ, USA, 2023; pp. 1–26.
37. Shaobo, X.; Qiankun, Z.; Xiaosong, H.; Yonggang, L.; Xianke, L. Battery sizing for plug-in hybrid electric buses considering variable route lengths. *Energy* **2021**, *226*, 120368. [CrossRef]
38. Carlsson, L.O.; Larsson, D.; Sauer, A.; Speckens, F.W. High-performance Battery for Premium Class Hybrid Vehicles. *MTZ Worldw.* **2021**, *82*, 28–35. [CrossRef]
39. Ellingsen, L.A.W.; Thorne, R.J.; Wind, J.; Figenbaum, E.; Romare, M.; Nordelöf, A. Life cycle assessment of battery electric buses. *Transp. Res. D Transp. Environ.* **2022**, *112*, 103498. [CrossRef]
40. A123 Systems Nanophosphate®High Power Lithium-Ion Cell. Available online: https://www.batteryspace.com/prod-specs/6610.pdf (accessed on 31 July 2023).
41. Rahantamialisoa, F.N.Z.; Zembi, J.; Miliozzi, A.; Sahranavardfard, N.; Battistoni, M. CFD simulations of under-expanded hydrogen jets under high-pressure injection conditions. *J. Phys. Conf. Ser.* **2022**, *2385*, 012051. [CrossRef]
42. Duronio, F.; Ranieri, S.; Di Mascio, A.; Vita, A. De Simulation of high pressure, direct injection processes of gaseous fuels by a density-based OpenFOAM solver. *Phys. Fluids* **2021**, *33*, 066104. [CrossRef]
43. Rahantamialisoa, F.; Battistoni, M.; Miliozzi, A.; Sahranavardfard, N.; Zembi, J. *Investigations on Hydrogen Injections Using a Real-Fluid Approach*; SAE Technical Papers 2023-01-0312; SAE: Warrendale, PA, USA, 2023. [CrossRef]
44. Panda, J.; Seasholtz, R.G. Measurement of shock structure and shock-vortex interaction in underexpanded jets using Rayleigh scattering. *Phys. Fluids* **1999**, *11*, 3761–3777. [CrossRef]
45. Hamzehloo, A.; Aleiferis, P.G. LES and RANS modelling of under-expanded jets with application to gaseous fuel direct injection for advanced propulsion systems. *Int. J. Heat Fluid Flow* **2019**, *76*, 309–334. [CrossRef]
46. Engine Combustion Network (ECN). Available online: https://ecn.sandia.gov/ (accessed on 31 July 2023).
47. Zembi, J.; Cruccolini, V.; Mariani, F.; Scarcelli, R.; Battistoni, M. Modeling of thermal and kinetic processes in non-equilibrium plasma ignition applied to a lean combustion engine. *Appl. Therm. Eng.* **2021**, *197*, 117377. [CrossRef]
48. Zembi, J.; Battistoni, M.; Mariani, F.; Irimescu, A.; Vaglieco, B.M.; Merola, S.S. Investigations on the impact of port water injection on soot formation in a DISI engine through CFD simulations and optical methods. *Fuel* **2023**, *337*, 127170. [CrossRef]
49. Salazar, V.; Kaiser, S. *Interaction of Intake-Induced Flow and Injection Jet in a Direct-Injection Hydrogen-Fueled Engine Measured by PIV*; SAE 2011 World Congress and Exhibition; Technical Paper 2011-01-0673; SAE: Warrendale, PA, USA, 2011. [CrossRef]
50. Ó Conaire, M.; Curran, H.J.; Simmie, J.M.; Pitz, W.J.; Westbrook, C.K. A comprehensive modeling study of hydrogen oxidation. *Int. J. Chem. Kinet.* **2004**, *36*, 603–622. [CrossRef]
51. Zembi, J.; Battistoni, M.; Mariani, F.; Irimescu, A.; Merola, S.S. *Pressure and Flow Field Effects on Arc Channel Characteristics for a J-Type Spark Plug*; SAE Technical Papers 2022-01-0436; SAE: Warrendale, PA, USA, 2022. [CrossRef]
52. Di Battista, D.; Di Bartolomeo, M.; Cipollone, R. Flow and thermal management of engine intake air for fuel and emissions saving. *Energy Convers. Manag.* **2018**, *173*, 46–55. [CrossRef]

53. Dickes, R.; Dumont, O.; Legros, A.; Quoilin, S.; Lemort, V. Analysis and comparison of different modeling approaches for the simulation of a micro-scale organic Rankine cycle power plant. In Proceedings of the ASME-ORC 2015, Brussels, Belgium, 12–14 October 2015.
54. Legros, A.; Guillaume, L.; Diny, M.; Lemort, V. Modelling, sizing and testing a scroll expander for a waste heat recovery application on a gasoline engine. In *IOP Conference Series: Materials Science and Engineering*; IOP Publishing: Bristol, UK, 2015; Volume 90. [CrossRef]
55. Fatigati, F.; Vittorini, D.; Coletta, A.; Cipollone, R. Assessment of the differential impact of scroll and sliding vane rotary expander permeability on the energy performance of a small-scale solar-ORC unit. *Energy Convers. Manag.* **2022**, *269*, 116169. [CrossRef]
56. Marchionni, M.; Fatigati, F.; Di Bartolomeo, M.; Di Battista, D.; Petrollese, M. Experimental and Numerical Dynamic Investigation of an ORC System for Waste Heat Recovery Applications in Transportation Sector. *Energies* **2022**, *15*, 9339. [CrossRef]
57. Di Battista, D.; Cipollone, R. Waste Energy Recovery and Valorization in Internal Combustion Engines for Transportation. *Energies* **2023**, *16*, 3503. [CrossRef]
58. Di Battista, D.; Carapellucci, R.; Cipollone, R. Integrated evaluation of Inverted Brayton cycle recovery unit bottomed to a turbocharged diesel engine. *Appl. Therm. Eng.* **2020**, *175*, 115353. [CrossRef]
59. Shi, M.; Wang, H.; Yang, C.; Wang, Y.; Niu, X. Experimental Research on the Matching Characteristics of the Compound VGT-STC System with a V-Type Diesel Engine. *Machines* **2022**, *10*, 788. [CrossRef]
60. Di Battista, D.; Di Bartolomeo, M.; Cipollone, R. Full energy recovery from exhaust gases in a turbocharged diesel engine. *Energy Convers. Manag.* **2022**, *271*, 116280. [CrossRef]
61. Di Bartolomeo, M.; Di Battista, D.; Fatigati, F.; Cau, G.; Cipollone, R. *Direct and Indirect Exhaust Heat Recovery from Turbocharged Heavy-Duty Engine*; SAE Technical Paper 2023-24-0122; SAE: Warrendale, PA, USA, 2023. [CrossRef]
62. Savva, Z.; Petallidou, K.C.; Damaskinos, C.M.; Olympiou, G.G.; Stathopoulos, V.N.; Efstathiou, A.M. $H_2$-SCR of NOx on low-SSA $CeO_2$-supported Pd: The effect of Pd particle size. *Appl. Catal. A Gen.* **2021**, *615*, 118062. [CrossRef]
63. Nanba, T.; Kohno, C.; Masukawa, S.; Uchisawa, J.; Nakayama, N.; Obuchi, A. Improvements in the $N_2$ selectivity of Pt catalysts in the NO-$H_2$-$O_2$ reaction at low temperatures. *Appl. Catal. B* **2003**, *46*, 353–364. [CrossRef]
64. Granger, P.; Dhainaut, F.; Pietrzik, S.; Malfoy, P.; Mamede, A.S.; Leclercq, L.; Leclercq, G. An overview: Comparative kinetic behaviour of Pt, Rh and Pd in the NO + CO and NO + $H_2$ reactions. *Top. Catal.* **2006**, *39*, 65–76. [CrossRef]
65. Zheng, L.; Casapu, M.; Stehle, M.; Deutschmann, O.; Grunwaldt, J.D. Selective Catalytic Reduction of NOx with Ammonia and Hydrocarbon Oxidation Over $V_2O_5$–$MoO_3$/$TiO_2$ and $V_2O_5$–$WO_3$/$TiO_2$ SCR Catalysts. *Top. Catal.* **2019**, *62*, 129–139. [CrossRef]
66. Lang, W.; Laing, P.; Cheng, Y.; Hubbard, C.; Harold, M.P. Co-oxidation of CO and propylene on Pd/$CeO_2$-$ZrO_2$ and Pd/$Al_2O_3$ monolith catalysts: A light-off, kinetics, and mechanistic study. *Appl. Catal. B* **2017**, *218*, 430–442. [CrossRef]
67. Sheng, L.; Ma, Z.; Chen, S.; Lou, J.; Li, C.; Li, S.; Zhang, Z.; Wang, Y.; Yang, H. Mechanistic insight into $N_2O$ formation during NO reduction by $NH_3$ over Pd/$CeO_2$ catalyst in the absence of $O_2$. *Chin. J. Catal.* **2019**, *40*, 1070–1077. [CrossRef]
68. Zhang, Q.; Geng, G.N.; Wang, S.W.; Richter, A.; He, K. Bin Satellite remote sensing of changes in NOx emissions over China during 1996–2010. *Chin. Sci. Bull.* **2012**, *57*, 2857–2864. [CrossRef]
69. Khoder, M.I. Atmospheric conversion of sulfur dioxide to particulate sulfate and nitrogen dioxide to particulate nitrate and gaseous nitric acid in an urban area. *Chemosphere* **2002**, *46*, 675–684. [CrossRef]
70. Millo, F.; Piano, A.; Rolando, L.; Accurso, F.; Gullino, F.; Roggio, S.; Bianco, A.; Pesce, F.; Vassallo, A.; Rossi, R. Synergetic Application of Zero-, One-, and Three-Dimensional Computational Fluid Dynamics Approaches for Hydrogen-Fuelled Spark Ignition Engine Simulation. *SAE Int. J. Engines* **2021**, *15*, 561–580. [CrossRef]
71. Liberzon, D. *Calculus of Variations and Optimal Control Theory: A Concise Introduction*; Princeton University Press: Princeton, NJ, USA, 2011.
72. Onori, S.; Serrao, L.; Rizzoni, G. *Hybrid Electric Vehicles: Energy Management Strategies*; Springer: Berlin/Heidelberg, Germany, 2016.
73. Donatantonio, F.; Ferrara, A.; Polverino, P.; Arsie, I.; Pianese, C. Novel Approaches for Energy Management Strategies of Hybrid Electric Vehicles and Comparison with Conventional Solutions. *Energies* **2022**, *15*, 1972. [CrossRef]


**Disclaimer/Publisher's Note:** The statements, opinions and data contained in all publications are solely those of the individual author(s) and contributor(s) and not of MDPI and/or the editor(s). MDPI and/or the editor(s) disclaim responsibility for any injury to people or property resulting from any ideas, methods, instructions or products referred to in the content.

energies

*Article*

# Retrofit of Diesel Engines with $H_2$ for Potential Decarbonization of Non-Electrified Railways: Assessment with Lifecycle Analysis and Advanced Numerical Modeling

**Mehrshad Kolahchian Tabrizi, Tarcisio Cerri, Davide Bonalumi *, Tommaso Lucchini and Morris Brenna**

Department of Energy, Politecnico di Milano, Via Lambruschini 4, 20156 Milano, Italy; mehrshad.kolahchian@polimi.it (M.K.T.); tarcisio.cerri@sursum-mi.com (T.C.); tommaso.lucchini@polimi.it (T.L.); morris.brenna@polimi.it (M.B.)
* Correspondence: davide.bonalumi@polimi.it; Tel.: +39-02-2399-3817

**Abstract:** The application of hydrogen in heavy-duty vehicles or trains has been suggested as a promising solution to decarbonize the transportation sector. In this study, a one-dimensional engine modeling is employed to evaluate the potential of hydrogen as a fuel for railway applications. A turbocharged diesel engine is simulated as the baseline unit, and the results are validated with experimental data. The same engine is converted to become compatible with hydrogen through some modifications in the turbocharger group and the injection and ignition systems to preserve the performance of the baseline configuration. The findings show that the engine traction power is reduced from 600 to 400 kW, indicating an inferior performance for the hydrogen-fueled engine. The energy consumption of the hydrogen-fueled engine on a real train mission profile is almost two times the diesel version. However, our Life Cycle Assessment analysis with a Well-to-Wheel system boundary shows a 56% reduction in equivalent $CO_2$ emissions for the engine fueled with photovoltaic-based green hydrogen. Substituting diesel with low-carbon hydrogen can decrease the train's carbon footprint from 4.27 to even less than 2 kg $CO_2$ eq./km, suggesting that moderately modified engines are a promising solution for decarbonizing non-feasibly electrified railway sections.

**Keywords:** hydrogen; trains; Life Cycle Assessment; transportation; green hydrogen; carbon footprint

**Citation:** Kolahchian Tabrizi, M.; Cerri, T.; Bonalumi, D.; Lucchini, T.; Brenna, M. Retrofit of Diesel Engines with $H_2$ for Potential Decarbonization of Non-Electrified Railways: Assessment with Lifecycle Analysis and Advanced Numerical Modeling. *Energies* **2024**, *17*, 996. https://doi.org/10.3390/en17050996

Academic Editors: Davide Lanni and Enzo Galloni

Received: 16 January 2024
Revised: 6 February 2024
Accepted: 8 February 2024
Published: 20 February 2024

**Copyright:** © 2024 by the authors. Licensee MDPI, Basel, Switzerland. This article is an open access article distributed under the terms and conditions of the Creative Commons Attribution (CC BY) license (https://creativecommons.org/licenses/by/4.0/).

## 1. Introduction

Hydrogen is becoming a potential solution for decarbonizing energy systems. According to net-zero emission (NZE) scenarios, hydrogen and $H_2$-based fuels will account for 10% of total final energy consumption in 2050 [1]. New applications like hard-to-abate industries, power generation, and transport will constitute 40% of the $H_2$ demand in 2030 [2]. For vehicles, the application of lithium-ion batteries in electric vehicles can be promising, specifically for smaller vehicles circulating in countries with lower electricity carbon intensity [3,4]. However, for decarbonizing heavy-duty vehicles or trains, hydrogen can become a more appropriate solution [5–7]. On the one hand, there has been a growing interest in introducing fuel cell-based hydrogen trains for use on railway lines that cannot be or are not conveniently electrified due to geometric constraints or low traffic characterizations [8,9]. On the other hand, many newly built diesel train fleets could benefit from the conversion of their internal combustion engine (ICE) to become compatible with hydrogen fuel, reducing their environmental impact with limited costs [10,11].

Hydrogen can be obtained using a wide range of methodologies, of which some of them can produce hydrogen with low environmental impacts, notably low $CO_2$ equivalent emissions. In 2022, only less than 1% of the produced hydrogen (less than 9.5 Mt) was obtained via a low-carbon route; however, the low-carbon $H_2$ production is expected to reach 450 Mt in 2050 [2]. High Technology Readiness Level (TRL) and the possibility of scaling up the production made the water electrolysis powered by renewable electricity

(Green $H_2$) and reforming of fossil fuels coupled with carbon capture, utilization, and storage (Blue $H_2$) the main routes for low-carbon $H_2$ production. Based on the NZE scenario defined by the International Energy Agency (IEA), by 2030, 40% of electricity will be generated by photovoltaic (PV) and wind sources with a subsequent increase to 70% by 2050 [1]. Therefore, wind and PV electricity will become the main power sources for electrolysis systems.

Life Cycle Assessment (LCA) can be employed to calculate the environmental impacts associated with products or services. Similarly, LCA can be applied to compute the carbon footprint of different $H_2$ production methods. Within an LCA analysis, a Well-to-Wheel system boundary can be used to compare the environmental performance of various fuels used for transportation purposes.

This work aims to evaluate the potential of hydrogen-fueled internal combustion engines for railway applications, featuring an LCA analysis to investigate the carbon footprint of these trains. Particularly, the study focuses on Diesel Multiple Units (DMUs) with electric traction motors, which are already in operation on non-electrified lines in several European countries. The diesel engine configuration is considered as a baseline. Numerical one-dimensional simulations using the Gasdyn code 2022 [12–16] are employed to estimate the fuel consumption of diesel or hydrogen-fueled railway vehicles on a real train mission profile. The simulation results for the diesel engine are validated with experimental data on power and fuel consumption in the operating points relevant to the train operation. The diesel engine unit is moderately converted to become compatible with hydrogen. The modifications are limited to the replacement of the combustion and injection systems and the turbocharger group to restrict the cost burdens. Finally, the climate profile of the baseline diesel train and the hydrogen-fueled train are analyzed.

## 2. Materials and Methods

In this section, the performance of the original diesel engine train is defined in Section 2.1. The methodology adopted for the conversion of diesel-based engines to hydrogen-fueled is explained in Section 2.2. The application of Life Cycle Assessment with the Well-to-Wheel system boundary is described in Section 2.3.

### 2.1. Performance of the Original Railway Vehicle

The investigated configuration consists of a single-decker articulated railway vehicle (also known as DMU—Diesel Multiple Unit) equipped with two identical diesel engines. Tables 1 and 2 summarize the main technical data of the vehicle and engine, respectively.

**Table 1.** Main characteristics of the railway vehicle under investigation.

| Parameter | Description |
|---|---|
| Diesel engine rated power [kW] | 2 × 365 |
| Vehicle mass [kg] | 68,000 |
| Vehicle maximum speed [km/h] | 140 |
| Maximum power at wheels [kW] | 600 |
| Starting tractive force [kN] | 80 |
| Maximum acceleration [m/s$^2$] | 1.0 |

**Table 2.** Main characteristics of the diesel engine.

| Parameter | Description |
|---|---|
| Engine arrangement [-] | 6 in line |
| Bore [cm] | 12.8 |
| Stroke [cm] | 16.6 |
| Total displacement [cm$^3$] | 12,816 |
| Compression ratio [-] | 18:1 |
| Emission compliance | Tier IIIA |

The schematic of the diesel–electric arrangement of the studied DMU is shown in Figure 1. The DMU under study has a diesel–electric arrangement. The power generated by the internal combustion engine is transferred to the wheels by decoupling the torque and speed parameters. In particular, the ICE can deliver constant power while the rotation speed and the delivered torque are kept constant, even as the torque and speed produced by the electric motors connected to the wheels vary.

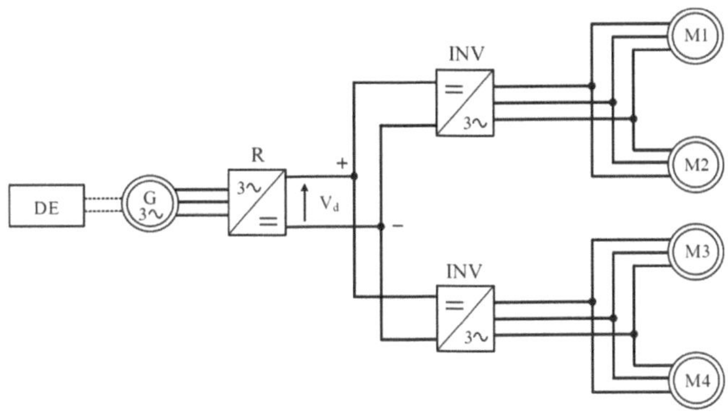


**Figure 1.** Diesel–electric arrangement for a DMU.

Figure 2 summarizes the performance of the railway vehicle operating with a diesel engine in terms of both maximum tractive force and the corresponding power at wheels as a function of the vehicle velocity. The maximum power supplied by two engines is higher than the one required by the wheels, where the difference is available for the onboard systems (electronic converters, interior lighting, air conditioning, etc.).

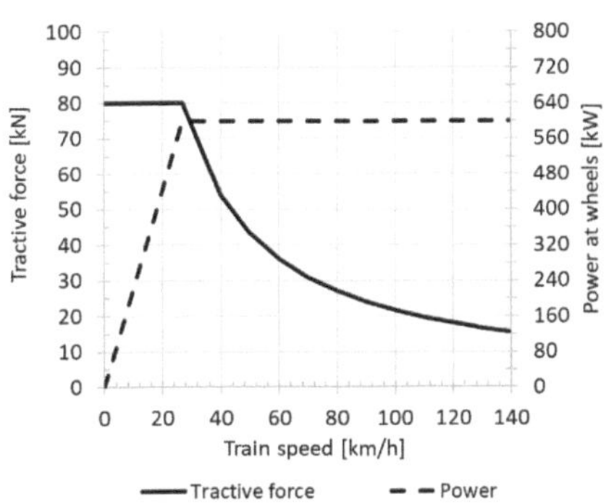


**Figure 2.** Tractive force (mechanical characteristic) and wheel power as a function of train speed when diesel thermal units work on the train.

Given the required engine constant power ($P$), the corresponding load ($L$), the duration for which the engine is switched on ($t$), and the revolution speed ($n$) at which power is delivered, it is possible to compute the fuel consumption for a specific mission profile using Equation (1):

$$m_{fuel} = P(n,\ L) \cdot BSFC(n, L) \cdot t_{switched\ on} \quad (1)$$

where $BSFC$ is the brake-specific fuel consumption for the corresponding operating point. One-dimensional numerical simulations are used to estimate $BSFC$. The Gasdyn code is employed in this work to simulate the investigated engine operating with diesel and hydrogen fuels. Gasdyn is a one-dimensional thermo-fluid dynamic model continuously developed and enhanced by the researchers at Politecnico di Milano. The main features of the numerical code are described in previous works [10–14]. The model is developed for the simulation of internal combustion engines coupled with their complete intake and exhaust systems, allowing for a detailed analysis of the fluid dynamic and thermal and chemical behaviors of the entire engine system. The Gasdyn code uses a set of different numerical solvers, including symmetric TVD finite difference techniques with second-order accuracy as described in [17]. Following a characteristic-based approach to the calculation of Riemann variables [18], boundary conditions and intra-pipe connections are developed for the representation of components such as turbochargers, multi-pipe junctions, abrupt area changes, inlet and outlet, etc. [18].

For compression ignition combustion, the model is designed to handle modern multi-pulse injection systems. Within the model, the combustion chamber is subdivided into three zones: fresh charge (air + residual gas), fuel (already vaporized), and burned gas. Pressure is assumed to be constant in the entire combustion chamber, directly ensuring the fulfillment of momentum conservation. Therefore, the solution of the mass and energy equations applied to the burned, unburned, and fuel zones allows for the computation of the pressure and temperature of the three zones. Each discrete injection event during the engine cycle is defined as a pulse, where up to 4 pulses can be considered. The burning rate of a single pulse is expressed as the sum of two stages, one describing the premixed phase and the other describing the diffusive phase. The rate of the burned mass fraction is calculated using the Wiebe equation, applied to the premixed and diffusive combustion processes. Distinct Wiebe coefficients are employed for each pulse (pre1, pre2, main, and post pulses) [19]. The entire engine layout is modeled, including the turbocharger, with a suitable model to predict the in-cylinder pressure development typical of compression ignition combustion.

Regarding the spark ignition combustion of the hydrogen-fueled engine, a two-zone, quasi-dimensional combustion model is used to predict the in-cylinder pressure development. The Zimont model [20] is employed to compute the fully developed turbulent flame speed, enabling the [20] evaluation of thermodynamic and chemical gas properties inside the combustion chamber [14,16]. In both combustion models, the heat transfer coefficient between the gas inside the cylinder and cylinder walls is calculated using the Woschni correlation [21].

Figure 3 compares the measured and predicted power and BSFC of the diesel engine configuration at full-load conditions. The operating point of the mounted thermal engine on DMU is specified at 1500 rpm. The predicted and estimated values are in good agreement. This alignment suggests that the proposed numerical approach can be used for a realistic estimation of diesel fuel consumption across different mission profiles.

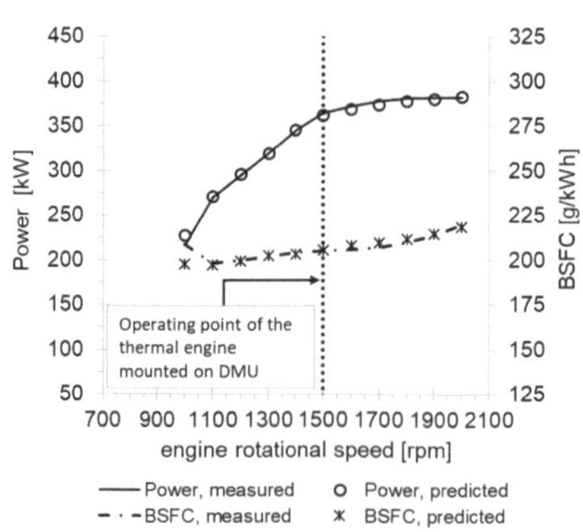


**Figure 3.** Comparison between measured and predicted performance of the diesel engine at full load conditions (the operating point of the thermal engine mounted on DMU is specified).

### 2.2. Engine Conversion to Hydrogen Fuel

The diesel engine unit is modified to become compatible with hydrogen fuel. The validated diesel engine model described in Section 2.1 is used for further simulations of the modified engine. To minimize the conversion costs, using consolidated components, the compression-ignition combustion system is replaced with the spark ignition, where hydrogen is delivered in the intake system via port-fuel injection (PFI). A different turbocharger is adopted due to the different air and exhaust mass flow rates. The hydrogen engine operates at the stoichiometric air/fuel ratio ($\lambda = 1$) to reach the maximum power at full load. Maximum allowed cylinder (during the combustion phase) and boost pressures are kept similar to the diesel configuration to avoid the increase in mechanical and thermal stresses. Table 3 summarizes the main modifications required to convert the diesel engine to operate with hydrogen fuel.

**Table 3.** Main modifications of the engine to be adequate for $H_2$ fueling.

| Parameter | Diesel Engine | Hydrogen Engine |
|---|---|---|
| Fuel injection system | Direct injection | Port injection |
| Ignition system | Compression ignition | Spark ignition |
| Compression ratio [-] | 18:1 | 12:1 |
| Air metering | Turbocharger (turbine diameter: 89.5 mm) | Turbocharger (turbine diameter: 70 mm) |
| Relative air/fuel ratio ($\lambda$) at rated power | 1.5 | 1.0 |

### 2.3. Well-to-Wheel Analysis

The process-based Life Cycle Assessment (LCA) method is one of the most analytical methods for evaluating the environmental profile of products (goods and services). LCA is regulated according to the principles defined in ISO 14040 and 14044 [22,23]. According to the application of LCA, different system boundaries can be considered. In transportation applications, a Well-to-Wheel (WTW) analysis is a common system boundary. Within the WTW analysis, the life cycle of a fuel is considered from its production up to its conversion into power at wheels. In this type of analysis, the impacts generated by the manufacturing of the transportation means (vehicle) are not included. In this study, only small modifications are applied to the engine; therefore, the diesel and $H_2$ trains are almost similar. In this case, a comparison based on WTW analysis becomes more relevant. The WTW emission of diesel used for transportation purposes in Europe is 89 g $CO_2$ eq./MJ diesel, and the low heating value (LHV) is 43.2 MJ/kg diesel [24].

In the literature, several studies using the LCA methodology calculated the Global Warming Potential (GWP) or the so-called carbon footprint of green $H_2$. The GWP values for PV-based $H_2$ via electrolysis can vary from 2.1 to 9.37 kg $CO_2$ eq./kg $H_2$ as a function of location, PV panel efficiency, and the database used [25]. The $H_2$ obtained through wind-powered electrolysis shows a lower carbon footprint compared to PV-based $H_2$ with a mean value equal to 1.29 kg $CO_2$ eq./kg $H_2$ [26]. Kolahchian et al. [25] estimated the carbon footprint of the production of PV-based $H_2$, taking into account the advancement of the PV industry and the current supply chain of PV panels in the Italian context. They compared their updated GWP values with the default values from the ecoinvent database [27,28]. They concluded that the ecoinvent database could not represent the carbon footprint of the production of green $H_2$ when state-of-the-art PV modules power the electrolysis system. Based on the real PV electricity production equivalent hours, calculated using the data from the Italian electricity transmission operator (Terna) [29], the GWP of the production of PV-based $H_2$ is 2.15 kg $CO_2$ eq./kg $H_2$. The annual potential of renewable electricity generation (PV, onshore and offshore wind, hydro, and geothermal) is around 715 TWh in Italy, where the majority, 50%, can be generated by PV systems [30]. Therefore, the choice of using PV-based $H_2$ can be justified. Also, blue hydrogen can play an important role during the transition period to a fully green $H_2$ market. The carbon footprint of blue $H_2$ for 1.85%

methane leakage and 90% carbon capture efficiency is around 3.5 kg $CO_2$ eq./kg $H_2$ (this value is calculated by averaging the results in Romano et al. [31]). Analyzing the impact of the origin of blue $H_2$ on the final GWP value of the train trip, a sensitivity analysis of the methane leakage rate in the gas supply chain is carried out.

Hydrogen should be compressed to increase its volume-based energy density, leading to a reduction in the storage size. The high-pressure storage for mobility applications offers pressures equal to 350 or 700 bars. In this work, hydrogen storage is assumed to be at 700 bars. Therefore, the produced hydrogen is compressed to pressures higher than the storage pressure, e.g., 800 bars. The energy requirement of the hydrogen multi-stage compression for reaching 800 bars is around 12% of hydrogen high-heating value (HHV), which means an additional power consumption equal to 4.7 kWh electricity/kg $H_2$ is needed [32,33]. Some studies reported an $H_2$ compression power consumption rate of less than 2 kWh/kg $H_2$ [34,35]. A manufacturer offers compressors with a power consumption of 2.7 kWh/kg $H_2$ for output pressures of up to 1000 bars [36]. In this work, the $H_2$ compression power requirement is assumed to be equal to 4 kWh/kg $H_2$ as an average value among the literature and manufacturer data. The Italian grid electricity with a carbon intensity of 0.371 kg $CO_2$ eq./kWh is considered for powering the compressors [37]. Also, for green $H_2$, the compression process powered by PV electricity is considered. The carbon footprint of PV electricity is extracted directly from the ecoinvent 3.5 database (cut-off method) using the default value for Italy (570 kWp open ground installation, multi-Si). Table 4 reports a summary of values considered within the Well-to-Wheel analysis.

**Table 4.** A summary of values considered in Well-to-Wheel analysis.

| Parameter | Description |
|---|---|
| GWP of green $H_2$ [kg $CO_2$ eq./kg $H_2$] | 2.15 |
| GWP of blue $H_2$ [kg $CO_2$ eq./kg $H_2$] | 3.5 |
| GWP of diesel [kg $CO_2$ eq./kg Diesel] | 3.85 |
| GWP of Italian grid electricity [kg $CO_2$ eq./kWh] | 0.371 |
| GWP of PV electricity for Italy [kg $CO_2$ eq./kWh] | 0.073 |
| Storage pressure [bar] | 700 |
| Compressor power consumption [kWh/kg $H_2$] | 4 |

## 3. Results and Discussion

The performance of the modified diesel engine fueled with hydrogen is analyzed and compared to the baseline diesel engine in Section 3.1. The consumption of both the baseline and modified diesel engine using the journey simulation is reported in Section 3.2. The results of the Well-to-Wheel analysis are presented in Section 3.3.

### 3.1. Performance of Modified Diesel Engine (Hydrogen-Fueled Engine)

The predicted performance of the modified diesel engine (hydrogen-fueled engine) is illustrated in Figure 4. According to the simulation results, the estimated performance of the hydrogen-fueled engine is sufficient to operate the railway vehicle, despite delivering a reduced amount of power compared to the original diesel engine. The performance reduction of the hydrogen-fueled engine is related to several factors:

- Reduced volumetric efficiency due to hydrogen PFI injection. The higher specific volume of hydrogen reduces the amount of air that can be trapped inside the cylinder, resulting in a reduction in the engine power compared to the diesel configuration.
- Lower compression ratio during engine operation to reduce knock risks. This factor clearly affects the efficiency and the maximum achievable power.
- Limited boost pressure. The reduced airflow rate restricts the increase in the boost pressure to achieve the same power as the original engine.

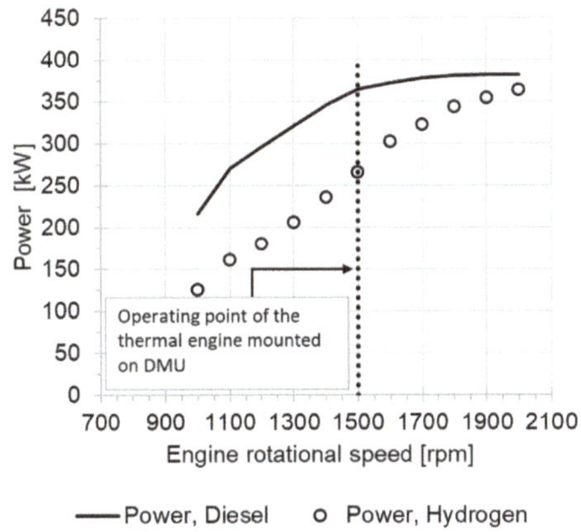


**Figure 4.** Comparison between the predicted effective power of the diesel and hydrogen engines at full-load operating points (the operating point of the thermal engine mounted on DMU is specified).

A more sophisticated and expensive upgrade, like the use of a direct high-pressure injection fuel system, is needed to achieve the previous level of power, leading to a rise in the total transformation costs [38,39]. Figure 5 shows the comparison between the fuel consumption of hydrogen and diesel engines as a function of engine rotational speed at full-load operating points. The mass-based hydrogen consumption is significantly lower than the diesel consumption. The notable difference between the two fuels is caused by the higher LHV of the hydrogen fuel compared to diesel fuel.

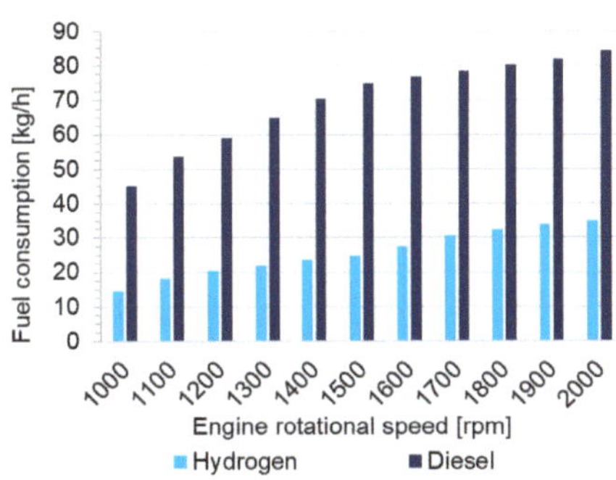


**Figure 5.** Comparison between predicted fuel consumption of hydrogen and diesel engines at full-load operating points.

Similar modifications applied to diesel engines to become compatible with natural gas are needed to obtain the hydrogen-fueled engine. However, the use of hydrogen entails higher costs for the injection system, as it must guarantee higher safety standards to avoid leaks and fires. According to the literature, the cost of the conversion of a diesel engine to become compatible with natural gas is in the range of EUR 8000 to EUR 12,000 for each engine [40]. The estimated cost includes the following stages:

- Engine disassembly;
- Revision and eventual replacement of some components;
- Replacement of pistons to reduce the compression ratio;
- Modification of the cylinder head to place the spark plugs;
- Installation of the phase sensor on the camshaft;
- Engine assembly;
- Installation of throttle valve for load control, ignition system, port-fuel injection system;

- Replacement of the turbocharger unit;
- New engine calibration.

Further costs are necessary for changing the after-treatment system required to reduce $NO_x$ emissions, updating the onboard safety systems, and installing high-pressure hydrogen tanks. The cost of a high-pressure hydrogen tank for storage at 700 bars is about 15 EUR/kWh (referred to the $H_2$ LHV) [41].

### 3.2. Train Round-Trip Consumption

Several simulations of a typical mission profile of the railway vehicle under investigation are conducted for both the diesel engine and hydrogen-fueled engine configurations. The altitude profile of the railway line under investigation, as a function of the distance from its departure station up to the arrival destination, is shown in Figure 6. The red vertical lines denote the intermediate stops.

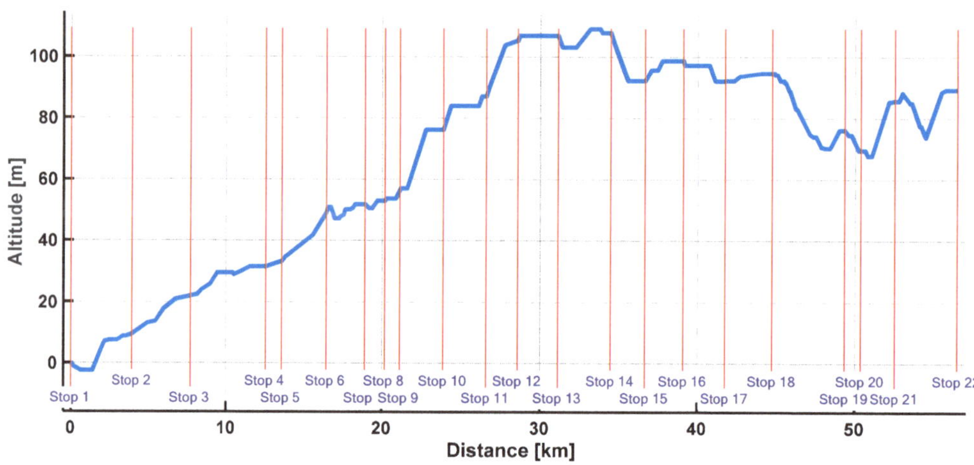


**Figure 6.** The altitude profile of the railway line under investigation (intermediate stops are denoted by red vertical lines).

In the baseline scenario, when the engine works with diesel, the simultaneous contribution of two onboard thermal engines can satisfy the 600 kW power request for the traction of the articulated railway vehicle. Meanwhile, the torque and speed of the electric motors are managed by inverters. This setup allows for the torque and rotational speed of the thermal unit to remain independent of the electric traction motor's performance. For the baseline diesel engine, according to Figure 2, the starting tractive force and the train base speed are 80 kN and 27 km/h, respectively. Using Equation (1), the mean fuel consumption for the railway vehicle traveling along the line under investigation is 1.11 kg/km, resulting in a round-trip diesel consumption equal to 127 kg.

As stated in Section 3.1, the hydrogen-fueled engine performs inferiorly compared to the original diesel unit, resulting in a power loss of about 100 kW. Figure 7 shows the new mechanical characteristics of the rail vehicle with hydrogen engines. Due to the lower performance of the hydrogen-fueled engine, the same railway vehicle has only 400 kW available for traction power instead of the original 600 kW.

Figure 8 compares the mechanical characteristics of a DMU in diesel and hydrogen configurations. For both the diesel and hydrogen versions, the motor drive can deliver the maximum starting tractive force. In the hydrogen version, having less power, the tractive force decreases at a lower speed, but for a suburban DMU, this did not prove to be a very penalizing factor. Figure 8 also illustrates that during the acceleration phase, the ICE operates for a large part of its characteristic curve at a constant power at the optimal operating point as the torque and speed parameters at the wheels are controlled by the electric motor drives.

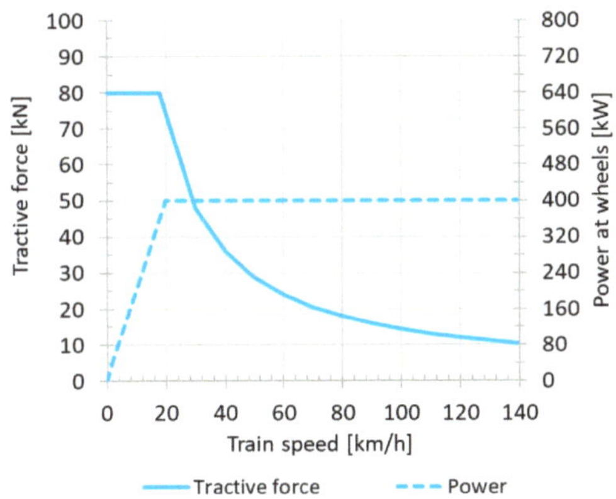


**Figure 7.** Tractive force (mechanical characteristic) and wheel power as a function of train speed when hydrogen thermal units work on the train.

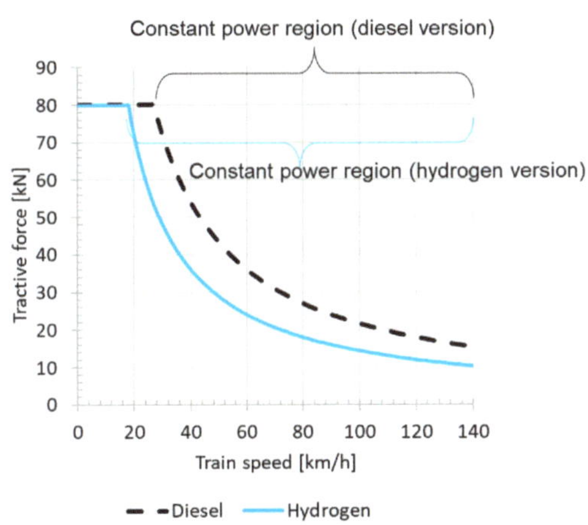


**Figure 8.** Mechanical characteristics for a DMU in diesel and hydrogen configurations.

Despite the reduced power of the hydrogen-fueled engine, the electric traction motors can still provide the maximum starting torque. Therefore, according to the simulation of the investigated suburban railway line equipped with the hydrogen-fueled engine, the traveling time does not increase substantially compared to the diesel engine scenario. The trip duration of the hydrogen train is predicted to be 5 min longer than the 100 min total travel time of the diesel version. The low permissible speeds and the proximity between the stops grant the limited time increase while, in the case of railway lines characterized by higher speeds and heavier rolling stokes like freight trains, performance problems might arise.

Based on the obtained mechanical characteristics curves of the hydrogen-fueled engine and assuming a similar onboard ancillary power requirement to the diesel train version, the performance of the hydrogen train along the same mission profile is predicted. The specific fuel consumption of the hydrogen engine for the operating point under consideration is equal to 90 g $H_2$/kWh. The simulation of the round trip of the train journey results in an overall hydrogen consumption of 87.7 kg, showing a 30% lower mass-based fuel consumption compared to the diesel version. Based on the LHV of the diesel and hydrogen, the energy content of the diesel consumed during a round trip is around 1500 kWh, which almost doubles and reaches around 2900 kWh for the hydrogen-fueled engine. This apparently high value of total hydrogen consumption compared to the diesel case (in terms of energy content) is due to the considerable difference in maximum power (about 200 kW lower for hydrogen-fueled engine), imposing the hydrogen version to run at the maximum power for a longer time in order to guarantee a nearly identical total trip time of the investigated mission profile.

### 3.3. Carbon Footprint

The carbon footprint of the train fueled with diesel, green, or blue $H_2$ is plotted in Figure 9. When diesel is used as the train's fuel, the WTW analysis shows a GWP value equal to 4.27 kg $CO_2$ eq./km. Green $H_2$ produced and compressed via PV electricity reduces the carbon footprint to 1.87 kg $CO_2$ eq./km, equivalent to a 56% decrease compared to the diesel train. Compressing the PV-based $H_2$ with the Italian grid electricity increases the GWP value to 2.79 kg $CO_2$ eq./km, corresponding to a 35% reduction compared to the conventional diesel train or a 48% increase with respect to the PV-based compression scenario. The latter point emphasizes the considerable GWP impact generated at the $H_2$ compression stage using the grid electricity. According to the assumptions made for blue $H_2$ in this work, the carbon mitigation in this scenario is minimized compared to the diesel-based train. The equivalent carbon emission of the train fueled with blue $H_2$ is almost 3.8 kg $CO_2$ eq. for each kilometer (only 11% less than the diesel-based train). The blue $H_2$ produced under a gas supply chain with a lower methane leakage like 0.2% (GWP = 1.7 kg $CO_2$ eq./kg $H_2$) can further reduce the train's carbon footprint to 2.44 kg $CO_2$ eq./km (43% lower than diesel), which even overperforms the green $H_2$ compressed by the Italian grid electricity. However, if blue $H_2$ is produced under the 3.5% methane leakage rate (GWP = 5.35 kg $CO_2$ eq./kg $H_2$), the advantage of a lower GWP associated with using hydrogen disappears, and the train emits 5.2 kg $CO_2$ eq./km (22% higher than diesel). The sensitivity analysis on the methane leakage rate for blue $H_2$ is depicted with an orange dashed rectangle.

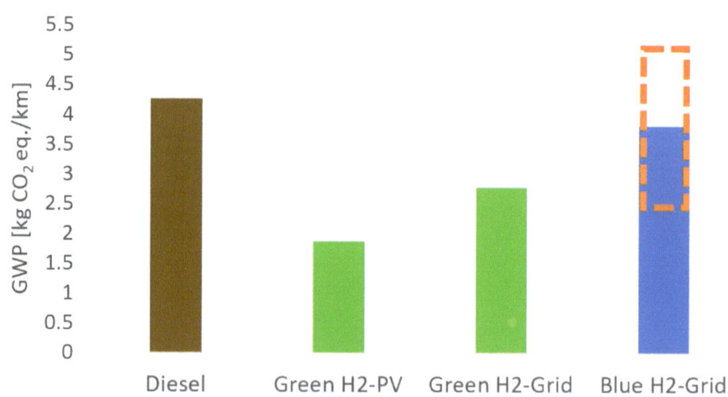


**Figure 9.** The GWP impact of the use phase of the studied train for different fuels per kilometer.

Figure 10 shows the GWP impact of the round trip of the train fueled with diesel or hydrogen. On each round trip, the conventional train results in a carbon footprint equal to 488 kg $CO_2$ eq. Utilizing hydrogen instead of diesel can bring a considerable amount of avoided emission, specifically in the case of the green $H_2$ powered and compressed via PV electricity. Compared to the diesel train, by using PV-based $H_2$, almost 274 or 170 kg $CO_2$ eq. is avoided for the PV or grid-compressed $H_2$, respectively. The average scenario for blue $H_2$ can result in a reduction of 52 kg $CO_2$ eq. for each train round trip. These reductions in equivalent $CO_2$ emissions suggest that carbon mitigation ranging from

56 to 11% can be obtained if diesel is substituted with hydrogen in a moderately modified diesel engine. Despite a considerable increase in the energy consumption of the hydrogen-fueled engine on a real train mission profile, the obtained reductions in equivalent $CO_2$ are promising, suggesting that moderately modified engines are viable candidates for decarbonizing railway sections that are not feasibly electrified.

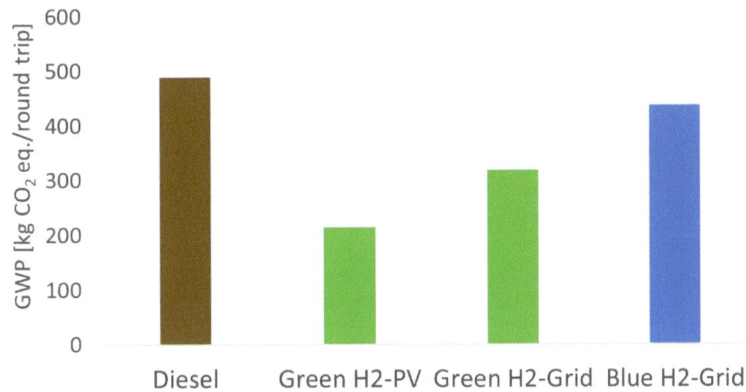


**Figure 10.** The GWP impact of the use phase of the studied train for different fuels for the round trip.

## 4. Conclusions

This paper demonstrates the potential of converting the internal combustion engines of a diesel–electric self-propelled train from diesel power to hydrogen power. A real single-decker articulated railway vehicle (Diesel Multiple Unit) currently operating on several non-electrified European railway lines is considered. This train is characterized by two or more diesel generator sets supplying power to the electric traction motors and onboard auxiliary services. The diesel engine is moderately modified to restrict the costs arising. A spark ignition system with hydrogen delivery in the intake system via port-fuel injection is utilized instead of the compression ignition combustion system. The turbocharger is replaced to handle different air and exhaust mass flow rates of the hydrogen-fueled engine. The baseline diesel engine is simulated using the Gasdyn code, and the model is validated through the measured performance values. The hydrogen-fueled engine is analyzed using the modified numerical model for spark ignition engines. Life Cycle Assessment analysis with a Well-to-Wheel system boundary is applied to evaluate the carbon footprint of the hydrogen train using both green and blue hydrogen. The main findings of this work are summarized in the following points:

- Switching the fuel from diesel to hydrogen causes a reduction in the power generated by the internal combustion engines while the electric traction motors still provide the maximum starting torque. The traction power of 600 kW for the original diesel engine is reduced to 400 kW for the modified engine working with hydrogen.
- In the case of the studied real railway line for suburban services, travel duration slightly increases from the initial 100 to 105 min for the hydrogen version. The low permissible speeds and the proximity between stops help to limit the time increase. However, for lines characterized by higher speeds and heavier rolling stokes, such as freight trains, performance problems might arise.
- The fuel consumption changes from 127 kg of diesel to 87.7 kg of hydrogen for the round trip of the mission profile. The considerable increase in the energy content of the requested fuel for the hydrogen engine is rooted in the engine's lower maximum power, forcing the hydrogen engine to run at the maximum power for a longer period to achieve nearly the same journey time as the baseline diesel engine.
- The Life Cycle Assessment analysis under a Well-to-Wheel system boundary shows that the carbon footprint of the studied train can significantly drop by utilizing low-

carbon hydrogen. Photovoltaic-based green hydrogen can reduce the equivalent $CO_2$ emissions by up to 56% compared to the conventional diesel train. Considering the optimistic methane leakage rate and the carbon capture efficiency for blue hydrogen production, the carbon footprint can decrease by up to 43% with respect to the baseline diesel train.

Evidencing the promising Global Warming Potential values by using low-carbon hydrogen in trains, the moderately modified engines can become a favorable candidate for decarbonizing non-feasibly or conveniently electrified railway sections.

**Author Contributions:** Conceptualization, D.B., T.L. and M.B.; Methodology, T.C., D.B. and T.L.; Software, M.K.T. and T.C.; Investigation, M.K.T. and T.C.; Data curation, M.K.T.; Writing—original draft, M.K.T. and T.C.; Writing—review & editing, D.B., T.L. and M.B.; Supervision, D.B., T.L. and M.B. All authors have read and agreed to the published version of the manuscript.

**Funding:** This research received no external funding.

**Data Availability Statement:** Data are contained within the article.

**Conflicts of Interest:** The authors declare no conflict of interest.

## References


1. International Energy Agency (IEA). *World Energy Outlook 2022*; International Energy Agency (IEA): Paris, France, 2022.
2. International Energy Agency (IEA). *Global Hydrogen Review 2023*; International Energy Agency (IEA): Paris, France, 2023.
3. Bonalumi, D.; Kolahchian Tabrizi, M. Re-Evaluation of the Global Warming Potential for the Production of Lithium-Ion Batteries with Nickel–Manganese–Cobalt Cathode Chemistries in China. *Energy Fuels* **2022**, *36*, 13753–13767. [CrossRef]
4. Kolahchian Tabrizi, M.; Bonalumi, D.; Lozza, G.G. Analyzing the Global Warming Potential of the Production and Utilization of Lithium-Ion Batteries with Nickel-Manganese-Cobalt Cathode Chemistries in European Gigafactories. *Energy* **2024**, *288*, 129622. [CrossRef]
5. Liu, F.; Mauzerall, D.L.; Zhao, F.; Hao, H. Deployment of Fuel Cell Vehicles in China: Greenhouse Gas Emission Reductions from Converting the Heavy-Duty Truck Fleet from Diesel and Natural Gas to Hydrogen. *Int. J. Hydrogen Energy* **2021**, *46*, 17982–17997. [CrossRef]
6. Haseli, Y.; Naterer, G.; Dincer, I. Comparative Assessment of Greenhouse Gas Mitigation of Hydrogen Passenger Trains. *Int. J. Hydrogen Energy* **2008**, *33*, 1788–1796. [CrossRef]
7. Piraino, F.; Genovese, M.; Fragiacomo, P. Towards a New Mobility Concept for Regional Trains and Hydrogen Infrastructure. *Energy Convers. Manag.* **2021**, *228*, 113650. [CrossRef]
8. Jafari Kaleybar, H.; Brenna, M.; Li, H.; Zaninelli, D. Fuel Cell Hybrid Locomotive with Modified Fuzzy Logic Based Energy Management System. *Sustainability* **2022**, *14*, 8336. [CrossRef]
9. Brenna, M.; Foiadelli, F.; Longo, M.; Zaninelli, D. Use of Fuel Cell Generators for Cell-Propelled Trains Renovation. In Proceedings of the 2018 IEEE International Conference on Environment and Electrical Engineering and 2018 IEEE Industrial and Commercial Power Systems Europe (EEEIC/I&CPS Europe), Palermo, Italy, 12–15 June 2018; pp. 1–5.
10. Bartolucci, L.; Cordiner, S.; Mulone, V.; Pasqualini, F.; Wancura, H. A $H_2$ Based Retrofit of Diesel Locomotives for $CO_2$ Emission Reductions: Design and Control Issues. *Int. J. Hydrogen Energy* **2022**, *47*, 32669–32681. [CrossRef]
11. Nqodi, A.; Mosetlhe, T.C.; Yusuff, A.A. Advances in Hydrogen-Powered Trains: A Brief Report. *Energies* **2023**, *16*, 6715. [CrossRef]
12. Montenegro, G.; Onorati, A.; Della Torre, A. The Prediction of Silencer Acoustical Performances by 1D, 1D–3D and Quasi-3D Non-Linear Approaches. *Comput. Fluids* **2013**, *71*, 208–223. [CrossRef]
13. Cornolti, L.; Onorati, A.; Cerri, T.; Montenegro, G.; Piscaglia, F. 1D Simulation of a Turbocharged Diesel Engine with Comparison of Short and Long EGR Route Solutions. *Appl. Energy* **2013**, *111*, 1–15. [CrossRef]
14. D'Errico, G.; Ferrari, G.; Onorati, A.; Cerri, T. Modeling the Pollutant Emissions from a S.I. Engine. *SAE Trans.* **2002**, *111*, 1–11.
15. D'Errico, G.; Onorati, A.; Ellgas, S. 1D Thermo-Fluid Dynamic Modelling of an S.I. Single-Cylinder $H_2$ Engine with Cryogenic Port Injection. *Int. J. Hydrogen Energy* **2008**, *33*, 5829–5841. [CrossRef]
16. D'Errico, G.; Cerri, T.; Lucchini, T. *Development and Application of S.I. Combustion Models for Emissions Prediction*; SAE Technical: Warrendale, PA, USA, 2006. [CrossRef]
17. Gascón, L.; Corberán, J.M. Construction of Second-Order TVD Schemes for Nonhomogeneous Hyperbolic Conservation Laws. *J. Comput. Phys.* **2001**, *172*, 261–297. [CrossRef]
18. Winterbone, D.E.; Pearson, R.J. *Theory of Engine Manifold Design*; Society of Automotive Engineers: Warrendale, PA, USA, 2000.
19. Gasdyn Software 2023. Theory Manual. Available online: https://www.polilink.polimi.it/en/casi-di-successo/gasdyn-software-2/ (accessed on 10 November 2023).
20. Lipatnikov, A.N.; Chomiak, J. Turbulent Flame Speed and Thickness: Phenomenology, Evaluation, and Application in Multi-Dimensional Simulations. *Prog. Energy Combust. Sci.* **2002**, *28*, 1–74. [CrossRef]

21. Heywood, J.B. *Internal Combustion Engine Fundamentals*; McGraw-Hill Education: New York, NY, USA, 2018; ISBN 1260116107.
22. *ISO 14040:2006*; Environmental Management—Life Cycle Assessment—Principles and Framework. International Organization for Standardization: Geneva, Switzerland, 2006.
23. *ISO 14044:2006*; Environmental Management—Life Cycle Assessment—Requirements and Guidelines. International Organization for Standardization: Geneva, Switzerland, 2006. Available online: https://www.iso.org/standard/38498.html (accessed on 5 July 2022).
24. Eriksson, M.; Ahlgren, S. *LCAs of Petrol and Diesel: A Literature Review*; Rapport (Institutionen för Energi Och Teknik, SLU): Uppsala, Sweden, 2013.
25. Kolahchian Tabrizi, M.; Famiglietti, J.; Bonalumi, D.; Campanari, S. The Carbon Footprint of Hydrogen Produced with State-of-the-Art Photovoltaic Electricity Using Life-Cycle Assessment Methodology. *Energies* **2023**, *16*, 5190. [CrossRef]
26. Cho, H.H.; Strezov, V.; Evans, T.J. A Review on Global Warming Potential, Challenges and Opportunities of Renewable Hydrogen Production Technologies. *Sustain. Mater. Technol.* **2023**, *35*, e00567. [CrossRef]
27. Jungbluth, N.; Stucki, M.; Frischknecht, R.; Buesser, S. *Photovoltaics, Ecoinvent Report No. 6 XII*; ESU-Services Ltd.: Uster, Switzerland, 2009.
28. Wernet, G.; Bauer, C.; Steubing, B.; Reinhard, J.; Moreno-Ruiz, E.; Weidema, B. The Ecoinvent Database Version 3 (Part I): Overview and Methodology. *Int. J. Life Cycle Assess.* **2016**, *21*, 1218–1230. [CrossRef]
29. Terna Electricity Data Archives and the Latest Electricity Report. Available online: https://www.terna.it/it/sistema-elettrico/statistiche/pubblicazioni-statistiche (accessed on 10 February 2023).
30. Colbertaldo, P.; Parolin, F.; Campanari, S. A Comprehensive Multi-Node Multi-Vector Multi-Sector Modelling Framework to Investigate Integrated Energy Systems and Assess Decarbonisation Needs. *Energy Convers. Manag.* **2023**, *291*, 117168. [CrossRef]
31. Romano, M.C.; Antonini, C.; Bardow, A.; Bertsch, V.; Brandon, N.P.; Brouwer, J.; Campanari, S.; Crema, L.; Dodds, P.E.; Gardarsdottir, S.; et al. Comment on "How Green Is Blue Hydrogen?". *Energy Sci. Eng.* **2022**, *10*, 1944–1954. [CrossRef]
32. Makridis, S.S. Hydrogen Storage and Compression. In *Methane and Hydrogen for Energy Storage*; Institution of Engineering and Technology: Stevenage, UK, 2016; pp. 1–28.
33. Tahan, M.-R. Recent Advances in Hydrogen Compressors for Use in Large-Scale Renewable Energy Integration. *Int. J. Hydrogen Energy* **2022**, *47*, 35275–35292. [CrossRef]
34. Zhang, J.Z.; Li, J.; Li, Y.; Zhao, Y. Established Methods Based on Compression and Cryogenics. In *Hydrogen Generation, Storage, and Utilization*; Wiley: Hoboken, NJ, USA, 2014; pp. 75–90.
35. Valente, A.; Iribarren, D.; Candelaresi, D.; Spazzafumo, G.; Dufour, J. Using Harmonised Life-Cycle Indicators to Explore the Role of Hydrogen in the Environmental Performance of Fuel Cell Electric Vehicles. *Int. J. Hydrogen Energy* **2020**, *45*, 25758–25765. [CrossRef]
36. Linde, A.G. The Ionic Compressor IC90 (90 MPa). Available online: https://www.boconline.co.uk/en/images/Datasheet_Ionic%20Fueler%20IC90_tcm410-410855.pdf (accessed on 10 December 2023).
37. Scarlat, N.; Prussi, M.; Padella, M. Quantification of the Carbon Intensity of Electricity Produced and Used in Europe. *Appl. Energy* **2022**, *305*, 117901. [CrossRef]
38. Munshi, S.; Garner, G.; Theissl, H.; Hofer, F.; Raser, B. Total Cost of Ownership (TCO) Analysis for Heavy Duty Hydrogen Fueled Powertrains. Available online: https://www.avl.com/documents/10138/2703324/20201225_Westport_AVL_Whitepaper_Hydrogen_HPDI_final.pdf/3f482167-d9ee-8a00-7e3a-ceb024373fe9 (accessed on 4 May 2021).
39. Jin, S.; Li, J.; Deng, L.; Wu, B. Effect of the HPDI and PPCI Combustion Modes of Direct-Injection Natural Gas Engine on Combustion and Emissions. *Energies* **2021**, *14*, 1957. [CrossRef]
40. Green Car Congress Omnitek Develops Diesel-to-Natural Gas Engine Conversion Kits for Mercedes OM904/OM906. Available online: https://www.greencarcongress.com/2012/04/omnitek-20120419.html (accessed on 20 December 2023).
41. James, B.D.; Houchins, C.; Huya-Kouadio, J.M.; DeSantis, D.A. *Final Report: Hydrogen Storage System Cost Analysis*; Strategic Analysis Inc.: Arlington, VA, USA, 2016.


**Disclaimer/Publisher's Note:** The statements, opinions and data contained in all publications are solely those of the individual author(s) and contributor(s) and not of MDPI and/or the editor(s). MDPI and/or the editor(s) disclaim responsibility for any injury to people or property resulting from any ideas, methods, instructions or products referred to in the content.

*energies*

MDPI

*Article*

# A Complete Assessment of the Emission Performance of an SI Engine Fueled with Methanol, Methane and Hydrogen

**Francesco Catapano, Silvana Di Iorio, Agnese Magno *, Paolo Sementa and Bianca Maria Vaglieco**

Institute of Science and Technology for Sustainable Energy and Mobility (STEMS)—CNR, Via G. Marconi 4, 80125 Naples, Italy; francesco.catapano@stems.cnr.it (F.C.); silvana.diiorio@stems.cnr.it (S.D.I.); paolo.sementa@stems.cnr.it (P.S.); biancamaria.vaglieco@stems.cnr.it (B.M.V.)

* Correspondence: agnese.magno@stems.cnr.it; Tel.: +39-0817177118

**Abstract:** This study explores the potentiality of low/zero carbon fuels such as methanol, methane and hydrogen for motor applications to pursue the goal of energy security and environmental sustainability. An experimental investigation was performed on a spark ignition engine equipped with both a port fuel and a direct injection system. Liquid fuels were injected into the intake manifold to benefit from a homogeneous charge formation. Gaseous fuels were injected in direct mode to enhance the efficiency and prevent abnormal combustion. Tests were realized at a fixed indicated mean effective pressure and at three different engine speeds. The experimental results highlighted the reduction of CO and $CO_2$ emissions for the alternative fuels to an extent depending on their properties. Methanol exhibited high THC and low $NO_x$ emissions compared to gasoline. Methane and, even more so, hydrogen, allowed for a reduction in THC emissions. With regard to the impact of gaseous fuels on the $NO_x$ emissions, this was strongly related to the operating conditions. A surprising result concerns the particle emissions that were affected not only by the fuel characteristics and the engine test point but also by the lubricating oil. The oil contribution was particularly evident for hydrogen fuel, which showed high particle emissions, although they did not contain carbon atoms.

**Keywords:** spark ignition engine; methanol; hydrogen; methane; particle emissions; lubricating oil

**Citation:** Catapano, F.; Di Iorio, S.; Magno, A.; Sementa, P.; Vaglieco, B.M. A Complete Assessment of the Emission Performance of an SI Engine Fueled with Methanol, Methane and Hydrogen. *Energies* **2024**, *17*, 1026. https://doi.org/10.3390/en17051026

Academic Editors: Davide Lanni and Enzo Galloni

Received: 19 January 2024
Revised: 16 February 2024
Accepted: 20 February 2024
Published: 22 February 2024

**Copyright:** © 2024 by the authors. Licensee MDPI, Basel, Switzerland. This article is an open access article distributed under the terms and conditions of the Creative Commons Attribution (CC BY) license (https://creativecommons.org/licenses/by/4.0/).

## 1. Introduction

European policy strives to achieve the challenging goal of zero carbon dioxide ($CO_2$) emissions by 2050 [1]. In this regard, the decarbonization of the transport sector plays a pivotal role. For this purpose, in the last few years, academic and industrial communities have devoted research efforts to the electrification of propulsion systems. Nevertheless, it is well recognized that electrically driven vehicles do not represent an immediate solution to climate change due to some barriers that still hinder their widespread use, such as the high purchase price of electric vehicles (EVs) limiting their usage to the wealthier social classes, the lack of charging infrastructure [2] and the potential environmental impacts of lithium mining [3]. Therefore, in parallel with the technological improvement of EVs, it is essential to continue to focus on internal combustion engines (ICEs) [4]. An effective way to guarantee a future for ICEs in the transition phase towards zero greenhouse gas emissions consists in the use of low/zero carbon fuels [5]—both liquid, such as methanol, and gaseous, such as methane and hydrogen. As known by the literature, the study of the effective impact of alternative fuels on environmental sustainability should not be limited to considering just the emissions at the vehicle exhaust but should also take into account all the factors spanning from fuel production to its utilization [6].

Methanol is a valid alternative for powering spark ignition (SI) engines thanks to its high octane quality and high heat of vaporization (HOV) resulting in improved engine thermal efficiency [7]. It can be produced from a lot of sources such as coal, natural gas, wood and waste [8]. A green method for producing methanol has also been developed. It consists of the use of captured carbon dioxide ($CO_2$) and hydrogen obtained from renewable sources such as solar and wind power [9]. Methanol can be used both pure and in blends at different mixture proportions. Several studies in the literature have investigated the emission characteristics of binary blends of methanol with gasoline as well as ternary blends of methanol, ethanol and gasoline [10–12]. They have shown reduced exhaust emissions of total hydrocarbons (THCs) and nitrogen oxides ($NO_x$) [13] and a blend-dependent behavior of the particulate matter (PM) [14]. Regarding the use of pure methanol in SI engines, the majority of the studies available in the literature were carried out on direct injection (DI) engines [15]. To the best of the authors' knowledge, there is a lack of investigations of port fuel injection (PFI) engines fueled by methanol, as also observed by Wei et al. [16] for heavy-duty applications. Therefore, a deeper comprehension of the benefits as well as the criticalities of the use of methanol for PFI systems is crucial. Moreover, a better insight into the effect of neat methanol on PM emissions is essential. Larsson et al. [17] observed that oxygenated fuels tend to decrease particle emissions thanks to their chemical structure, which inhibits the formation of soot precursors and increases post-oxidation. Nevertheless, the formation of particles also depends on the physical properties of the fuel such as the volatility. Low volatility, in fact, can cause pool fires where particles are formed. On the other hand, Svensson et al. [18] observed that methanol has a non-sooting behavior and, hence, particles measured at the exhaust of a pure methanol-fueled engine could be due to sources other than the fuel such as oil leakage [17].

Methane and hydrogen are also excellent candidates to replace gasoline in SI engines thanks to their clean-burn characteristics [19,20]. The existing literature on the effect of gaseous fuels on engine performance and emissions is mainly focused on the PFI strategy [21,22]. This technique allows engines to benefit from a homogeneous charge formation, but, on the other hand, it can induce abnormal combustion and power drop [23]. These issues could be overcome through the DI strategy [24]. This method can lead to high $NO_x$ emissions, thus requiring the lean mode operation. As observed by Thawko et al. [25], the DI of gaseous fuel is a promising method to improve engine performance and avoid PFI drawbacks. However, some knowledge gaps exist on the application of this technology to gaseous fuels, and one of them regards particle emissions. Methane contains carbon in its molecular structure, so it can emit fuel-derived particles. However, it was found that in some operating conditions, the contribution of the lubricant oil to particle formation is predominant with respect to that of the fuel [26]. The impact of lubricant oil on particle emissions is even more evident for hydrogen, which is a carbon-free fuel and, hence, should not emit fuel-derived particles.

This study tries to contribute to the knowledge of the effect of carbon-neutral fuels for SI engines, and its novelty consists in a comparative analysis of methanol, methane and hydrogen. The investigation was carried out on a low-displacement, single-cylinder SI engine. Different injection strategies were adopted depending on the fuel properties. Liquid fuels were, in fact, injected in PFI mode, while gaseous ones were injected directly into the chamber. The engine was operated in stationary conditions at three test points included in the urban and extra-urban cycles. For a comparative analysis among the different fuels, experiments were realized at a fixed indicated mean effective pressure (imep). Indicative data were acquired to analyze the combustion behavior. Carbon monoxide (CO), $CO_2$, THC and $NO_x$ were measured at raw exhaust. Particles were characterized in terms of number and size at diluted exhaust.

## 2. Materials and Methods

### 2.1. Experimental Apparatus

Figure 1 illustrates the schematic diagram of the experimental setup. Tests were carried out on a four-stroke, single-cylinder SI engine equipped with a naturally aspirated four-valve head. The engine crankshaft was connected to an electrical dynamometer. A list of engine technical specifications is reported in Table 1.

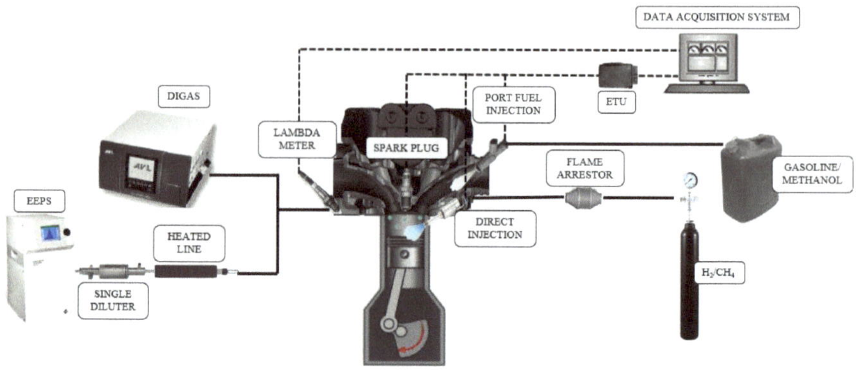


**Figure 1.** Experimental setup.

**Table 1.** Engine specification.

| Engine | Spark Ignition | |
|---|---|---|
| Number of Cylinders | 1 | |
| Bore [mm] | 72 | |
| Stroke [mm] | 60 | |
| Displacement [cm$^3$] | 244.3 | |
| Compression Ratio | 11.5:1 | |
| Max. Power [kW] | 16 @ 8000 rpm | |
| Max. Torque [Nm] | 20 @ 5500 | |
| Intake | Turbo-charged | |
| Injection System | Gaseous fuels—DI | Liquid fuels—PFI |
| Number of Nozzle Holes | 6 | 3 |
| $p_{inj}$ [bar] | 5 | 2.5 |

The ignition system consisted of a spark plug mounted in the center of the cylinder head. The engine was equipped with two different injection systems. In its marked version, it was fitted with a DI system characterized by a side-mounted, six-hole injector located between the intake valves. The DI injection system was adapted to also operate with gaseous fuels through an air strata injector typically used for the simultaneous injection of air and gasoline. It was set in the cylinder head with an ad hoc designed adaptor [27]. Moreover, the PFI system was implemented through a three-hole injector for liquid fuels placed in the intake manifold.

For the liquid fuels, the fuel supply system consisted of a fuel tank, an electric, low-pressure pump and a fuel filter. On the other hand, the gaseous fuels were stored in a pressurized bottle at 200 bar, decompressed to 6 bar by a regulator and then supplied to the cylinder. A flame arrestor and a pneumatic-actuated valve were installed in the gaseous fuel supply line to avoid any risk of backfire.

Engine parameters, such as the spark and injection timing and throttle valve opening to adjust the airflow, were controlled by an AVL engine timing unit (ETU).

The operating excess air ratio (λ) was measured using a wideband oxygen sensor (LSU 4.9 Bosch) located in the exhaust manifold and monitored using a lambda meter (LA4, ETAS).

In-cylinder pressure was measured with a high-speed piezoelectric pressure sensor (GH12D) flush mounted on the cylinder head. The charge output from this transducer was converted to an amplified voltage and sent to the high-speed data acquisition system AVL Indimodul for combustion analysis. Engine crank angle detection was performed by an AVL 365C encoder with a multiplier at a resolution of 0.1 degrees.

Regulated exhaust emission species such as CO, $CO_2$, THC and $NO_x$ were measured by an AVL DiGas 4000, whose details are shown in Table 2, with uncertainties of ±0.2%, ±0.15%, ±0.2% and ±1%, respectively [28].

Engine exhaust particle number and size distribution were measured through a TSI Engine Exhaust Particle Spectrometer (EEPS) 3090 whose main specifications are listed in Table 3. The instrument is based on the electrical mobility principle described in [29] and is capable of detecting particles in the range of 5.6–560 nm at 10 Hz frequency. The particle number distributions were calculated from the electrometer currents after adjusting for time delay, multiple charges and other factors by a proprietary inversion matrix [30].

Before measuring the particles, the exhaust gas sample was passed through a 150 °C heated probe to prevent water condensation and then it was diluted at a ratio of 1:9 by a single diluter (SD).

**Table 2.** DiGas specifications [31].

| Species | Principle Measurement | Measurement Range | Resolution |
|---|---|---|---|
| CO | Non-dispersive infrared (NDIR) | 0–10% vol. | 0.01%vol. |
| $CO_2$ | Non-dispersive infrared (NDIR) | 0–20% vol. | 0.1%vol. |
| THC | Non-dispersive infrared (NDIR) | 0–20,000 ppm vol. | 1 ppm |
| $NO_x$ | Chemiluminescent detector (CLD) | 0–5000 ppm vol. | 1 ppm |

**Table 3.** EEPS specifications [29].

| Properties | Range |
|---|---|
| Particle Size Range | 5.6–560 nm |
| Particle Size Resolution | 16 channels per decade (32 total) |
| Electrometer Channels | 22 |
| Maximum Data Rate | 10 size distribution per second |
| Inlet Aerosol Temperature | 10–52 °C |
| Storage Temperature | −20 to 50 °C |

### 2.2. Procedure

In this study, three different alternative fuels, methanol, methane and hydrogen, were tested. They were used in pure form. Performance and pollutant emissions were analyzed and compared to those of conventional gasoline. Table 4 shows the main physical–chemical properties of the fuels.

**Table 4.** Fuel properties.

| Properties | Gasoline | Methanol | Methane | Hydrogen |
|---|---|---|---|---|
| Chemical Formula | $C_4$–$C_{12}$ | $CH_3OH$ | $CH_4$ | $H_2$ |
| Density at 15 °C [kg/L] | 746 | 796 | 0.67 | 0.08 |
| Boiling Point [°C] | 27–225 | 64 | −161.4 | −252.9 |
| LHV [MJ/kg] | 42.94 | 20.1 | 50 | 120 |
| $AFR_{st}$ | 14.5 | 6.4 | 17.24 | 34.20 |
| RON | 95.0 | 108.6 | >120 | >130 |
| Flammability Limits [vol %] | 1.4/7.6 | 6.7/36 | 5.3/14 | 4/75 |
| C [% mass] | 85.9 | 38.0 | 75 | - |
| H [% mass] | 13.4 | 12.1 | 25 | 100 |
| O [% mass] | 0.6 | 50.0 | - | - |

Methanol, methane and hydrogen represent three high-octane-number (ON) fuels. As with gasoline, methanol can be stored in liquid form at a standard temperature and pressure. The chemical structure of the alcohol fuel, with its lower carbon content and higher amount of oxygen with respect to gasoline, contributes to determining its properties and combustion characteristics. Methanol has, in fact, a lower stoichiometric air–fuel ratio ($AFR_{st}$) and a low heating value (LHV) which is almost half that of gasoline.

Gaseous fuels are characterized by a higher LHV than liquid ones, even if the thermal values per volume of methane and even more so of hydrogen are lower because of their low density. Methane shows poor lean-burn capability. On the other hand, the hydrogen–air mixture combusts in the range of 4–75% air by volume, ensuring that combustion is achieved even with a poor mixture.

Two different approaches were adopted for the experiments with liquid and gaseous fuels to take into account their different properties. To exploit the intrinsic nature of the fuels, the liquid ones were injected in PFI mode, thus allowing for a better homogeneous charge formation. On the other hand, DI technology was chosen for methane and hydrogen to improve the volumetric efficiency. To allow for comparison among the tested fuels, experiments were carried out at the same imep, that is, 6.0 bar. To obtain data sufficiently meaningful and worthwhile, three engine speeds were selected, 2000, 2500 and 3000 rpm, which are representative of the operating points of a typical commercial vehicle.

A synthesis of the investigated condition is shown in Table 5. The start of injection (SOI) and start of spark (SOS) were modified to guarantee a stable combustion operation. Tests with liquid fuels were carried out at throttled conditions, the throttle position (TP) was at 4% and the duration of injection (DOI) was properly set to run under stoichiometric mode at all engine speeds. This allowed us to maximize the performance while maintaining low emissions.

**Table 5.** Investigated operating conditions.

| Engine Speed [rpm]/Imep [bar] | Fuel | Injection Mode | SOI [cad BTDC] | SOS [cad ATDC] | TP [%] | DOI [cad] |
|---|---|---|---|---|---|---|
| 2000/6 | gasoline | PFI | 315 | 37.0 | 4 | 64 |
| | methanol | PFI | 360 | 40.0 | 4 | 155 |
| | $CH_4$ | DI | 305 | 22.5 | 8 | 175 |
| | $H_2$ | DI | 310 | 9.6 | 95 | 200 |
| 2500/6 | gasoline | PFI | 315 | 39.0 | 4 | 80 |
| | methanol | PFI | 360 | 40.0 | 4 | 170 |
| | $CH_4$ | DI | 315 | 24.3 | 14 | 238 |
| | $H_2$ | DI | 330 | 6.5 | 95 | 320 |
| 3000/6 | gasoline | PFI | 315 | 44.0 | 4 | 89 |
| | methanol | PFI | 360 | 45.0 | 4 | 175 |
| | $CH_4$ | DI | 315 | 28.5 | 10 | 270 |
| | $H_2$ | DI | 352 | 10.0 | 95 | 245 |

The $\lambda$ values at which the engine was operated for the tested fuels at all investigated conditions are shown in Figure 2. Experiments with gaseous fuels were realized at leaner conditions to reduce the $NO_x$ emissions.

When the engine was fueled with methane, $\lambda$ was kept at 1.15 at all tested engine speeds by adjusting the TP. This $\lambda$ was the maximum value achievable to guarantee the desired imep with a stable operation because of the narrow flammability limits of methane. Hydrogen was tested under wide open throttle (WOT) conditions to benefit from its wide flammability range. In order to improve the efficiency and at the same time reduce the $NO_x$ emissions, the engine was run with $\lambda$ values equal to 1.5, 1.4 and 1.6 at 2000, 2500 and 3000 rm, respectively.

The engine ran at steady and stable conditions for at least 5 min before starting the measurements at each engine speed. Each acquisition was repeated three times. The average values of the three repetitions are shown in the experimental results section.

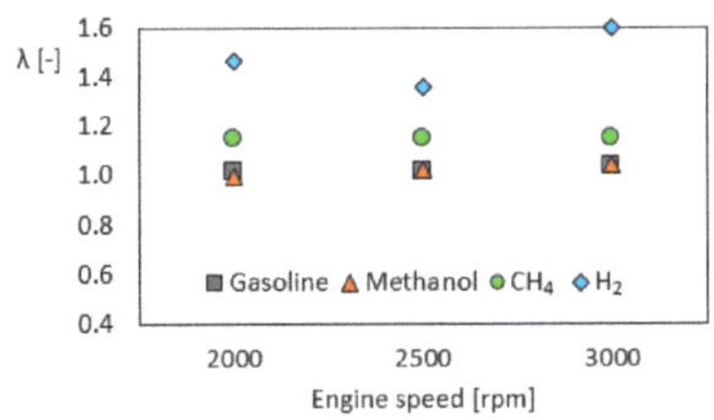


**Figure 2.** λ values for gasoline, methanol, methane and hydrogen at all investigated conditions.

## 3. Results

### *3.1. Combustion Analysis*

In order to achieve a better comprehension of the impact of the tested alternative fuels on the engine performance and emissions, the combustion behavior was investigated through the analysis of the indicated parameters. The in-cylinder pressure was detected and, for an effective comparative analysis among the fuels, the maximum pressure and crank angle degree (cad) at which it occurred was calculated as shown in Figure 3.

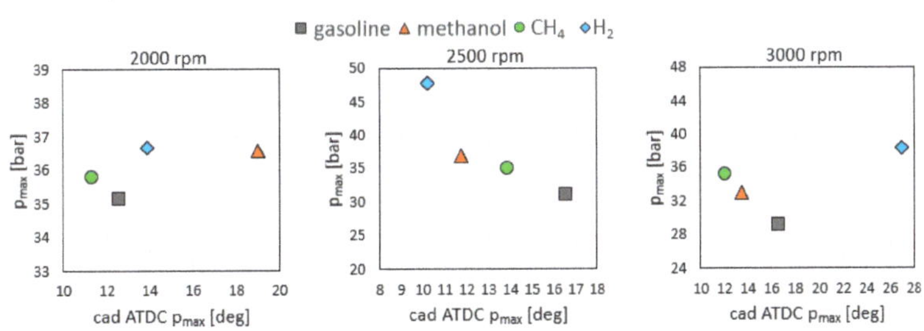


**Figure 3.** $p_{max}$ versus cad ATDC at $p_{max}$ for gasoline, methanol, methane and hydrogen at 2000, 2500, and 3000 rpm, 6.0 bar of imep.

As mentioned above, to take advantage of the different fuel properties, different injection strategies—PFI and DI—were applied for liquid and gaseous fuels, respectively. Therefore, the differences in the combustion evolution depended on both the injection mode and the fuel type.

At 2500 and 3000 rpm, as concerns the liquid fuels, methanol had a higher and more advanced peak of pressure compared to gasoline, up to 5 cad at 2500 rpm. Since a similar SOS was set for both fuels at these engine speeds, the earlier combustion of the alcohol fuel could be ascribed to the faster combustion rate owing to the oxygen content of methanol, and the higher maximum pressure was due to the higher ON. At 2000 rpm, instead, the SOS was 3 cads retarded for methanol, thus limiting the effect of the fuel properties and leading to a delayed but always higher maximum pressure with respect to gasoline.

Regarding the gaseous fuels, a definite trend was not distinguishable since the combustion phasing was affected by different factors such as the injection mode, the control parameters and the fuel properties. The maximum pressure phasing of hydrogen with respect to methane changed because of the combination of the SOS applied for the different fuels and the higher flame speed of the hydrogen. In general, a higher pressure peak could be observed for hydrogen at all engine speeds. To justify this trend, it is important to take into account, beyond its properties, that WOT was implemented for hydrogen to benefit from its wide flammability limits entailing that a larger amount of air entered the cylinder with respect to methane tested in throttle operation.

Interesting information about combustion evolution can be drawn from the rate of heat release (ROHR), from which the peak and combustion duration could be determined. The latter was calculated as the difference between the cads for 90 and 10% of the mass fraction burned (MBF90 and MBF10, respectively), representative of the end and start of combustion. These parameters are reported in Figure 4. Regarding the liquid fuels, methanol showed a lower combustion duration than gasoline thanks to its higher flame velocity and higher peak of heat release thanks to its larger oxygen content that guaranteed more complete combustion.

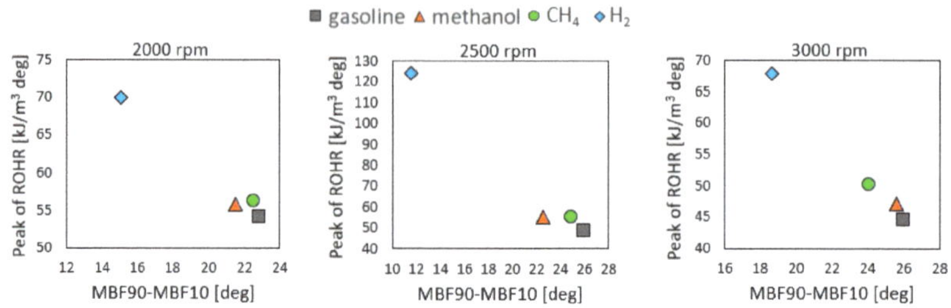


**Figure 4.** Peak of ROHR versus combustion duration for gasoline, methanol, methane and hydrogen at 2000, 2500 and 3000 rpm, 6.0 bar of imep.

As concerns the gaseous fuels, in general, methane—compared to gasoline—showed faster combustion characterized by greater heat released. This trend could be correlated to the DI of the fuel that, improving the turbulence and, hence, the fuel–air mixing, resulted in increased flame propagation. This phenomenon for hydrogen was amplified by its intrinsic higher flame speed.

## 3.2. Exhaust Emission Analysis

To better analyze the environmental impact of the tested alternative fuels, gaseous emissions were measured at the exhaust, as shown in Figure 5. Emission concentrations were converted to mass flow rate by determining the inlet air and fuel mass flow rate. Therefore, the pollutant mass flow rates were expressed as grams of pollutant per kWh by referring them to the indicated power.

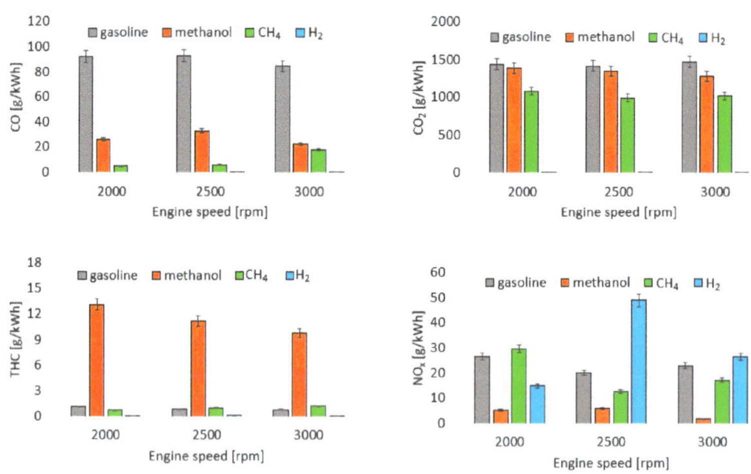


**Figure 5.** CO, $CO_2$, THC and $NO_x$ emissions for gasoline, methanol, methane and hydrogen at 2000, 2500 and 3000 rpm, 6.0 bar of imep.

Regarding the liquid fuels, methanol showed a reduction of CO emissions of about 70% with respect to gasoline and quite similar $CO_2$ emissions. It is well known that the mechanisms of CO and $CO_2$ formation depend on the C/H ratio and the oxygen content in the fuel. Therefore, the lower amount of carbon atoms in the methanol resulted in reduced CO emissions. On the other hand, the oxygenated fuel entailed a large oxygen presence in the rich areas of the chamber, thus promoting the conversion reactions of CO to $CO_2$. Methanol was characterized by THC emissions about one order of magnitude higher with respect to gasoline. This trend could be ascribed to the lower volatility of the alcohol fuel, which can result in worse fuel evaporation, and the formation of liquid fuel pools near the intake valves [32]. $NO_x$ emissions were decreased compared to gasoline, with values up to 90% lower at 3000 rpm because of the higher HOV of the fuel, which reduced the temperature in the combustion chamber, thus inhibiting the $NO_x$ formation reaction, although there was a leaning effect of the charge due to the larger oxygen presence in the methanol.

With regard to the gaseous fuels, methane allowed to reduce CO and $CO_2$ emissions compared to the liquid ones. The emissions of THC and $NO_x$ were instead affected by both the fuel properties and the operating conditions. Methane combustion was characterized by lower THC than gasoline at 2000 rpm and slightly higher THC at 2500 and 3000 rpm. Conversely, $NO_x$ emissions were higher at 2000 rpm compared to gasoline and lower at higher speeds. At 2000 rpm, because of the typical low temperature, the evaporation of the gasoline was worsened, resulting in less efficient combustion and thus in higher THC emissions. As the speed increased, the temperature increased, resulting in better evaporation. The slight increase in the THC emissions of methane could be ascribed to the lower temperature that is typical of its combustion as well as the lower quenching distance of methane when it was direct-injected. Compared to methanol, methane emitted less THC at each tested condition because of the worst evaporation that is typical of liquid fuel. The cooling of the charge also contributed to producing a lower amount of $NO_x$.

Concerning hydrogen, since it does not contain any carbon atoms in its molecules, the formation of $CO_2$, CO and THC does not occur, and these species could just be produced by the lubricating oil. In the tested conditions, the level of emissions of these gases was too low to be detected by the available equipment. On the other hand, hydrogen showed high emissions of $NO_x$, especially at 2500 and 3000 rpm. The high combustion velocity led to an increased rate of heat release (Figure 4) and, consequently, to a higher in-cylinder temperature that promoted $NO_x$ formation. Thanks to its high flammability limits, hydrogen can be burnt in lean conditions, thus reducing the flame temperature and, hence, preventing $NO_x$ reactions. This phenomenon was amplified in operating conditions with a lower engine speed, 2000 rpm, which was characterized by lower temperature.

Figure 6 shows the contour plot depicting the particle number and diameter over time for hydrogen fueling in operating conditions with the higher engine speed, 3000 rpm.

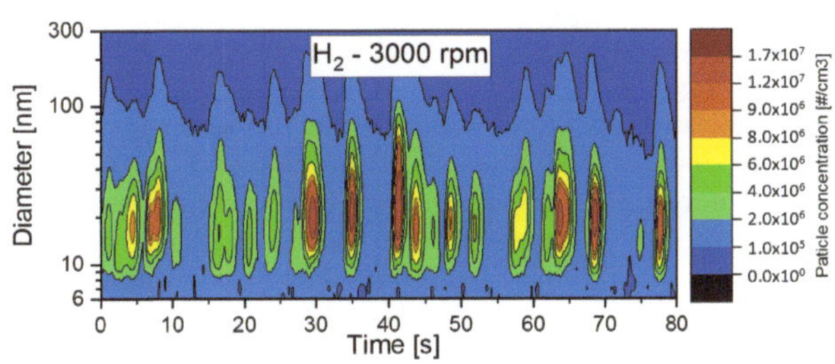


**Figure 6.** Time-resolved particle concentration and size distributions for hydrogen at 3000 rpm.

Particles ranged between 6 and 300 nm, with a peak concentration of order of magnitude around $10^7$ #/$cm^3$ in the size range of 10–50 nm. Since hydrogen does not contain any carbon atoms in its molecular structure, these measured particles were not ascribable to the fuel but rather to other sources such as the lubricant oil. This result was also confirmed by previous research studies of the authors [33,34] through simultaneous physical and chemical analyses of the particles emitted by a hydrogen-fueled engine.

An interesting outcome arising from the time-resolved particle concentration and size distributions in Figure 6 was the oscillation behavior of the particle concentration. Although the tests were carried out in fixed operating conditions in stationary mode, the emission of particles was not continuous but exhibited variations in the concentrations. These particle fluctuations were also observed by Thawko et al. [35], who characterized the particles emitted by a DI ICE fed with a hydrogen-rich reformate. This trend could be due to the entrainment of vapor oil in the combustion chamber. This oil accumulates during engine running and, periodically, at a frequency depending also on the operating conditions, burns when the self-ignition temperature is reached.

In order to compare the particle emissions from the different fuels, the average PSDs were calculated at each operating condition, as shown in Figure 7. The data are plotted on a y-log scale to better highlight the differences among the tested fuels. The intensity and the shape of the PSDs depended on the operating conditions and fuel type.

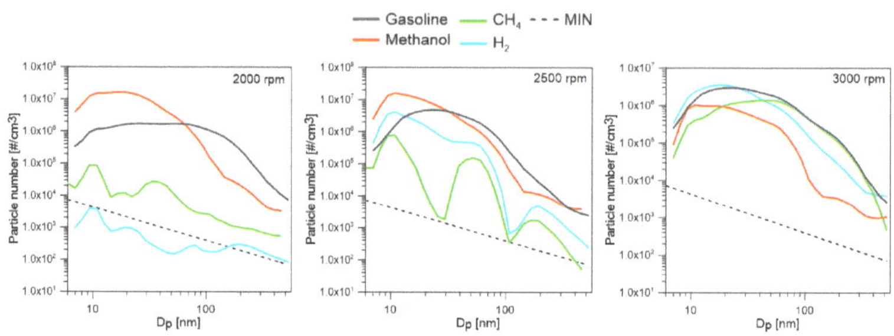


**Figure 7.** PSDs for gasoline, methanol, methane and hydrogen at 2000, 2500 and 3000 rpm, 6.0 bar of imep.

Regarding the liquid fuels, at all investigated engine speeds, the PSDs of methanol were shifted towards a smaller diameter with respect to gasoline fuel, with a distinct nucleation mode (NM) at around 10 nm and a small inflection point of the accumulation mode (AM) centered at 200 nm. Gasoline has a broader distribution, with a little hump typical of the NM at 10 nm and a more pronounced tail of the accumulated particles. At 2000 and 2500 rpm, methanol had a higher concentration of nucleated particles with respect to gasoline. It is known from the literature that the volatile species typical of the NM can originate from the partial combustion of the fuel as well as from the lubricating oil [36]. It is known from the literature that alcohol fuels can induce oil dilution, thus contributing to oil-derived particle formation [37]. Oxygenated fuels were expected to reduce particle emissions due to their oxygen content, which inhibits soot precursor formation and improves post-oxidation. The higher volatility of the alcohol fuels can cause pool fires near the intake valves where the nuclei particles are formed [32]. This result is also consistent with the higher THC emissions [14,38] (Figure 5). At 3000 rpm, instead, the PSD of methanol dropped down compared to gasoline because of the higher temperature reached at high engine speeds, which increases the rates of evaporation and oxidation. The better fuel evaporation and charge formation, in fact, reduced the mechanism of particle formation. Moreover, the increased particle post-oxidation led to smaller-particle emissions.

The PSDs of the gaseous fuels were strongly affected by the engine test point because of the in-cylinder environmental condition that impacted the participation of the lubricating oil in combustion. At 2000 rpm, methane and hydrogen emitted little or no particle emissions, respectively. The PSD of methane, in fact, was higher than the lower limit of the spectrometer, while, for hydrogen fueling, it was not distinguishable due to noise.

At 2500 rpm, the gaseous fuels showed definite PSDs with lower concentrations than the liquid ones. Both methane and hydrogen presented two bumps at 10 and 60 nm and a lower tail peaked at 200 nm. In particular, hydrogen had a higher PSD with respect to methane. At high engine speeds, the PSDs of the gaseous fuels were comparable to those of the liquid fuels. Methane exhibited a little peak of the NM at 10 nm and a wide AM overlapping the gasoline one. Hydrogen showed a pronounced NM with concentration values even higher than those of gasoline. As mentioned above, the main source of particles when gaseous fuels are burnt is the lubricating oil that enters the cylinder because of the piston ring dynamics [39]. The partial combustion of high-molecular-weight hydrocarbons in the oil, in fact, leads to the formation of nanoparticles that then grow and coagulate to form larger aggregates.

To better highlight the effect of fuel and operating conditions on particle emissions, the total particle number (Np) and mean diameter (Dm) were calculated and are depicted in the bubble graph in Figure 8. Np values are reported on the y-axis while Dm is represented by the bubble dimension. Methanol allowed for a reduction in particle size at all engine speeds thanks to the embedded oxygen that promoted the oxidation of the particles. At 2000 rpm, the alcohol fuel emitted a large number of particles due to the higher HOV of the fuel that, together with the lower temperature reached at low engine speeds, led to poor atomization and the mixing of the oxygenated fuels, thus promoting the formation of fuel-rich zones were nuclei particles originated from. With the increase in engine speed to 2500 and 3000 rpm, the in-cylinder temperature increased, overcoming the charge cooling effect typical of methanol and, hence, reducing the sources of particles.

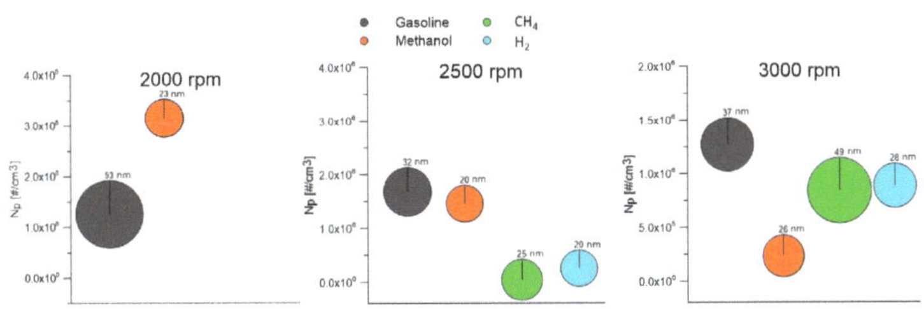


**Figure 8.** Particle number and diameter for gasoline, methanol, methane and hydrogen at 2000, 2500 and 3000 rpm, 6.0 bar of imep.

When gaseous fuels were used, the particle number and size decreased. At 2000 rpm, the bubbles representative of the particles emitted from methane and hydrogen are not shown since their values were too low to be well detected by the used spectrometer. As mentioned above, for the gaseous fuels, the main role in particle formation is played by the lubricating oil. It can be observed that the particle number and size of the particles emitted by both methane and hydrogen increased with the engine speed. With increases in the engine speed, the flame temperature also increased. Therefore, a major amount of the oil film deposited on the cylinder surface participated in combustion, leading to particle formation. No significant differences in terms of number and size were observed at 2500 rpm, evidencing the role of the lubricating oil. On the other hand, at high engine speeds, methane exhibited a higher number of particles of larger diameter with respect to hydrogen. This trend can be ascribed to the higher DOI of methane resulting in a longer interaction of the fuel jet with the cylinder surface and, hence, dragging large amounts of oil into the combustion. Moreover, another aspect to consider is that methane combustion evolves at a lower temperature than hydrogen, thus facilitating the process of particle growth and coagulation.

This experimental study has highlighted the advantages in terms of pollutant emission reduction obtained at different extents from the use of alternative fuels—both liquid and gaseous. However, for a complete evaluation of the effective environmental benefit of using alternative fuels, the entire process of fuel production, transportation and distribution should be taken into account—the well-to-tank pathway (WtT). In this regard, the engine indicated power ($P_{ind}$) was related to the WtT global warming potential (GWP 100a) in kg $CO_{2eq}$ obtained by the literature [6] as shown in Figure 9. Data were expressed in % with respect to gasoline. All the fuels showed positive values, indicating an advantage in terms of total $CO_2$ reduction compared to gasoline. Moreover, the best behavior was shown by methanol, followed by methane and then by hydrogen.

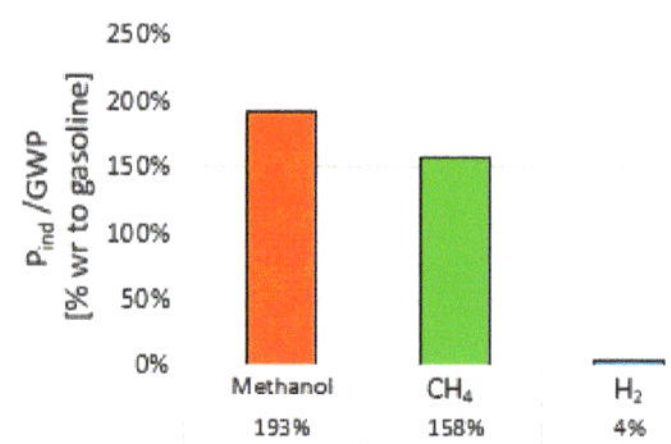


**Figure 9.** $P_{ind}$/GWP ratio in % with respect to gasoline for methanol, methane and hydrogen at 3000 rpm.

## 4. Conclusions

This study investigated the combustion and emission characteristics of methanol, methane and hydrogen in a small displacement SI engine. Fuels were injected in different modes depending on their nature—PFI for the liquids and DI for the gaseous ones—to better exploit their potential. Tests were carried out at different engine speeds and at the same imep. The following conclusions are drawn from the research activity:

Combustion analysis:

- The indicative data revealed that methanol is characterized by advanced—up to 5 cad—and faster combustion with a higher pressure peak—up to 18%—and heat release compared to the reference fuel.
- Gaseous fuels are characterized by a lower combustion duration and higher heat release than gasoline. This result could be in part due to the injection configuration, highlighting the positive effect of the DI technology for gaseous fuels.
- For hydrogen fueling, the effect of the higher flame speed was more evident, resulting in the fastest combustion—up to 11 cad of combustion duration—at 2500 rpm.

Emission characteristics:

- All the alternative fuels guarantee a reduction of CO and $CO_2$ emissions.
- Methanol shows high THC emissions, with a maximum value of 13 g/kWh at 2000 rpm, due to the less efficient fuel vaporization. $NO_x$ emissions are, instead, decreased. Regarding the particles, methanol shows not negligible emissions. In general, smaller size particles were measured, while their number depended on the engine test point. The nature of these particles is not clear. An important factor affecting their formation is, in fact, the lubricating oil.
- For gaseous fuels, the emission levels also depended on the operating conditions that affect the in-cylinder temperature. When gaseous fuels were used, the effect of oil on the particle emissions was clear, especially for hydrogen, which should not emit particles on its own. Lower particle emissions with a larger diameter were measured for methane fueling. For hydrogen, an increase in particle emissions was observed with increases in engine speed.
- All the tested fuels contributed to sustainable mobility thanks to the benefit in terms of total $CO_2$ reductions with respect to gasoline, with the best performance achieved by methanol.

Therefore, this study has highlighted that there is considerable merit in continuing to investigate alternative fuels as potential future automotive energy vectors since the objective of $CO_2$ abatement cannot be achieved by focusing on a single technology but rather by diversifying the available solutions. From this perspective, this study has pointed out that a proper injection strategy and control could help to make the most of fuel characteristics.

Moreover, continuing the study of particle emissions due to lubricating oil when the engine is fueled with alcohol is important to determine the nature of such particles, if fuel- or oil-derived, and better understand how to control their emissions. This type of analysis is also relevant when gaseous fuels are used for the proper optimization of engine oil.

**Author Contributions:** F.C.: Methodology, Investigation. S.D.I.: Methodology, Investigation, Writing—original draft, Visualization. A.M.: Methodology, Investigation, Writing—original draft, Visualization. P.S.: Methodology, Investigation. B.M.V.: Supervision. All authors have read and agreed to the published version of the manuscript.

**Funding:** This research has been partially supported by the European Union—NextGenerationEU—National Sustainable Mobility Center CN00000023, Italian Ministry of University and Research Decree n. 1033—17/06/2022, Spoke 12, CUP B43C22000440001.

**Data Availability Statement:** The data presented in this study are available on request from the corresponding author. The data are not publicly available due to privacy.

**Acknowledgments:** The authors thank Carlo Rossi and Bruno Sgammato for the engine assessment and support with the experiments.

**Conflicts of Interest:** The authors declare no conflicts of interest.

## Abbreviations

| | |
|---|---|
| $AFR_{st}$ | Stoichiometric Air/Fuel Ratio |
| AM | Accumulation Mode |
| ATDC | After Top Dead Center |
| cad | Crank Angle Degree |
| CLD | Chemiluminescent Detector |
| CO | Carbon Monoxide |
| $CO_2$ | Carbon Dioxide |
| DI | Direct Injection |
| Dm | Mean Diameter |
| DOI | Duration of Injection |
| EEPS | Engine Exhaust Particle Spectrometer |
| ETU | Engine Timing Unit |
| EV | Electric Vehicle |
| GWP | Global Warming Potential |
| HOV | Heat of Vaporization |
| ICE | Internal Combustion Engine |
| Imep | Indicated Mean Effective Pressure |
| $\lambda$ | Excess Air Ratio |
| LHV | Low Heating Value |
| MBF10 | cad for 10% of Mass Fraction Burned |
| MBF90 | cad for 90% of Mass Fraction Burned |
| NDIR | Non-Dispersive Infrared |
| NM | Nucleation Mode |
| $NO_x$ | Nitrogen Oxides |
| Np | Particle Number |
| ON | Octane Number |
| PFI | Port Fuel Injection |
| $P_{ind}$ | Indicated Power |
| PM | Particulate Matter |
| PSD | Particle Size Distribution |
| ROHR | Rate of Heat Release |
| SD | Single Diluter |
| SI | Spark Ignition |
| SOI | Start of Injection |
| SOS | Start of Spark |
| THC | Total Hydrocarbons |
| TP | Throttle Position |
| WOT | Wide Open Throttle |
| WtT | Well-to-Tank |

## References


1. 2050 Long-Term Strategy. n.d. Available online: https://climate.ec.europa.eu/eu-action/climate-strategies-targets/2050-long-term-strategy_en (accessed on 27 March 2023).
2. Adhikari, M.; Ghimire, L.P.; Kim, Y.; Aryal, P.; Khadka, S.B. Identification and analysis of barriers against electric vehicle use. *Sustainability* **2020**, *12*, 4850. [CrossRef]
3. Kaunda, R.B. Potential environmental impacts of lithium mining. *J. Energy Nat. Resour. Law* **2020**, *38*, 237–244. [CrossRef]
4. Serrano, J.R.; Novella, R.; Piqueras, P. Why the development of internal combustion engines is still necessary to fight against global climate change from the perspective of transportation. *Appl. Sci.* **2019**, *9*, 4597. [CrossRef]
5. Karczewski, M.; Chojnowski, J.; Szamrej, G. A review of low-$CO_2$ emission fuels for a dual-fuel RCCI engine. *Energies* **2021**, *14*, 5067. [CrossRef]
6. Breuer, J.L.; Scholten, J.; Koj, J.C.; Schorn, F.; Fiebrandt, M.; Samsun, R.C.; Albus, R.; Görner, K.; Stolten, D.; Peters, R. An Overview of Promising Alternative Fuels for Road, Rail, Air, and Inland Waterway Transport in Germany. *Energies* **2022**, *15*, 1443. [CrossRef]
7. Bae, C.; Kim, J. Alternative fuels for internal combustion engines. *Proc. Combust. Inst.* **2017**, *36*, 3389–3413. [CrossRef]
8. Verhelst, S.; Turner, J.W.; Sileghem, L.; Vancoillie, J. Methanol as a fuel for internal combustion engines. *Prog. Energy Combust. Sci.* **2019**, *70*, 43–88. [CrossRef]
9. Boretti, A. Renewable hydrogen to recycle $CO_2$ to methanol. *Int. J. Hydrogen Energy* **2013**, *38*, 1806–1812. [CrossRef]
10. Di Iorio, S.; Catapano, F.; Magno, A.; Sementa, P.; Vaglieco, B.M. The Potential of Ethanol/Methanol Blends as Renewable Fuels for DI SI Engines. *Energies* **2023**, *16*, 2791. [CrossRef]
11. Turner, J.W.; Lewis, A.G.; Akehurst, S.; Brace, C.J.; Verhelst, S.; Vancoillie, J.; Sileghem, L.; Leach, F.; Edwards, P.P. Alcohol fuels for spark-ignition engines: Performance, efficiency and emission effects at mid to high blend rates for binary mixtures and pure components. *Proc. Inst. Mech. Eng. Part D J. Automob. Eng.* **2018**, *232*, 36–56. [CrossRef]
12. Turner, J.W.; Lewis, A.G.; Akehurst, S.; Brace, C.J.; Verhelst, S.; Vancoillie, J.; Sileghem, L.; Leach, F.; Edwards, P.P. Alcohol Fuels for Spark-Ignition Engines: Performance, Efficiency, and Emission Effects at Mid to High Blend Rates for Ternary Mixtures. *Energies* **2020**, *13*, 6390. [CrossRef]
13. Zhang, Z.; Wen, M.; Cui, Y.; Ming, Z.; Wang, T.; Zhang, C.; Ampah, J.D.; Jin, C.; Huang, H.; Liu, H. Effects of methanol application on carbon emissions and pollutant emissions using a passenger vehicle. *Processes* **2022**, *10*, 525. [CrossRef]
14. Geng, P.; Zhang, H.; Yang, S. Experimental investigation on the combustion and particulate matter (PM) emissions from a port-fuel injection (PFI) gasoline engine fueled with methanol–ultralow sulfur gasoline blends. *Fuel* **2015**, *145*, 221–227. [CrossRef]
15. Ho, C.S.; Peng, J.; Yun, U.; Zhang, Q.; Mao, H. Impacts of methanol fuel on vehicular emissions: A review. *Front. Environ. Sci. Eng.* **2022**, *16*, 121. [CrossRef]
16. Wei, Y.; Zhu, Z.; Liu, S.; Liu, H.; Shi, Z.; Zeng, Z. Investigation on injection strategy affecting the mixture formation and combustion of a heavy-duty spark-ignition methanol engine. *Fuel* **2023**, *334*, 126680. [CrossRef]
17. Larsson, T.; Mahendar, S.K.; Christiansen-Erlandsson, A.; Olofsson, U. The effect of pure oxygenated biofuels on efficiency and emissions in a gasoline optimised disi engine. *Energies* **2021**, *14*, 3908. [CrossRef]
18. Svensson, E.; Li, C.; Shamun, S.; Johansson, B.; Tuner, M.; Perlman, C.; Lehtiniemi, H.; Mauss, F. Potential Levels of Soot, $NO_x$, HC and CO for Methanol Combustion. SAE Tec Pap 2016-01-0887. 2016. Available online: https://saemobilus.sae.org/content/2016-01-0887/ (accessed on 18 January 2024).
19. Cho, H.M.; He, B.-Q. Spark ignition natural gas engines—A review. *Energy Convers. Manag.* **2007**, *48*, 608–618. [CrossRef]
20. Akal, D.; Öztuna, S.; Büyükakın, M.K. A review of hydrogen usage in internal combustion engines (gasoline-Lpg-diesel) from combustion performance aspect. *Int. J. Hydrogen Energy* **2020**, *45*, 35257–35268. [CrossRef]
21. Duc, K.N.; Duy, V.N.; Hoang-Dinh, L.; Viet, T.N.; Le-Anh, T. Performance and emission characteristics of a port fuel injected, spark ignition engine fueled by compressed natural gas. *Sustain. Energy Technol. Assess.* **2019**, *31*, 383–389. [CrossRef]
22. Ma, F.; Liu, H.; Wang, Y.; Li, Y.; Wang, J.; Zhao, S. Combustion and emission characteristics of a port-injection HCNG engine under various ignition timings. *Int. J. Hydrogen Energy* **2008**, *33*, 816–822. [CrossRef]
23. Ortiz-Imedio, R.; Ortiz, A.; Ortiz, I. Comprehensive analysis of the combustion of low carbon fuels (hydrogen, methane and coke oven gas) in a spark ignition engine through CFD modeling. *Energy Convers. Manag.* **2021**, *251*, 114918. [CrossRef]
24. Yosri, M.; Palulli, R.; Talei, M.; Mortimer, J.; Poursadegh, F.; Yang, Y.; Brear, M. Numerical investigation of a large bore, direct injection, spark ignition, hydrogen-fuelled engine. *Int. J. Hydrogen Energy* **2023**, *48*, 17689–17702. [CrossRef]
25. Thawko, A.; Eyal, A.; Tartakovsky, L. Experimental comparison of performance and emissions of a direct-injection engine fed with alternative gaseous fuels. *Energy Convers. Manag.* **2021**, *251*, 114988. [CrossRef]
26. Thawko, A.; Tartakovsky, L. The Mechanism of Particle Formation in Non-Premixed Hydrogen Combustion in a Direct-Injection Internal Combustion Engine. *Fuel* **2022**, *327*, 125187. [CrossRef]
27. Di Iorio, S.; Sementa, P.; Vaglieco, B.; Catapano, F. An Experimental Investigation on Combustion and Engine Performance and Emissions of a Methane-Gasoline Dual-Fuel Optical Engine. SAE Technical Paper 2014-01-1329. 2014. Available online: https://saemobilus.sae.org/content/2014-01-1329/ (accessed on 18 January 2024).
28. Holman, J.P. *Experimental Methods for Engineers*; McGraw-Hill: New York, NY, USA, 2012.
29. TSI Incorporated. Engine Exhaust Particle Sizer TM Spectrometer Model 3090. Available online: https://tsi.com/products/particle-sizers/fast-particle-sizer-spectrometers/engine-exhaust-particle-sizer-(eeps)-3090/ (accessed on 30 March 2015).

30. TSI. Updated Inversion Matrices for Engine Exhaust Particle Sizer TM (Eeps TM) Spectrometer Model 3090—Application Note EEPS-005 (A4). 2015. Available online: https://tsi.com/getmedia/22bf0106-13d9-4503-b179-bc76cb55e100/Updated_Inversion_Matrices_EEPS-005-A4-web?ext=.pdf (accessed on 18 January 2024).
31. AVL Emission Tester Series 4000. 2007. Available online: https://www.yumpu.com/en/document/view/7031754/avl-emission-tester-series-4000-product-brochure-avl-ditest (accessed on 18 January 2024).
32. Catapano, F.; Di Iorio, S.; Luise, L.; Sementa, P.; Vaglieco, B.M. Influence of ethanol blended and dual fueled with gasoline on soot formation and particulate matter emissions in a small displacement spark ignition engine. *Fuel* **2019**, *245*, 253–262. [CrossRef]
33. Apicella, B.; Catapano, F.; Di Iorio, S.; Magno, A.; Russo, C.; Sementa, P.; Tregrossi, A.; Vaglieco, B.M. Comprehensive analysis on the effect of lube oil on particle emissions through gas exhaust measurement and chemical characterization of condensed exhaust from a DI SI engine fueled with hydrogen. *Int. J. Hydrogen Energy* **2023**, *48*, 22277–22287. [CrossRef]
34. Apicella, B.; Catapano, F.; Di Iorio, S.; Magno, A.; Russo, C.; Sementa, P.; Tregrossi, A.; Vaglieco, B.M. Comprehensive analysis on the effect of lube oil on particle emissions through gas exhaust measurement and chemical characterization of condensed exhaust from a DI SI engine fueled with hydrogen. Part 2: Effect of operating conditions. *Int. J. Hydrogen Energy* **2024**, *49*, 968–979. [CrossRef]
35. Thawko, A.; Yadav, H.; Eyal, A.; Shapiro, M.; Tartakovsky, L. Particle emissions of direct injection internal combustion engine fed with a hydrogen-rich reformate. *Int. J. Hydrogen Energy* **2019**, *44*, 28342–28356. [CrossRef]
36. Raza, M.; Chen, L.; Leach, F.; Ding, S. A Review of particulate number (PN) emissions from gasoline direct injection (gdi) engines and their control techniques. *Energies* **2018**, *11*, 1417. [CrossRef]
37. Larsson, T.; Olofsson, U.; Erlandsson, A.C. Undiluted measurement of the particle size distribution of different oxygenated biofuels in a gasoline-optimised disi engine. *Atmosphere* **2021**, *12*, 1493. [CrossRef]
38. Lou, D.; Wang, T.; Fang, L.; Tan, P.; Hu, Z.; Zhang, Y.; Xu, Z.; Cheng, C.; Wang, S.; Zhang, Y. Investigation of the combustion and particle emission characteristics of a GDI engine with a 50 MPa injection system. *Fuel* **2022**, *315*, 123079. [CrossRef]
39. Hora, T.S.; Shukla, P.C.; Agarwal, A.K. Particulate emissions from hydrogen enriched compressed natural gas engine. *Fuel* **2016**, *166*, 574–580. [CrossRef]

**Disclaimer/Publisher's Note:** The statements, opinions and data contained in all publications are solely those of the individual author(s) and contributor(s) and not of MDPI and/or the editor(s). MDPI and/or the editor(s) disclaim responsibility for any injury to people or property resulting from any ideas, methods, instructions or products referred to in the content.

energies

MDPI

Article

# A New Method of Failure Mode and Severity Effects Analysis for Hydrogen-Fueled Combustion Systems

**Adrian Gill, Ireneusz Pielecha * and Filip Szwajca**

Faculty of Civil and Transport Engineering, Poznan University of Technology, pl. Marii Skłodowskiej-Curie 5, 60-965 Poznań, Poland; adrian.gill@put.poznan.pl (A.G.); filip.szwajca@put.poznan.pl (F.S.)
* Correspondence: ireneusz.pielecha@put.poznan.pl; Tel.: +48-61-224-4502

**Abstract:** This article aims to align its content with current trends in hybrid risk analysis methods while utilizing experimental research. This paper presents a hybrid methodology for analyzing the failure severity of a two-stage hydrogen-powered combustion system and details its implementation. This methodology assumes the use of the original FMESA method (Failure Mode and Effects Severity Analysis) with dedicated tabular scales of the failure severity. Obtaining results under the FMESA using experimental research is intended to reduce epistemic uncertainty, which is an important component of hazard severity or risk models. Its essence is to change the way of obtaining the results of the basic components of known methods such as FMEA/FMECA (Failure Mode and Effect Analysis/Failure Mode, Effects and Criticality Analysis). Experimental research was also used to develop the original failure severity scales for a two-stage hydrogen-fueled combustion system. The article presents a review of the literature on methods for identifying and analyzing hazards in hydrogen systems, the FMESA method with its mathematical model, results in the form of tabular scales of the failure severity, results of selected experimental tests, and quantitative results of a severity analysis of eleven failure modes of a two-stage hydrogen-fueled combustion system for a selected engine operating point.

**Keywords:** hydrogen combustion; combustion indexing; failure severity analysis

**Citation:** Gill, A.; Pielecha, I.; Szwajca, F. A New Method of Failure Mode and Severity Effects Analysis for Hydrogen-Fueled Combustion Systems. *Energies* **2024**, *17*, 4802. https://doi.org/10.3390/en17194802

Academic Editors: Enzo Galloni and Davide Lanni

Received: 29 June 2024
Revised: 19 September 2024
Accepted: 23 September 2024
Published: 25 September 2024

**Copyright:** © 2024 by the authors. Licensee MDPI, Basel, Switzerland. This article is an open access article distributed under the terms and conditions of the Creative Commons Attribution (CC BY) license (https://creativecommons.org/licenses/by/4.0/).

## 1. Introduction

Evolving the performance characteristics of an internal combustion engine's fueling system components poses unique challenges for safety engineering. This is particularly true for hydrogen fueling systems, whose inherent properties can generate risks, leading to significant failure or losses. Studies of hydrogen fueling systems often highlight the consequences of fire or explosion incidents. This emphasis stems from both the intrinsic properties of hydrogen itself, such as its low ignition energy and wide flammability range, and the characteristics of hydrogen fires, which make them difficult to detect and extinguish (hydrogen flames are colorless or nearly colorless, and often the only effective approach is to cut off the fuel source [1]). Furthermore, burning hydrogen in an internal combustion engine with an excessively low air–fuel ratio can lead to knocking combustion [2,3]. This undesirable combustion process can ultimately lead to the failure of engine components [4]. However, despite being improper combustion, various methods exist for its control [5,6] or minimization [7,8].

The exploitation of hydrogen is becoming increasingly common. Environmental analyses indicate various possibilities for hydrogen extraction [9,10], and predictions for the cost of $H_2$ production in the coming years appear to be decreasing. Forecasts suggest that within a few years, hydrogen production costs could be more than halved by 2025 [9]. Hydrogen may not be the safest option for fuel, but other fuels used today have also been seen as unsafe during different times of production and utilization [10].

While safety incidents like fire or explosion are major concerns in hydrogen-fueled systems, changes in the operating characteristics of mixture formation and combustion systems components also significantly impact combustion quality. In two-stage combustion systems, the use of passive or active pre-combustion chambers fundamentally alters the combustion process by enabling diverse control strategies [5,11,12]. Importantly, these changes in operating characteristics are much more likely to occur compared to safety events and can even be considered typical for such systems.

Safety engineering provides powerful tools to control hazards and rationally manage processes while considering their inherent risks. In the case of hydrogen-operated systems, methods like FMEA/FMECA (Failure Mode and Effect Analysis/Failure Mode, Effects and Criticality Analysis) or HAZOP (Hazard and Operability Study) are highly recommended. These methods are not only valuable in hydrogen sectors but are even included in quality management system standards for other sectors, such as IRIS (International Railway Industry Standard). As the authors of a paper [13] point out, "for the risk analysis of hydrogen fuel cell vehicles, the existing research mainly adopts the method of theoretical analysis, and the common methods are the FMECA, the FFMEA, the FTA (Fault Tree Analysis), and the HAZOP".

HAZOP and FMEA/FMECA methods are particularly valuable during the system design stage, as the prevention of undesirable conditions, such as leaks, malfunctions, or component failures, offers the greatest safety benefits. While HAZOP focuses on identifying potential hazards and their operational causes, FMEA delves into the failure modes of each system component and qualitatively assesses the probability and severity of their consequences [13–15]. Correa-Jullian and Groth's application of FMEA to a liquid hydrogen storage system [14] exemplifies this approach, identifying critical failure scenarios such as pressure relief valve malfunctions, air-operated valve failures, and evaporator ruptures. Their qualitative assessment of failure types and severity levels highlights the importance of preventive measures in mitigating hydrogen-related risks.

Cui et al. [13], by contrast, conducted a fire hazard analysis and explosion simulation for hydrogen leakage in fuel cell vehicles. They employed FMEA analysis in a traditional manner, focusing on identifying the potential types of failure. The authors of another study [16] addressed a similar analytical problem but utilized FTA and ETA (Event Tree Analysis) methods.

Another publication [15], a comprehensive study by the U.S. Department of Transportation's National Highway Traffic Safety Administration (NHTSA), analyzes the causes and consequences of hydrogen fuel cell vehicle failures. This NHTSA report aimed to evaluate safety issues in hydrogen-powered vehicles and identify areas for developing federal safety standards. A key finding highlights the criticality of redundancy in high-pressure components. Without redundancy in the compressed-hydrogen fuel system (container, pressure relief device (PRD), or first valve), a single-point failure can cause large-scale hydrogen release or venting. Additionally, container failures may release mechanical energy. While smaller hydrogen releases and ruptures of other components can also be dangerous, they typically have less severe consequences compared to large leaks and tank ruptures, as noted in [15]. The study employed a qualitative scale to assess the potential severity of consequences.

The HAZOP method finds frequent application in the context of hydrogen systems, primarily for identifying scenarios that could lead to adverse events. This use of the method is well supported by publications such as [13,17–19]. A clear example is the work of [19], where HAZOP was employed to identify accident scenarios caused by internal factors during the evaluation of organic-hydride-hydrogen-refueling stations utilizing methylcyclohexane. Similarly, Ehrhart et al. [17] used HAZOP to assess risk and model ventilation performance for hydrogen releases within auto repair shops. Jung et al. [18] demonstrate a slightly different application of HAZOP, conducting an analysis of hydrogen behavior under cryogenic adsorption conditions.

FMEA's application faces criticism due to shortcomings and, in some cases, even its debatable value for certain risk assessments [20]. To address these limitations, various authors have proposed extensions of FMEA or created hybrid methods incorporating

FMEA or FMECA as a core component. Examples of extension methods are found in works by [20,21], while [14,22,23] showcase hybrid methods. These hybrid methods combine established safety engineering techniques, most commonly FMEA, HAZOP, and FTA [23].

FMEA/FMECA risk assessments, not only for hydrogen systems, typically focus on the most severe consequences of failure in terms of loss magnitude. The severity level is usually assigned subjectively based on worst-case scenarios [24]. However, such scenarios often have a very low probability of occurring. Impacts with lesser losses are much more likely, potentially leading to reduced system functionality. Therefore, neglecting the criticality (risk) of failure with average severity is unreasonable, especially since these events could still result in unacceptable risk levels. While numerous studies analyze criticality in hydrogen systems, there is a dearth of research focusing on this type of failure, particularly in two-stage combustion systems.

A literature review and the authors' experience in this current article show that there is a lack of studies on the failure criticality associated with the quality of combustion processes in hydrogen-fueled systems. More so, such studies have not been conducted for two-stage hydrogen combustion systems. In addition, traditional methodologies (such as FMEA/FMECA) assessments of damage criticality do not include scales based on measurable parameters that characterize the systems. On the one hand, this is a welcome generalization, but, on the other hand, it is a simplification that introduces additional inaccuracy and subjectivity into the assessments. Based on the literature review, the areas of ignorance that are the subject of the article were identified. The purpose of this study is to present an original method of damage severity assessment—FMESA (Failure Mode and Effects Severity Analysis)—and to indicate the results of its application to a two-stage hydrogen combustion system. The method is part of the current trends in hybrid risk analysis, combining various well-known methods of this type of analysis. Its essence is a specific approach to identifying the components of the analysis, supported by an original tabular damage severity scale developed for a two-stage hydrogen combustion system at its typical operating point.

Section 2 details the method's concept and describes the test stand. Section 3 presents the analysis' results, including the developed failure severity scale for the selected engine operating point and experimental data for selected engine operation indicators, along with examples of critical failure to components (e.g., intake, combustion, and exhaust systems), as identified in [16]. Section 4 discusses the results and analyzes the proposed method's applicability. Finally, Section 5 summarizes the entire work.

## 2. Materials and Methods

### *2.1. Description of the Test Stand*

The combustion process analysis was conducted on a single-cylinder AVL 5804 engine (AVL factory, Graz, Austria) with a displacement of 0.5 $dm^3$ and a TJI (turbulent jet ignition) combustion system. The study employed both passive and active combustion chambers. The engine was fueled with technical hydrogen (99.999% purity) at a pressure of 7 bar. The characteristics of the technical hydrogen fuel are provided in Table 1. An AMKASYN electrospinning asynchronous brake served as the engine load. A scheme of the test stand is shown in Figure 1.

The operation of an engine fueled by gaseous fuels makes it possible to use higher excess air ratio values (lean mixtures). This means that the efficiency of the engine can be increased. At the same time, higher compression ratios also can be applied (up to the limit of knock combustion). Using the TJI system makes it possible to take advantage of ultra-lean combustion, where a nearly stoichiometric mixture is aimed at the prechamber, and a very lean mixture is aimed at the main chamber. The escaping jets from the prechamber ignite the lean charge in the main chamber. Thus, combining gaseous fuel with a TJI system and a high compression ratio results in increased engine efficiency and effectiveness. With such solutions, the efficiency gap between internal combustion engines and fuel cells is reduced [10].

**Table 1.** Typical physiochemical properties of hydrogen [25].

| Species | Hydrogen |
|---|---|
| Chemical formula | $H_2$ |
| LHV [MJ/kg] | 120 |
| Laminar burning velocity at $\lambda = 1$ [m/s] | 3.51 |
| Auto-ignition temperature [K] | 773–850 |
| Research octane number | >100 |
| Flammability limit in air [vol %] | 4.7–75 |
| Quench distance [mm] | 0.64 |
| Absolute minimum ignition energy [mJ] | 0.02 |
| Latent heat of vaporization [kJ/kg] | 461 |

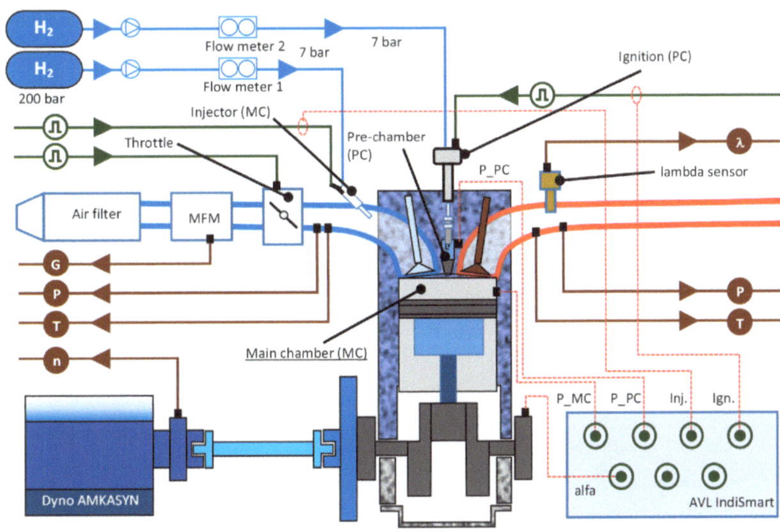


**Figure 1.** Test stand diagram for a two-stage hydrogen combustion system.

Hydrogen consumption was recorded using two independent mass flowmeters: a Micro Motion ELITE CMFS010M (Coriolis phenomenon, by Emerson) for the main chamber (MC) and a EL_FLOW flowmeter (by Bronkhorst, AK Ruurlo, The Netherlands) for the prechamber (PC). Air consumption was measured with a Sensycon Sensyflow thermal mass flowmeter from ABB (Zurich, Switzerland). An open-loop Engine Management Unit (EMU Black controller) from Ecumaster (Cracov, Poland) controlled the throttle and ignition. The active pre-chamber utilized a Beru pencil coil ignition system. This system allowed for setting both the coil charging time (fixed at 3 ms) and the ignition advance time using an M10 spark plug. Finally, the excess air ratio was determined with an LSU 4.9 broadband probe (by Bosch) interfaced with an LCP80 controller (by IMFsoft, Ostrava, Czech Republic).

The combustion chamber was equipped with a pre-chamber (6 holes—radially), whose volume was 2.29 cm$^3$. This value accounted for 0.45% of the total engine stroke volume (at the BDC) and 6.6% of the capacity (at the TDC).

### *2.2. Description of Analysis Methodology and FMESA Analysis Method*

The FMECA method, also known as E-FMECA (empirical FMECA), takes a different approach to identifying essential components compared to traditional FMECA. While traditional FMECA relies on inductive reasoning (forward identification) for this purpose [26], E-FMECA leverages experimental studies to determine these critical components.

FMESA, moreover, can be categorized as a hybrid risk analysis method, as it uses various methods and approaches to identify undesirable system states. Among other things, it assumes the use of the HAZOP methodology to conduct the process of identifying the so-called local effects of system failure (local effects). The idea and the individual steps of FMESA are shown in Figure 2.

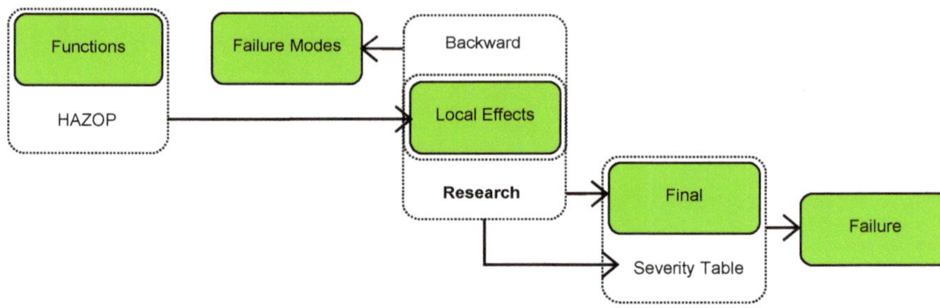


**Figure 2.** FMESA steps with FMEA/FMECA components (green color).

The core of the FMESA (Figure 2) consists of FMEA/FMECA components, while their identification is realized with the support of HAZOP and experimental studies. These studies are used, on the one hand, to precisely formulate the core component of FMESA (i.e., final effects), and, on the other hand, they enable the realization of a quantitative procedure for indicating effects on the basis of measured values of key system/installation parameters (see Section 3.1). The individual components of the FMESA results sheet, with sample records, are shown in Table 2.

**Table 2.** The column headings of the FMESA worksheet with their origin and labels.

| FMESA Column Header Name | Symbol | Description |
|---|---|---|
| Element | E | Pressure reducer |
| Function | F | Maintaining constant pressure behind the main valve |
| Guide Word | GW | more |
| Deviation/Local Effect | D | Fuel reduction |
| Failure Mode | FM | Reducer pin jammed |
| Measurable Parameters | MP | Lambda value = 1, 2. . . Pmx = . . .; Tex = . . . |
| Final Effect | FE | Engine power drop |
| Failure Severity | FS | Max[FS(MP)] = FS(λ) = H1.2 |

The first step in FMECA is to identify the functions performed by the system elements. This is in line with the types of FMEA analyses defined, for example, by [27] as PF-FMEA (Product Function Failure Modes and Effects Analysis) or PI-FMEA (Product Inter-face Failure Modes and Effects Analysis). PF-FMEA is used to evaluate the effects at the customer level following the failure of each of the top-level product functions. This is a powerful tool for understanding customer needs, as each function is examined with respect to its intended use [27].

Local effect (LE) analysis, a crucial component of FMECA, utilizes the HAZOP approach. This method employs guidewords to systematically question the achievement of design intentions and operational conditions for processes, procedures, or systems. By applying these guidewords, the goal is to identify deviations from the intended process flow, known as irregularities. Notably, this aligns perfectly with the objective of this article—analyzing combustion process quality. While a detailed description of HAZOP is available in [28], this work focuses on the key components relevant to achieving the article's specific goals. Table 3 presents a set of guidewords tailored for the FMECA method, along with explanations and examples for their application.

**Table 3.** A set of guide words adopted for the E-FMECA method. Own compilation based on [19,28,29].

| Guide Word | Interpretation | Examples |
| --- | --- | --- |
| NO | No part of intended result or intention is achieved; task not completed | No flow; no data or control signal passed; operator does not take action |
| MORE | Quantitative increase in output or in the operating condition; do more than or more of the required action | Larger quantity handled; valves opened more than required; data are passed at a higher rate than intended |
| LESS | Quantitative decrease, e.g., lower temperature; do less of the required action | Smaller quantity handled; not all valves opened in a step; data are passed at a lower rate than intended |
| AS WELL AS | Quantitative increase, but, e.g., with additional material, impurities present; do something in addition to the required task; simultaneous execution of another operation/step | Some additional or spurious signal is present; additional material handled; open additional valve |
| PART OF | Quantitative decrease (e.g., only one or two components in a mixture); not all task in an action carried out; only some of the intention is achieved, i.e., only part of intended fluid transfer takes place | The data or control signal are incorrect; action within a step omitted |
| REVERSE | Opposite (e.g., backflow); do the opposite of the required action; covers reverse flow in pipe | Closes valves instead of opens; needs to reverse previous action; normally not relevant |
| OTHER THAN | No part of the intention is achieved, something completely different happens (e.g., flow or wrong material); a result other than the original intention is achieved, i.e., transfer of wrong material | Acts on wrong valve; incorrect material handled; the data or control signals are incorrect |
| EARLY | Something happens early relative to clock time, e.g., cooling or filtration | The signals arrive too early with reference to clock time |
| LATE | Something happens late relative to clock time, e.g., cooling or filtration | The signals arrive too late with reference to clock time |
| BEFORE | Something happens too early in a sequence, carry out the action after the time specified | The signals arrive earlier than intended within a sequence |
| AFTER | Something happens too late in a sequence, carry out the action after the time specified | The signals arrive later than intended within a sequence; changes order of steps; takes action too slowly |
| GW applied to parameters | Physical properties of a material or process; physical conditions (temperature, speed); specified intention of a component of a system or design (e.g., information transfer); operational aspects | Increase in combustion temperature; increase in the $\lambda$-value; increasing CoC |

In FMECA, deviations (Ds) encompass any departure from an element's intended function. This includes negative fulfillment, qualitative or quantitative changes, substitutions with alternative actions, and disruptions in the timing or sequence of execution. Table 2 exemplifies a qualitative change in function realization.

The FMESA method defines local effects (LEs) as the consequences of a failure mode (FM) on the function, performance, or state of the analyzed component [24]. Identified deviations (Ds) are assumed to correspond to LEs. In the context of power systems, LEs typically manifest as changes in component functionality, leading to variations in the recorded operating parameters of the entire system or plant. An example of an LE is excessive pressure limitation caused by a malfunctioning reducing valve. Additionally, LEs can encompass observable events or operating states during experiments, such as the bending of a pipe in a hydrogen delivery system.

Leveraging the established definitions of local effects (LEs) and the principles of HAZOP methodology, FMECA employs a deductive approach, also known as backward reasoning or "worksback". This approach analyzes the identified deviations (Ds) or formulated LEs and systematically reasons backward to identify the potential types of component failure (FMs) that could cause these deviations or LEs. In essence, we are working backward to understand how the component can fail and lead to the observed effects.

The FMESA methodology focuses on identifying measurable parameters (MPs) to quantify the final effects of a component malfunction in a power system. These MPs capture the changes in operating parameters caused by the malfunction. They are vital for both quantitative assessment of the effects (detailed in Section 2.3) and evaluation of the criticality of the failure (discussed in Section 3.1). Table 2 showcases examples of MPs (Measurable Parameters, e.g., lambda value and maximum pressure), while their specific values are presented in Section 2.3.

The FMESA method does not rely solely on theoretical analysis for final effects (FEs). Instead, it leverages controlled experiments on a test bench. Previously identified local effects (LEs) are simulated, and their actual impacts are measured and recorded using available equipment and sensors. This approach bridges the gap between theoretical predictions and real-world behavior.

The experimental findings culminated in the development of severity tables for a two-stage hydrogen-fueled combustion system. These tables are presented in Section 3, embodying the research objectives outlined in this article.

The severity tables developed in Section 3.1 rely on identifying operating parameters that reflect how failure impacts the system's operation. For instance, excess air factor effectively captures the consequences of excessive pressure reduction by the control valve. By monitoring changes in this parameter within its designated limits (low parameter limit [LPL] and high parameter limit [HPL]), we can establish a spectrum of potential consequences, ranging from decreased power output and increased toxic emissions to engine failure.

Thus, if by FM we denote the set of possible failure forms of plant elements, i.e., FM = $\{FM_i\}$ (i = 1, 2, ..., m), it is assumed that it is possible to assign to the i-th failure form a set of operating parameters characterizing it, i.e.,

$$\bigvee_{i=1}^{n} FM_i \rightarrow MP_i = \{MP_{ij}\}; j = 1, 2, \ldots, n, \quad (1)$$

where $MP_i = \{MP_{ij}\}$ (j = 1, 2, ..., n) is a finite set of possible measurable parameters of system operation.

The measurable operating parameters of the system are the following:

- Exhaust temperature ($T_{ex}$)//$MP_{i1}$;
- Shift in the center of combustion (CoC)//$MP_{i2}$;
- Exhaust outlet pressure ($P_{sp}$)//$MP_{i3}$;
- Boost pressure ($P_{in}$)//$MP_{i4}$;
- Excess air ratio ($\lambda$)//$MP_{i5}$;
- Amount of nitrogen oxides in the exhaust gasses ($NO_x$)//$MP_{i6}$.

For example, the set of parameters for fault form FM1 formulated as "locked gate in closed position" takes the form MP1 = $\{P_{ex}; NO_x; T_{ex}\}$.

Each final effect is characterized by a corresponding level of severity. Thus, let further FS be a finite set of ratings, FSk (k = 1, 2, ..., l), such that l = 5 and that

$$\bigvee_{i=1}^{m} \bigvee_{j=1}^{n} FS = \{VL, L, M, H, VH\}, \quad (2)$$

where VL, L, M, H, and VH are symbols of ratings/levels of severity in ascending order, understood as very low, low, medium, high, and very high, respectively.

The severity (level of "do-ness") of $xMP_{ij}$ values, representing the measurable parameter i for failure mode j, is determined using predefined scales. These scales are created by plotting the assigned severity levels against the corresponding parameter values. Examples of such scales can be found in Tables 4–9.

**Table 4.** Severity scale of damage to a two-stage hydrogen-fueled combustion system for the exhaust gas temperature parameter.

| Measurable Parameter (MP) | MP Value ($xMP_{i1}$) | Classification Description/Final Effect (FE) | Classification Term or Name | Classification No. ($FS_k = \varphi(z)$) |
|---|---|---|---|---|
| $MP_{i1}$—Exhaust temperature (Tsp) | 40–220 | Misfiring; loss of power; lack of fuel delivery | Very High | VH |
| | 220–240 | Large $\lambda$; power loss, chronic combustion | High | H |
| | 240–280 | Lambda in the region of 2.0–3.0 | Medium | M |
| (LPL;HPL) | 280–340 | Slight increase in $\lambda$ | Low | L |
| = (40;400) deg C | 340–400 | Knocking combustion, threatens to destroy the engine | Very High | VH |

**Table 5.** Severity scale of damage to a two-stage hydrogen-fueled combustion system for the center of the combustion change parameter.

| Measurable Parameter (MP) | MP Value ($xMP_{i2}$) | Classification Description/Final Effect (FE) | Classification Term or Name | Classification No. ($FS_k = \varphi(z)$) |
|---|---|---|---|---|
| $MP_{i2}$—Changing the center of combustion (CoC) | 0–4 | Damage to the engine, decrease in mechanical efficiency | Very High | VH |
| | 4–7 | Proper engine operation | High | H |
| | 7–10 | Reduction in $NO_x$, decrease in engine efficiency | Medium | M |
| (LPL;HPL) | 10–14 | Increased $T_{ex}$; lower $NO_x$ | Low | L |
| = (0;20) | 14–20 | Engine overheating, $NO_x$ decrease | High | H |

**Table 6.** Severity scale of damage to a two-stage hydrogen-fueled combustion system for the parameter change in boost pressure.

| Measurable Parameter (MP) | MP Value ($xMP_{i3}$) | Classification Description/Final Effect (FE) | Classification Term or Name | Classification No. ($FS_k = \varphi(z)$) |
|---|---|---|---|---|
| $MP_{i3}$—Changing boost pressure | 0.0–0.8 | Leak in intake system; turbo failure; no combustion; ignition loss | Very High | VH |
| | 0.8–1.0 | Operation at low $\lambda$; knock combustion; | High | H |
| (LPL;HPL) | 1.0–1.2 | Specific work; $\lambda$ in the range 0.95–3 | Medium | M |
| = (0;2) | 1.2–1.6 | Overcharging; ignition loss; power loss | Low | L |
| | 1.6–2.0 | Lack of fuel pressure; faulty fuel supply line; frequent ignition loss | VL | VL |

**Table 7.** Severity scale of damage to a two-stage hydrogen-fueled combustion system for the parameter change in cylinder pressure.

| Measurable Parameter (MP) | MP Value ($xMP_{i4}$) | Classification Description/Final Effect (FE) | Classification Term or Name | Classification No. ($FS_k = \varphi(z)$) |
|---|---|---|---|---|
| $MP_{i4}$—Change in cylinder pressure | 0.0–15 | No compression; defective engine | Very High | VH |
| | 15–40 | Low load; proper working conditions | Very low | VL |
| | 40–60 | Increased load | Low | L |
| (LPL;HPL) = (0;100 bar) | 60–80 | High engine load; possible knocking combustion | High | H |
| | 80–100 | Knocking combustion; air intake damage, fuel supply jammed in excess position; possible engine damage | Very high | VH |

**Table 8.** Severity scale of damage to a two-stage hydrogen-fueled combustion system for the lambda change parameter.

| Measurable Parameter (MP) | MP Value ($xMP_{i5}$) | Classification Description/Final Effect (FE) | Classification Term or Name | Classification No. ($FS_k = \varphi(z)$) |
|---|---|---|---|---|
| $MP_{i5}$—Lambda value change | 0.0–0.8 | No combustion; no fuel supply; collapsed fuel supply | High | H |
| | 0.8–1.0 | Knock combustion; very high $NO_x$; | Very High | VH |
| (LPL;HPL) | 1.0–2.0 | Possible knock; large $NO_x$ | Medium | M |
| = (0;5 bar) | 2.0–3.0 | Proper combustion; mid-range $NO_x$ | Low | L |
| | 3.0–5.0 | not enough fuel; ignition loss; low $T_{ex}$; near zero $NO_x$ | Very Low | VL |

**Table 9.** Severity scale of damage to a two-stage hydrogen-fueled combustion system for the parameter of the amount of nitrogen oxides in the exhaust gas.

| Measurable Parameter (MP) | MP Value ($xMP_{i6}$) | Classification Description/Final Effect (FE) | Classification Term or Name | Classification No. ($FS_k = \varphi(z)$) |
|---|---|---|---|---|
| $MP_{i6}$—amount of | 0–20 | Large lambda, ignition loss, power loss | Very High | VH |
| nitrogen oxides in | 20–50 | High lambda, proper combustion | Very Low | VL |
| the exhaust gas ($NO_x$) | 50–100 | Lambda in the region of 1.5–2.0; normal operation; passive chamber operation possible | Very Low | VL |
| (LPL;HPL) | 100–300 | High load; low lambda; possible knocking combustion | Medium | M |
| = (0;500 ppm) | 300–500 | Knocking combustion; pre-empted ignition; possible engine damage | Very High | VH |

For linear ascending severity scales taking values in the interval <0;1>, when any j-th parameter of plant operation takes values in the interval ($LPL_j$; $HPL_j$), the generalized form of the severity function can be expressed by the following formula:

$$\varphi\left(xMP_{ij}\right) = \begin{cases} VH & \text{when} \quad f^{-1}(0.80) \le xMP_{ij} \le HPL_j \\ H & \text{when} \quad f^{-1}(0.60) \le xMP_{ij} < f^{-1}(0.80) \\ M & \text{when} \quad f^{-1}(0.40) \le xMP_{ij} < f^{-1}(0.60) \\ L & \text{when} \quad f^{-1}(0.20) \le xMP_{ij} < f^{-1}(0.40) \\ VL & \text{when} \quad LPL_j \le xMP_{ij} < f^{-1}(0.20) \end{cases} \tag{3}$$

where $xMP_{ij}$ denotes the measured value of the j-th parameter associated with the i-th failure mode.

The function f(.) is a mapping of the beginnings and ends of the intervals of xMPij values (Tables 4–9, col. 2) to the interval <0;1>, but it can be any other subset of the ordinal numbers. The essence, however, is the division of this subset resulting from severity levels. Suggesting typical risk matrices, five severity levels were assumed for the FMESA method, equally dividing the counter-domain of the f(.) function, i.e., the interval <0;1>. Successive beginnings and ends of severity levels thus correspond to the numbers <0;0.2), <0.2;0.4), <0.4;0.6), <0.6;0.8), and <0.8;1.0>. The function f(.) was introduced primarily to write down the procedure for obtaining grades or severity levels in the clearest (mathematical) way possible. These values, however, can also be used in the case of the need to determine numerical and non-linguistic severity values.

The effects of function (3) can be represented as severity matrices. An example of a failure severity matrix for a two-stage hydrogen-powered combustion system regarding exhaust gas temperature ($T_{ex}$) is shown in Figure 3.

As can be seen, among other things, from the matrix (Figure 3), the form of the function f(.) is usually not linear; moreover, it can have a specific, empirical character. To create it, it is necessary to use expert knowledge of the system's functioning at different values of measurable parameters.

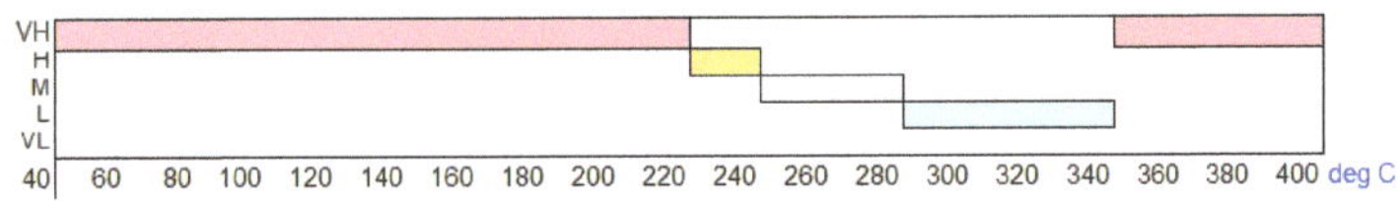


**Figure 3.** An example of a severity matrix for the $T_{ex}$ parameter.

Building on Equation (3), we can determine a severity rating from the set FS for each measured parameter $MP_{ij}$. However, Equation (1) establishes that each failure mode ($FM_i$) is characterized by a set of parameters ($MP_i$). Therefore, each failure mode also receives a set of severity ratings denoted by $\Omega$. These ratings are derived from the individual severity scores calculated for each parameter within the set $MP_i$.

$$\bigvee_{i=1}^{n} FM_i \to \Omega_i = \{\omega_{ij}\} : \ \omega_{ij} = FS_k \to MP_{ij}. \tag{4}$$

The final level/estimation of $FS_i$—severity for the i-th form of failure is determined as follows:

$$FS_i = \max_j \{\omega_{ij}\}; j = 1, 2, \ldots, n \tag{5}$$

*2.3. Analysis of the Internal Combustion Engine Operating Conditions*

This failure analysis employed a single-cylinder research engine fueled by hydrogen and equipped with a two-stage TJI combustion system. To investigate various failure scenarios, different operating conditions were implemented. These conditions included combustion in both passive (no prechamber fuel supply) and active (fuel supplied to both prechamber and main chamber) TJI configurations. Additionally, varying excess air ratios were used to explore knocking combustion, which typically occurs at low air-to-fuel ratios ($\lambda$). Changes in $\lambda$ also reflected varying engine loads, as the tests maintained a constant fuel dose to the main chamber. Finally, increasing the prechamber fuel dose was another control variable, demonstrating the possibility of full engine control. Collectively, these test conditions aimed to encompass a broad spectrum of potential failure modes in a hydrogen-fueled TJI engine.

Figure 4 presents pressure waveforms within the cylinder's main chamber for various operating conditions. These conditions include using the passive or active chamber, along with varying excess air ratio ($\lambda$) and prechamber fuel dose. The data clearly demonstrate that increasing $\lambda$ corresponds to a higher engine load. Additionally, the active chamber's configuration exhibits a more significant influence on pressure waveforms. Notably, a high global excess air ratio negatively impacts combustion quality in the passive chamber. This is evident from the significant deviation of the combustion pressure line in the passive chamber case compared to the active chamber at high $\lambda$ values, indicating a less optimal combustion process.

Thermodynamic analyses were performed using AVL Concerto V4.5 software (AVL, Graz, Austria). Analysis of cylinder pressure waveforms (Figure 4) revealed the impact of the prechamber fuel dose on thermodynamic indicators as excess air ratio ($\lambda$) changes. Regardless of the initial $\lambda$ value (passive chamber), increasing the prechamber dose causes a slight decrease in $\lambda$ (Figure 5a). Notably, the ignition angle remained constant throughout the tests (Figure 5b). This control method, however, led to a significant increase in the combustion center angle (CA) when using the passive chamber (no prechamber dose). This occurs because $\lambda$ cannot be controlled within the prechamber, resulting in slow combustion and embers slowly flowing into the main chamber. Such a slow process is reflected in the rate and amount of heat release. Conversely, increasing the prechamber dose significantly stabilizes combustion, substantially shortening the initial combustion phase. The Center of Combustion (CoC) angle is accelerated by approximately 4 degrees. Interestingly, at $\lambda = 3$, this acceleration is doubled (8 degrees) compared to other $\lambda$ values (Figure 5d). Overall, increasing the prechamber dose stabilizes CoC while shortening the combustion angle, indicating faster prechamber combustion. As expected, increasing $\lambda$ generally leads to longer combustion times, with a significant increase in the combustion angle only observed at high λ-values (Figure 5c).

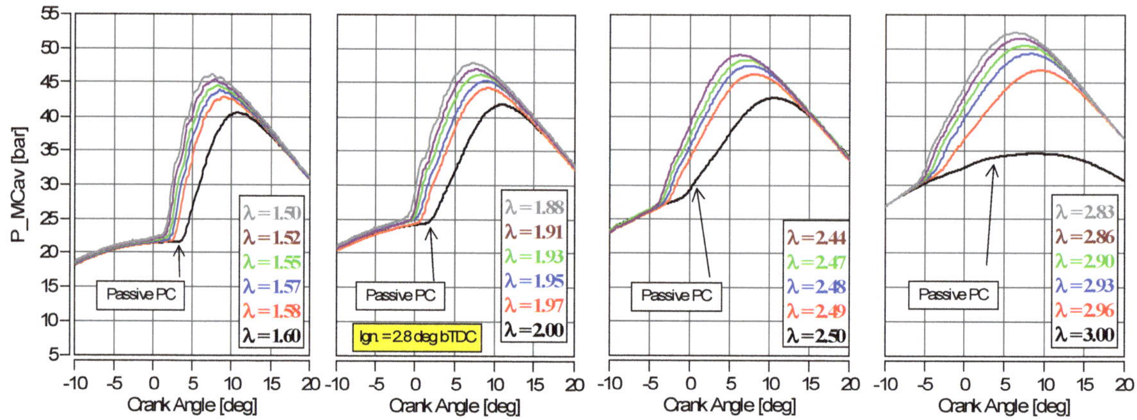


**Figure 4.** Cylinder pressure for varying values of excess air ratio (passive chamber—no fuel supply to the pre-chamber, n = 1500 rpm, qo_MC = 3 mg/inj).

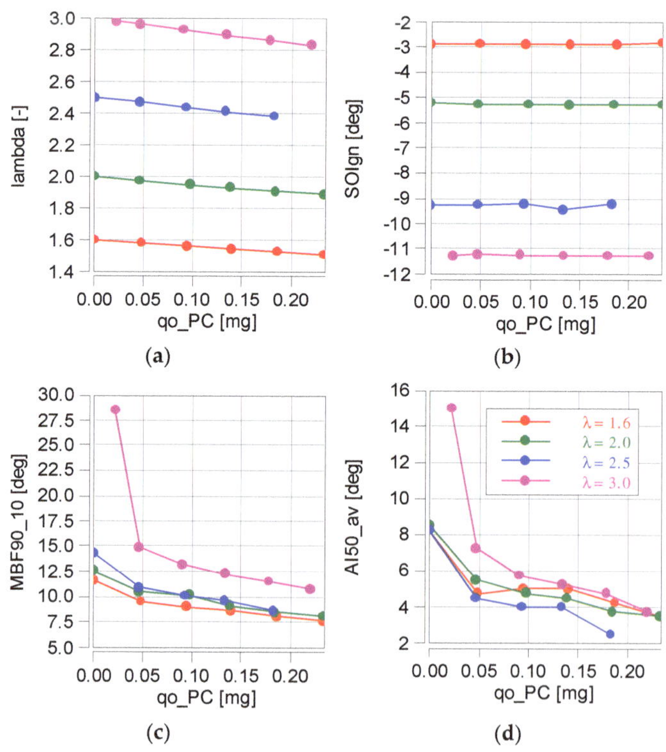


**Figure 5.** Indicators of engine adjustments when changing the pre-chamber fuel dose: (**a**) change in excess air ratio (decrease of λ when increasing the pre-chamber dose); (**b**) constancy of ignition timing (no change in angle causes the center of combustion to change); (**c**) change in combustion duration (reduction in combustion duration as a result of increasing the pre-chamber fuel dose); and (**d**) change in combustion center (at a constant SOI value, acceleration of the combustion center is observed).

The subsequent section explores the engine's operating ratios. Altering the excess air ratio (λ) necessitates a corresponding change in boost pressure (Figure 6a). The analysis revealed a linear relationship; increasing λ requires a higher engine boost pressure. Conversely, as λ increases, both the in-cylinder and exhaust gas temperatures decrease. This trend is confirmed by this study's findings, which demonstrate a linear decrease in temperature with increasing λ (Figure 6b).

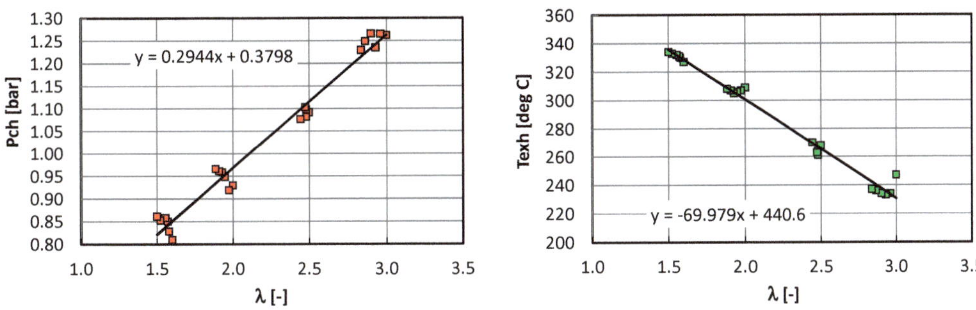


(a) (b)

**Figure 6.** Change in engine adjustment indicators depending on the excess air ratio: (**a**) boost pressure; (**b**) exhaust gas temperature.

As shown in Figure 7a, increasing boost pressure alongside a rising excess air ratio (λ) leads to a corresponding increase in maximum cylinder pressure. Notably, the peak pressure in the main chamber is roughly 1 bar higher compared to the prechamber. This relationship appears independent of the specific λ value. However, it is important to acknowledge that a larger diameter for the channels connecting the prechamber to the main chamber could decrease this pressure differential.

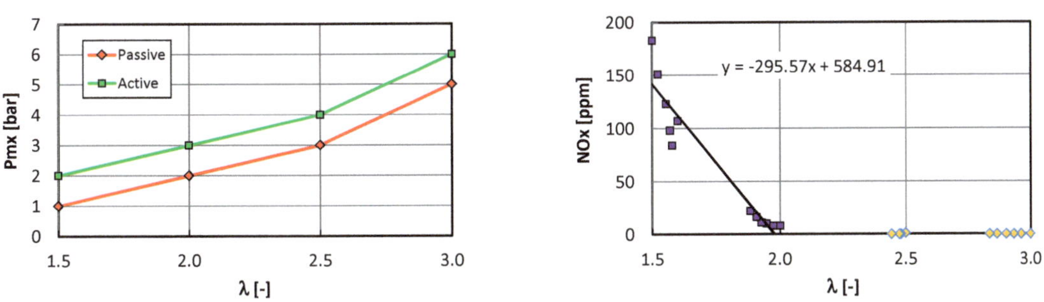


(a) (b)

**Figure 7.** Change in engine performance indicators: (**a**) maximum cylinder pressure in relation to the passive and active pre-chamber system; (**b**) concentration of nitrogen oxides in the exhaust gasses.

Increasing the excess air ratio (λ) dilutes the air–fuel mixture, leading to a decrease in nitrogen oxide ($NO_x$) concentration. This reduction is primarily driven by the lower combustion chamber temperatures associated with higher λ values. Notably, $NO_x$ concentration remains significant up to λ = 2. Beyond this point, minimum $NO_x$ values close to zero are observed (yellow points in Figure 7b). Consequently, for the purpose of this analysis, the relationship between λ and $NO_x$ is assumed to be linear up to λ = 2 (Figure 7b).

## 3. Results

### 3.1. Tabulated Severity Scale of Failure to Two-Stage Hydrogen-Fueled Combustion System

Severity is defined as "significance of grading of the failure-mode's effect on item operation, on the item surrounding, or on the item operator" [30], to be estimated with respect to the defined boundaries of the system under analysis. Typically, the severity scale is presented in tabular form, including the names of the severity categories, their description, and rank numbers. A useful example of a severity scale was given, for example, by Crowe in his paper [27] and Dhillon in his paper [24]. Example notations they used to describe severity categories are "No real impact on system performance, and the customer may not even notice the failure"; "The failure affects safe item operation, involves non-compliance with government rules and regulations"; "High customer dissatisfaction due to the nature of the failure, such as a major system (e.g., automobile engine) function being inoperative"; and "The nature of failure causes only slight annoyance. The customer will probably only notice a slight deterioration of the performance".

Building upon established severity assessment concepts, original severity scales were developed for one point of operation in a hydrogen-fueled two-stage combustion system. These scales not only provide qualitative descriptions of severity categories but also introduce quantification based on measured system/installation performance parameters (Tables 4–9).

It should be added that the adopted ranges of measured parameters (MP value; Tables 4–9, col. 2, for an example engine operating point) of a hydrogen-fueled engine (1500 rpm; IMEP = 0.36 MPa at $\lambda = 1.6$ and IMEP = 0.42 MPa at $\lambda = 3.0$). A more precise range can be obtained after analyzing varying engine speeds and loads.

### 3.2. FMESA: Selected Results of Failure Analysis by the FMESA Method

Table 10 presents the results of the failure analysis and severity assessment for the hydrogen-fueled two-stage combustion system. The selected failures represent those commonly encountered in hydrogen combustion systems and exhibit the highest FS (failure severity) scores. In cases where a wide range of measured parameter values indicated two severity levels, the more critical option was chosen.

Table 10 details the analysis of system elements and their potential failures. The first column identifies the analyzed elements. For each element, its failure form is determined based on its function (col. F) and a corresponding keyword (col. GW). This approach helps identify deviations or abnormalities in how the element fulfills its function (col. D). Notably, one element's function can have multiple potential deviations (e.g., an ignition coil). These deviations or non-functions are considered local effects within FMESA (detailed in Section 2.2 of the methodology). Associated with these local effects are the damage characteristics, the root causes of the deviations, listed in the (FM) column. To assess damage severity following the FMESA methodology, measurable parameters (col. MP) are selected and recorded. These parameters reflect changes in system operation caused by the specific damage. Using these parameters and the criticality tables (Section 3.1), a severity level is assigned in the form of a "Classification No." Since the FMESA method is hybrid (quantitative–qualitative) and severity is estimated based on measured parameters, we present the parameter value alongside the "Classification No." (col. FS). Finally, this Table includes descriptions of the final effects (col. FE) for informational purposes; these do not impact the overall results of the method.

**Table 10.** Results of failure analysis of a two-stage hydrogen-fueled combustion system by FMESA for the selected engine operating point.

| Element | Function | Guide Word | Deviation/Local Effect | Failure Mode | Values of Measurable Parameters | Final Effect | Failure Severity |
|---|---|---|---|---|---|---|---|
| **E** | **F** | **GW** | **D** | **FM** | **MP** | **FE** | **FS** |
| Fuel pressure regulator | Pressure reduction after the main valve | more | Reduction in fuel dosage | Reducer pin jam | $\lambda > 3 \rightarrow$ VL<br>15 < Pcyl = <20 bar $\rightarrow$ VL<br>Tex < 220 deg C $\rightarrow$ VH | Decrease in engine power | VH<br>(Tex < 220) |
| Supply line (fuel lines) | Fuel transport | less | Reduction in fuel dosage | Tube rupture | $\lambda > 3.;\rightarrow$ VL<br>15 < Pcyl = <30 bar $\rightarrow$ VL<br>200 < Tex < 320 deg C $\rightarrow$ H | Decrease in engine power | H<br>(Tex 200–320) |
| Prechamber valve | Fuel delivery | no | No fuel dosage for prechamber | Pc valve jamming | CoC > 10 deg $\rightarrow$ H<br>15 < Pcyl = Pcyl—5 bar $\rightarrow$ VL<br>$\lambda$ unchanged | Decrease in engine power | H<br>(CoC > 10) |
| MC injector | Fuel delivery | no | Drastic reduction in the dose to the main chamber | Injector mechanical damage | $\lambda > 3 \rightarrow$ VL<br>15 < Pcyl = < 20 bar $\rightarrow$ VL<br>Tex < 220 deg C $\rightarrow$ VH | No engine operation | VH<br>(Tex < 220) |
| Ignition coil | Energy generation to discharge on the spark plug | no | Misfire | Contact corrosion | 15 < Pcyl = < 30 bar $\rightarrow$ VL<br>Tex < 220 deg C $\rightarrow$ VH | Uneven engine operation | VH<br>(Tex < 220) |
| Air intake in front of the cylinder | Limited airflow into the cylinder | | Power limitation | Leakage, for example, at the throttle | CoC = 3–5 deg $\rightarrow$ VH<br>$\lambda$ = 1–2 or less $\rightarrow$ VH | Decrease in engine power; knocking combustion | VH<br>(CoC = 5) |
| Exhaust system | Flue gas discharge in the installation | other than | Escape of exhaust fumes into the environment | Leakage/breakage | Tex < 220 deg C $\rightarrow$ VH<br>0 < NOx < 50 ppm $\rightarrow$ VL(>2 0) | A significant amount of fumes in the environment | VH<br>(Tex < 220) |
| Exhaust system | Exhaust to the installation | no | No exhaust | Locked slider in closed position | Tex = 300 deg C $\rightarrow$ L | Limited engine power | L<br>(Tex = 300) |
| Hydrogen inlet line | Fuel transportation | less | Limited hydrogen flow | Cable bend | $\lambda > 2 \rightarrow$ L<br>15 < Pcyl = < 30 bar $\rightarrow$ VL | Decrease in engine power | L<br>($\lambda > 2$) |
| Ignition coil | Energy generation to discharge on the spark plug | less | Too large an ignition advance | Intake pressure too low, map sensor | CoC < 2–3 deg $\rightarrow$ VH<br>Tex > 300 $\rightarrow$ VH<br>NOx > 300 ppm $\rightarrow$ VH | Knock combustion | VH<br>(CoC = 2–3) |
| Ignition coil | Energy generation to discharge on the spark plug | more | Too large an ignition delay | Intake pressure too high, map sensor | CoC > 15 deg $\rightarrow$ H<br>Tex average $\rightarrow$H level $\rightarrow$ H<br>50 < NOx < 150 ppm $\rightarrow$ M | Chronic combustion | H<br>(CoC = 15) |

## 4. Discussion

The core of FMESA consists of FMEA/FMECA components, while their identification is realized with the support of HAZOP and experimental studies. The rationale for such an approach was primarily the desire to provide a methodical and comprehensive analysis of any system. This is because both FMEA and HAZOP assume a structured and systematic form of analysis (of both systems and designed or existing products, processes, procedures, organizational changes, or draft legal agreements [29]). However, the FMEA process begins with obtaining the results of identifying the types of defects (according to [30]), which, in our opinion, does not properly stimulate the analyst/expert's reasoning in this identification (and which is often the assumption). Much better in this aspect is HAZOP, in which the identification of damage types is an intermediate step but obtained according to deductive reasoning (or so-called backward reasoning [26]). The team conducting the HAZOP analysis takes undesired results and deviations from the intended results and process conditions and looks backward for their possible causes and damage types [28]. This gives less uncertainty to the conclusion of the inference, the reason for which has been described in works [31,32], among others. On the other hand, it can be generally said that backward reasoning refers to the situation in which, for a known conclusion, one tries to find justifying premises. Understood in this way, reasoning involves finding (unknown) causes (expansions) for (known) effects (expansions). With regard to technical systems, it can be said to be applied in determining how a certain system condition occurred, usually involving the failure of system components. The specifics of the implementation of HAZOP presented here give the rationale for its use within FMESA.

Obtaining FMESA results using experimental studies is intended to reduce epistemic uncertainty, a crucial component of hazard and risk models (e.g., [33–35]). An essential part of the method is the implementation of experimental studies, which provide information on key FMESA elements and are used to develop damage severity scales. We propose presenting these scales in tabular form, a common and user-friendly approach. However, graphical representations (Figure 3) are also possible, showing severity changes as a function of measurable parameters. Since we propose developing such scales in a unified form for combustion systems (independent of individual parameter ranges), we call them "severity matrices of failure to hydrogen-fueled combustion systems". However, it is important to remember that these scales and matrices were developed for a two-stage combustion system (for the selected engine operating point) and require calibration for different systems or installations (accurate classification metrics ranges can be obtained after analyzing varying engine speeds and loads).

The results for the final effect (FE) component and the final failure severity level are identified through controlled experimental studies, which avoid operating the plant to destruction. In this case, identification relies on expert prediction based on trends in system operation parameters. This aligns with the modern safety management approach, especially in high-risk domains like those addressed by resilience engineering [36], which emphasizes proactive decision making based on system behavior rather than waiting for a failure to analyze it retrospectively.

The method offers several promising extensions. Firstly, it can incorporate nonlinear function mappings or fuzzy sets within the classification metric, potentially expanding its applicability. These functions could belong to known classes, further enhancing universality. Ideally, identifying dedicated functions for specific plant types would eliminate the need for extensive research by the method's user. Additionally, developing databases containing failure severity functions as characteristics of hydrogen plant components presents an interesting and valuable long-term goal.

When presenting the method's results, we propose a slight modification to the typical notation for degrees of failure severity. Traditionally, verbal expressions or their acronyms are used. However, because our method is hybrid (severity is estimated based on measured parameter values), we believe it is more informative to present the parameter value alongside the assigned "Classification No." (Table 5, column FS).

The method's results directly contribute to risk assessment. One of the key components of risk models is the level of impact associated with events. Our method's severity levels can be easily integrated into these models. Furthermore, by incorporating a probabilistic measure, such as failure mode frequency, the severity measure (FS) can be transformed into a criticality measure. The model's future expansion could also involve estimating confidence intervals for measurable parameter values.

This method is particularly valuable during plant design or modification stages. Identifying critical installation elements and the most impactful failure modes is crucial for analyzing systems in general. Such analysis can significantly influence decisions about allowing a system to operate. Additionally, FMESA's severity assessments can be used to rank plant variants, informing decisions like selection.

Our failure severity tables offer a unique two-dimensional approach. Firstly, severity levels are assigned quantitatively based on measured values of critical system parameters. Secondly, these tables codify the relationship between parameter values and specific, concrete failure effects within the installation.

## 5. Final Remarks

In application areas, evaluating systems from the perspective of their failure severity should become the cornerstone of systems management. This analysis can determine whether a system is authorized for operation.

Hydrogen's unique properties necessitate a rigorous analysis of failure consequences (often addressed in FMEA-based risk assessments). While these assessments typically focus on worst-case scenarios like fire or explosion (which have a low probability), less severe, functionality-related effects are more likely to occur. Ignoring the criticality of these average-severity events is unreasonable, as they can significantly contribute to an unacceptable overall risk level. Existing studies on hydrogen plant severity assessment lack a focus on such moderate consequences, particularly for two-stage combustion systems.

This article aligns with current trends in hybrid risk analysis, which combine various analytical methods. It moves a step further by incorporating experimental studies to validate the results. The core objective is to present the methodology and implementation of a novel FMESA approach specifically tailored for two-stage hydrogen-fueled combustion systems. This includes the development of dedicated tabular scales to assess failure severity within this system type.

This method complements, rather than replaces, existing system management tools. The mathematical model is designed for flexibility, allowing for easy adaptation through different function forms or integration of fuzzy sets for linguistic variable mapping.

This work demonstrates the effectiveness of the FMESA method in quantitatively and qualitatively assessing failure severity for components in hydrogen-fueled combustion systems for an example engine operating point. FMESA evaluates the impact of failure on both system operation and overall combustion quality. We showcase this capability by presenting failure severity estimations for a two-stage hydrogen-fueled combustion system. The core contribution of this article lies in the introduction of the novel FMESA method for failure severity analysis. Furthermore, by applying FMESA to a two-stage combustion system, we address a critical gap in existing research.

**Author Contributions:** Conceptualization, A.G., I.P. and F.S.; methodology, A.G., I.P. and F.S.; software, A.G. and F.S.; validation, A.G. and I.P.; formal analysis, A.G., I.P. and F.S.; investigation, A.G., I.P. and F.S.; resources, I.P. and F.S.; data curation, A.G. and F.S.; writing—original draft preparation, A.G. and I.P.; writing—review and editing, A.G., I.P. and F.S.; visualization, I.P. and A.G.; supervision, A.G., I.P. and F.S.; project administration, A.G., I.P. and F.S.; funding acquisition, I.P. and F.S. All authors have read and agreed to the published version of the manuscript.

**Funding:** This work was supported by the Poznan University of Technology [interdisciplinary grant 0415/SIGR/7286].

**Data Availability Statement:** The original contributions presented in the study are included in the article, further inquiries can be directed to the corresponding author.

**Conflicts of Interest:** The authors declare no conflicts of interest.

## Abbreviations

| | |
|---|---|
| BDC | bottom dead center |
| CoC | center of combustion |
| E-FMECA | empirical FMECA |
| ETA | event tree analysis |
| FE | final effects |
| FM | failure mode |
| FMEA/FMECA | failure mode and effect analysis/failure mode, effects and criticality analysis |
| FMESA | failure mode and effects severity analysis |
| FS | failure severity |
| FTA | fault tree analysis |
| HAZOP | hazard and operability study |
| HPL | high parameter limit |
| LE | local effect |
| LHV | lower heat value |
| LPL | low parameter limit |
| MC | main chamber |
| MPs | measurable parameters |
| NHTSA | national highway traffic safety administration |
| P | pressure |
| PC | pre-chamber |
| PFI | port fuel injection |
| PF-FMEA | product function failure modes and effects analysis |
| PI-FMEA | product inter-face failure modes and effects analysis |
| PRD | pressure relief device |
| qo | fuel dose |
| SOI | start of combustion |
| TDC | top dead center |
| TJI | turbulent jet ignition |
| $\lambda$ | excess air coefficient |

## References


1. Safe Use of Hydrogen as a Fuel in Commercial Industrial Applications. Polish Register of Shipping S.A., Gdansk 2021. Available online: https://prs.pl/en/prs-en/ (accessed on 20 May 2024). (In Polish)
2. Corrigan, D.J.; Di Blasio, G.; Ianniello, R.; Silvestri, N.; Breda, S.; Fontanesi, S.; Beatrice, C. Engine knock detection methods for spark ignition and prechamber combustion systems in a high-performance gasoline direct injection engine. *SAE Int. J. Engines* **2022**, *15*, 883–897. [CrossRef]
3. Pielecha, I.; Szwajca, F. Combustion characteristics of a hydrogen-fueled TJI engine under knocking conditions. *Energies* **2024**, *17*, 1324. [CrossRef]
4. Kalghatgi, G.; Algunaibet, I.; Morganti, K. On knock intensity and superknock in SI engines. *SAE Int. J. Engines* **2017**, *10*, 1051–1063. [CrossRef]
5. Pielecha, I.; Szwajca, F.; Skobiej, K. Experimental investigation on knock characteristics from pre-chamber gas engine fueled by hydrogen. *Energies* **2024**, *17*, 937. [CrossRef]
6. Aramburu, A.; Guido, C.; Bares, P.; Pla, B.; Napolitano, P.; Beatrice, C. Knock detection in spark ignited heavy duty engines: An application of machine learning techniques with various knock sensor locations. *Measurement* **2024**, *224*, 113860. [CrossRef]
7. Pan, M.; Wei, H.; Feng, D. Effects of exhaust gas recirculation on knock intensity of a downsized gasoline spark ignition engine. *J. Energy Res. Technol.* **2019**, *141*, 011101. [CrossRef]
8. Wang, K.; Zhang, Z.; Sun, B.; Zhang, S.; Lai, F.; Ma, N.; Ju, X.; Luo, Q.; Bao, L.-Z. Experimental investigation of the working boundary limited by abnormal combustion and the combustion characteristics of a turbocharged direct injection hydrogen engine. *Energy Convers. Manag.* **2024**, *299*, 117861. [CrossRef]
9. Noor, W.B.; Amin, T. Towards sustainable energy: A comprehensive review on hydrogen integration in renewable energy systems. *Future Energy* **2024**, *3*, 1–17. [CrossRef]
10. Locke, K. The urgency of hydrogen: Environmental issues and the need for change. *Future Sustain.* **2024**, *2*, 46–58. [CrossRef]

11. Yip, H.L.; Srna, A.; Yuen, A.C.Y.; Kook, S.; Taylor, R.A.; Yeoh, G.H.; Medwell, P.R.; Chan, Q.N. A review of hydrogen direct injection for internal combustion engines: Towards carbon-free combustion. *Appl. Sci.* **2019**, *9*, 4842. [CrossRef]
12. Liu, F.; Zhou, L.; Hua, J.; Liu, C.; Wei, H. Effects of pre-chamber jet ignition on knock and combustion characteristics in a spark ignition engine fueled with kerosene. *Fuel* **2021**, *293*, 120278. [CrossRef]
13. Cui, S.; Zhu, G.; He, L.; Wang, X.; Zhang, X. Analysis of the fire hazard and leakage explosion simulation of hydrogen fuel cell vehicles. *Therm. Sci. Eng. Prog.* **2023**, *4*, 101754. [CrossRef]
14. Correa-Jullian, C.; Groth, K.M. Liquid hydrogen storage system FMEA and data requirements for risk analysis. In Proceedings of the 30th European Safety and Reliability Conference and 15th Probabilistic Safety Assessment and Management Conference, Venice, Italy, 1–5 November 2020; pp. 973–980. [CrossRef]
15. Stephens, D.R.; Rose, S.E.; Flamberg, S.A.; Ricci, S.M.; George, P.E. *Failure Modes and Effects Analysis for Hydrogen Fuel Cell Vehicles—Subtask 1*; National Highway Traffic Safety Administration: Washington, DC, USA, 2009; p. 170. Available online: https://www.nhtsa.gov/sites/nhtsa.dot.gov/files/811075_0.pdf (accessed on 20 June 2024).
16. Rodionov, A.; Wilkening, H.; Moretto, P. Risk assessment of hydrogen explosion for private car with hydrogen-driven engine. *Int. J. Hydrogen Energy* **2011**, *36*, 2398–2406. [CrossRef]
17. Ehrhart, B.D.; Harris, S.R.; Blaylock, M.L.; Muna, A.B.; Quong, S. Risk assessment and ventilation modeling for hydrogen releases in vehicle repair garages. *Int. J. Hydrogen Energy* **2021**, *46*, 12429–12438. [CrossRef]
18. Jung, W.-C.; Jung, P.-K.; Moon, H.-M.; Chang, M.-H.; Yun, S.-H.; Lee, H.-G.; Hwang, M.-W.; Woo, I.S. Hydrogen breakthrough behaviors for cryogenic adsorption and HAZOP study. *Fusion Eng. Des.* **2018**, *134*, 123–127. [CrossRef]
19. Suzuki, T.; Izato, Y.; Miyake, A. Identification of accident scenarios caused by internal factors using HAZOP to assess an organic hydride hydrogen refueling station involving methylcyclohexane. *J. Loss Prevent. Proc.* **2021**, *71*, 104479. [CrossRef]
20. Fu, Y.; Qin, Y.; Wang, W.; Liu, X.; Jia, L. An extended FMEA model based on cumulative prospect theory and type-2 intuitionistic fuzzy VIKOR for the railway train risk prioritization. *Entropy* **2020**, *22*, 1418. [CrossRef]
21. Macura, D.; Laketić, M.; Pamučar, D.; Marinković, D. Risk analysis model with interval type-2 fuzzy FMEA—Case study of railway infrastructure projects in the Republic of Serbia. *Acta Polytech. Hung.* **2022**, *19*, 103–118. [CrossRef]
22. Appoh, F.; Yunusa-Kaltungo, A.; Sinha, J.K.; Kidd, M. Practical demonstration of a hybrid model for optimising the reliability, risk, and maintenance of rolling stock subsystem. *Urban Rail Transit* **2021**, *7*, 139–157. [CrossRef]
23. Casamirra, M.; Castiglia, F.; Giardina, M.; Lombardo, C. Safety studies of a hydrogen refuelling station: Determination of the occurrence frequency of the accidental scenarios. *Int. J Hydrogen Energy* **2009**, *34*, 5846–5854. [CrossRef]
24. Dhillon, B.S. *Design Reliability: Fundamentals and Applications*; CRC Press: Boca Raton, FL, USA; London, UK; New York, NY, USA; Washington, DC, USA, 1999.
25. Jin, Y.; Luo, H.; Zhang, G.; Zhai, C.; Ogata, Y.; Matsumara, Y.; Ichikawa, T.; Nakashimada, Y.; Kim, W.; Nishida, K. Ignition timing effect on the combustion performance of hydrogen addition in methane fermentation gas in a local energy system. *Fuel* **2022**, *324*, 124714. [CrossRef]
26. Gill, A. *Layered Models of Safety Systems for Rail Transport Applications*; Poznan University of Technology Publishing House: Poznan, Poland, 2018; ISBN 978-83-7775-517-4.
27. Crowe, D.; Feinberg, A. *Design for Reliability*, 1st ed.; CRC Press LLC: Boca Raton, FL, USA, 2001. [CrossRef]
28. International Electrotechnical Commission (IEC). *Hazard and Operability Studies (HAZOP Studies)—Application Guide*, 2nd ed.; IEC: Geneva, Switzerland, 2016.
29. *ISO 31010*; Risk Management—Risk Assessment Techniques. ISO: Geneve, Switzerland, 2009.
30. *IEC 60812*; Analysis Techniques for System Reliability—Procedure for Failure Mode and Effects Analysis (FMEA). International Electrotechnical Commission: Geneva, Switzerland, 2006.
31. Gill, A.; Kadziński, A. Hazard identification model. In Proceedings of the 20th International Scientific Conference, Transport Means, Kaunas University of Technology, Juodkrante, Lithuania, 5–7 October 2016; pp. 885–890.
32. Urbański, M. *Abductive Reasoning. Models and Procedures*; Adam Mickiewicz University Scientific Publishing House: Poznan, Poland, 2009.
33. Aven, T. *Risk Analysis*, 2nd ed.; John Wiley & Sons: Hoboken, NJ, USA, 2015. [CrossRef]
34. SRA. Society for Risk Analysis Glossary, Society for Risk Analysis (SRA). 2018. Available online: https://www.sra.org/wp-content/uploads/2020/04/SRA-Glossary-FINAL.pdf (accessed on 21 June 2024).
35. Thekdi, S.; Aven, T. Characterization of biases and their impact on the integrity of a risk study. *Saf. Sci.* **2024**, *170*, 106376. [CrossRef]
36. Azadeh, A.; Salehi, V.; Arvan, M.; Dolatkhah, M. Assessment of resilience engineering factors in high-risk environments by fuzzy cognitive maps: A petrochemical plant. *Saf. Sci.* **2014**, *68*, 99–107. [CrossRef]


**Disclaimer/Publisher's Note:** The statements, opinions and data contained in all publications are solely those of the individual author(s) and contributor(s) and not of MDPI and/or the editor(s). MDPI and/or the editor(s) disclaim responsibility for any injury to people or property resulting from any ideas, methods, instructions or products referred to in the content.

energies

MDPI

*Article*

# Turpentine as an Additive for Diesel Engines: Experimental Study on Pollutant Emissions and Engine Performance

**Robert Mădălin Chivu [1,2,*], Jorge Martins [2], Florin Popescu [1], Krisztina Uzuneanu [1], Ion V. Ion [1], Margarida Goncalves [3], Teodor-Cezar Codău [4], Elena Onofrei [5] and Francisco P. Brito [2,*]**

1 Department of Thermal Systems and Automotive Engineering, "Dunarea de Jos" University, 800001 Galati, Romania; krisztina.uzuneanu@ugal.ro (K.U.); ion.ion@ugal.ro (I.V.I.)
2 Mechanical Engineering and Resource Sustainability Center (MEtRICs), Department of Mechanical Engineering, School of Engineering, Campus Azurém, University of Minho, 4800-058 Guimarães, Portugal; jmartins@dem.uminho.pt
3 Mechanical Engineering and Resource Sustainability Center (MEtRICs), Department of Science and Technology of Biomass, Faculty of Science and Technology, NOVA University of Lisbon, 2829-516 Caparica, Portugal; mmpg@fct.unl.pt
4 Fibrenamics–Institute for Innovation in Fiber-Based Materials and Composites, Center for Textile Science and Technology, University of Minho, Campus de Azurém, 4800-058 Guimarães, Portugal
5 Faculty of Industrial Design and Bbusiness Management, Technical University "Gheorghe Asachi" of Iasi, 700050 Iași, Romania; elena.codau@academic.tuiasi.ro
* Correspondence: robert.chivu@ugal.ro (R.M.C.); francisco@dem.uminho.pt (F.P.B.)

**Abstract:** The need for reducing fossil fuel consumption and greenhouse gas (GHG) emissions in internal combustion engines has raised the opportunity for the use of renewable energy sources. For the progressive replacement of fossil fuels like diesel, those derived from the sustainable management of forest resources may be a good option. In Portugal, pine trees (pinus pinaster) are among the most widely cultivated tree species. Turpentine can be extracted from their sap without harming the tree. Turpentine is known to be a good fuel with a lower viscosity than regular diesel but with a comparable caloric value, boiling point and ignition characteristics, although it is not widely used as a compression ignition fuel. Moreover, recent research has highlighted the possibility of substantially increasing the turpentine yield through biotechnology, bringing it closer to economic viability. The present study investigates the performance, pollutant emissions and fuel consumption of a 1.6 L four-cylinder direct-injection diesel engine operating with several blends of commercial diesel fuel and turpentine obtained from pine trees. The aim of this study was to assess whether it would be possible to maintain or even improve the performance, fuel consumption and GHG and pollutant emissions (HC, $NO_x$, CO and PM) of the engine with the partial incorporation of this biofuel. Turpentine blends of up to 30% in substitution of regular diesel fuel were tested. The main novelties of the present work are related to (i) the careful testing of a still-insufficiently studied fuel that could gain economical attractiveness with the recent developments in yield improvement through biotechnology and (ii) the tests conducted under fixed engine load positions typical of road and highway conditions. The addition of this biofuel only slightly impacted the engine performance parameters. However, a slightly positive effect was observed in terms of torque, with an increase of up to 7.9% at low load for the 15T85D mixture and 6.8% at high load being observed. Power registered an increase of 9% for the 15T85D mixture at low speed and an increase of 5% for the 30T70D mixture at high speed when compared to the reference fuel (commercial diesel fuel). While the efficiency and fossil GHG emissions were improved with the incorporation of turpentine, it had a mixed effect on polluting emissions such as unburned hydrocarbons (HC) and smoke (PM) and a negative effect on nitrogen oxides ($NO_x$). $NO_x$ emissions increased by 30% for high loads and 20% for low loads, mainly as an indirect effect of the improvement in the engine performance and not so much as a consequence of the marginally higher oxygen content of turpentine relative to commercial diesel fuel.

**Keywords:** turpentine fuel; diesel substitution; engine testing; terpene biofuels

**Citation:** Chivu, R.M.; Martins, J.; Popescu, F.; Uzuneanu, K.; Ion, I.V.; Goncalves, M.; Codău, T.-C.; Onofrei, E.; Brito, F.P. Turpentine as an Additive for Diesel Engines: Experimental Study on Pollutant Emissions and Engine Performance. *Energies* **2023**, *16*, 5150. https://doi.org/10.3390/en16135150

Academic Editors: Davide Lanni and Enzo Galloni

Received: 31 May 2023
Revised: 22 June 2023
Accepted: 28 June 2023
Published: 4 July 2023

**Copyright:** © 2023 by the authors. Licensee MDPI, Basel, Switzerland. This article is an open access article distributed under the terms and conditions of the Creative Commons Attribution (CC BY) license (https://creativecommons.org/licenses/by/4.0/).

## 1. Introduction

In the last 30 years, the main energy source for the field of transport has been represented by fossil fuels. They are still needed today for internal combustion engines (ICEs). Fossil fuels are not a renewable source of energy, and that is why alternative energy sources such as biofuels have been produced from various sources [1] and blended with fossil fuels. Biofuels are an important alternative to fossil fuels as they produce fewer greenhouse gases (GHG) on a life-cycle basis and can be produced from renewable sources such as agricultural crops, forest products and food waste. They also aim to improve the quality of life by significantly reducing harmful gases. Biofuels can be used in a variety of ways, from powering vehicles to producing electricity. Because the standards of the European Union are becoming increasingly challenging in terms of the pollutant emissions of internal combustion engines [2], emissions harmful to humans and the environment must be drastically reduced. The transesterification of plant/animal fats with methanol is the basis for producing biodiesel, an environmentally beneficial and renewable fuel that is primarily made up of high-fatty-acid methyl esters. In addition to being non-toxic, flammable, non-explosive and biodegradable, biodiesel also boasts a large supply of raw materials and supply security [3–5].

Biodiesel production in Europe is a rapidly emerging industry, as the European Union has set ambitious targets for biofuel production in the continent. To meet these targets, Europe has invested heavily in research and development (R&D) for biodiesel production, which has resulted in the development of numerous biodiesel plants across the continent. Biodiesel is produced by combining vegetable oils and animal fats with methanol and a catalyst to produce a fuel that is substantially similar to diesel fuel. It is then blended with diesel fuel to create biodiesel blends, which can be used in any diesel engine. In addition to being a renewable source of energy, biodiesel also has several environmental benefits. It emits fewer pollutants than traditional diesel, except for nitrogen oxides. The European biodiesel industry is highly competitive, with many major players competing for market share. The largest biodiesel producers in the region are Germany, France and the Netherlands. These countries have invested heavily in R&D for biodiesel production, leading to the development of several plants across the continent. In addition, several smaller countries are beginning to enter the market, such as Austria, the Czech Republic and Denmark [6,7].

Despite its popularity as an alternative biofuel to diesel fuel, new alternatives to biodiesel are being developed. These alternatives do not rely on the intensive cultuvation of energy crops but rather on sustainable forest management, in which the extraction of energy does not require cutting down trees. Turpentine is such an example. It is a volatile organic compound derived from the resin of certain coniferous trees such as Pinus Pinaster, the most abundant conifer species in Portugal, and it is obtained without the need to cut down the tree. It is a powerful solvent and cleaning agent that is derived from pine trees. It has been used for centuries as a medicine, disinfectant and paint thinner. Turpentine is a clear, colourless liquid with a slight pungent odour and flavour. It is composed mainly of a mixture of terpenes and terpenoids, and its main component is alpha-pinene. It is also used in making varnishes, adhesives and inks. It can also be used to make solvents, which can be used to dissolve oils and waxes. Turpentine has several medicinal uses. Turpentine can be irritating, and it is toxic if ingested or inhaled in large quantities [8].

In recent years, turpentine has been gaining attention as a potential biofuel. It is a renewable, sustainable, carbon-neutral fuel source. Turpentine can be produced from sustainably harvested pine trees or from the waste products of sawmills and paper mills. Turpentine can also be made from vegetable oils and other plant-based materials. Turpentine has an energy density that can surpass that of diesel fuel and can be used, to a certain degree, in diesel engines. It has a low sulphur content and generally does not produce as many pollutants as traditional fuels, except for $NO_x$. Turpentine can also be blended with other biofuels, such as ethanol and biodiesel, to increase the global incorporation of renewable fuels into engine fuel blends and thus reduce fossil fuel consumption and GHG emissions. This is valid for both diesel and gasoline engine fuel blends [9–12].

Mixing turpentine with diesel provides many beneficial properties that can improve the performance of diesel engines. Turpentine also has a lower viscosity than diesel, which can improve the fuel's ability to flow through an engine's fuel system. In addition, it can also reduce the amount of sulphur dioxide produced by diesel engines, helping to reduce air pollution and extending the life of aftertreatment systems. Being a biofuel that emits $CO_2$ with its combustion, it can be considered as carbon-neutral or even carbon-negative because of the $CO_2$ absorption by the plant during its whole lifetime [13]. Finally, turpentine can be used to help improve the fuel economy of a diesel engine. This is related to factors such as its higher heating value, its improved combustion due to oxygen in the molecule and its lower viscosity, which facilitates injection and fuel atomisation [14].

Although turpentine has been used for a long time in a lot of applications, its research and use as a biofuel is still not widespread. There are some notable studies on this fuel being incorporated into diesel engines [9,12,15–17].

Some of the studies involving turpentine or pine oil reported a lower efficiency and higher fuel consumption. This was the case with [12], which assessed blends of turpentine and diesel, and [18], which assessed blends of jatropha oil and turpentine. However, there were some cases where a lower consumption was recorded during the driving cycle [12], reporting a lower fuel consumption for pine oil blends than diesel or jatropha blends [19]. Thus, it seems that further exploration of the impact of turpentine blends on consumption is still needed.

The authors of [20] reported that pollutant emissions such as HC, CO and smoke decreased by 65%, 30% and 70%, respectively, compared to diesel at a high engine load. However, at maximum load, pine oil had up to 25% higher $NO_x$ production than the reference fuel.

Torque and power are parameters that are important in assessing vehicle performance. Often, authors either report results on turpentine blends for a full engine load [21] or as a function of several engine loads. However, not only is a full engine load an engine setting that is rarely used in regular driving but also many studies do not even highlight the torque/power corresponding to a given engine load. This makes it difficult to assess the usefulness of the results for real driving scenarios.

From the abovementioned literature review, it is apparent that this fuel is still insufficiently explored. The recent advances in biotechnologies for increasing turpentine yields have raise the potential economic interest of this fuel, especially in a country such as Portugal, in which the pine industry has such a strong economical role. Additionally, the existing studies on turpentine fuel do not present the results in a way that is practical for assessing vehicle performance under engine loads that are typical of road driving and highway driving. Under these circumstances, it seems important to carefully test this fuel that is still insufficiently studied now that its economical relevance might increase.

The present study performs a series of tests using mixtures of turpentine with diesel in different proportions in a light-duty four-cylinder 1.6 L direct-injection engine.

The mixtures tested were 5T95D (turpentine 5% and diesel 95%), 10T90D, 15T85D, 20T80D and 30T70D. The performance of these mixtures was compared under four combinations of speed and load that are typical of different driving scenarios: 1700 RPM and 2250 RPM, roughly corresponding to a speed of around 90 km/h and 120 km/h for a light duty car, respectively, and two different loads. The variation in the degree of turbocharging was avoided by fixing it at 0.5 bar. The performance characteristics (torque power, consumption and efficiency) and pollutant emissions were analysed and compared.

## 2. Data and Methodology

In this study, commercially available turpentine was combined with diesel in order to evaluate the influence of the additive on the combustion behaviour, engine performance and exhaust emissions. Details about the fuel preparation process that is typically performed to obtain this fuel, as well its main characteristics, are presented in the following section.

## 2.1. Fuel Preparation

Biofuel extraction involves the steam distillation of pine resin, as shown in Figure 1.

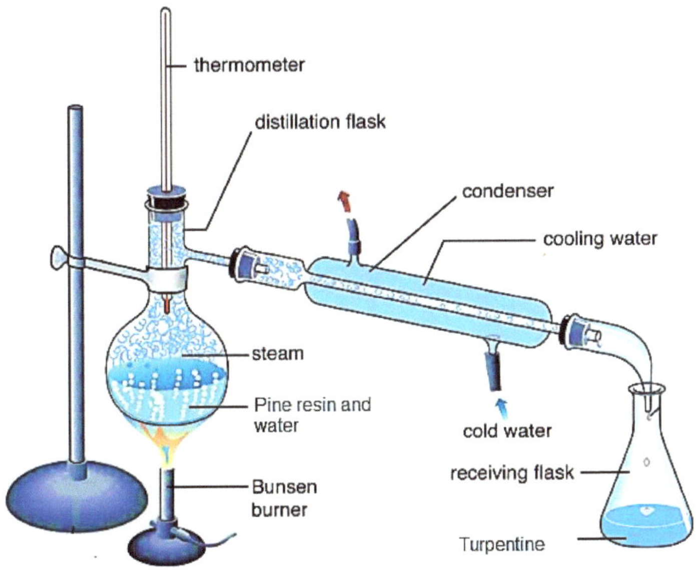


**Figure 1.** Schematic representation of the distillation process for the extraction of turpentine, adapted from [22].

The typical methodology of pine sap extraction can be found in [23,24]. Pine sap is usually added to a distillation flask together with a quantity of distilled water; considering that the high viscosity of this sap depends on its temperature, it is necessary to add a part of water. After the mixture is exposed to a heat source, the mixture homogenises, and the sap from the trees changes its viscosity and colour, becoming a liquid. Depending on the intensity of the heat source and the amount added to the flask, the duration until the mixture reaches its boiling point is approximately 15 min. As the mixture boils, the content in the flask decreases. This is due to the decrease in the level of water content in the mixture; therefore, a quantity of water is gradually added to extract as much of the substance as possible. Water vapour and turpentine vapour are condensed to reach a liquid state, and they drain into a container. The newly formed mixture is left to settle for 24 h; due to the different density of turpentine compared to water, it separates from the water and can be collected. In general, it is commonly observed that the majority of fuels that are derived through the process of steam distillation and have not undergone transesterification are regarded as fuels characterised by their low viscosity and a lower cetane number [25,26]. In the present study, the authors only performed the blending and some tests to check the solubility of turpentine in diesel and in water. Figure 2 illustrates the decantation process of water that was purposefully added by the authors to turpentine in a test tube to confirm the full phase separation of the fuel and water.

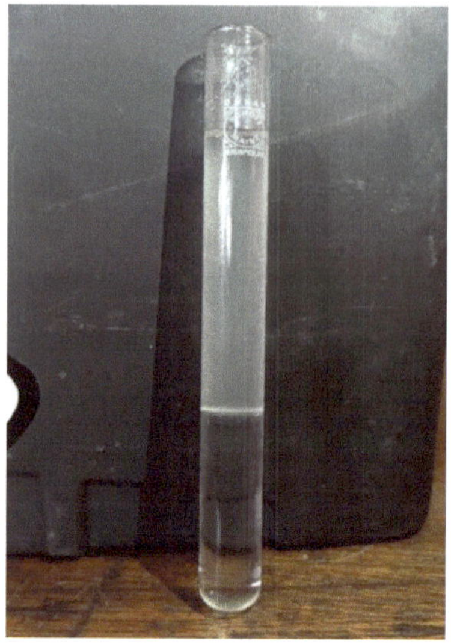

**Figure 2.** Separation of turpentine from water by decantation.

### 2.2. Fuel Properties

In the paper [24], Merghini et al. conducted a study regarding the main chemical compounds of turpentine. Twenty-four chemical compounds were identified in this substance, and the dominant compounds were monoterpene hydrocarbons with 81.4% α-pinene and 6.47% β-pinene; the amounts of the chemical compounds in turpentine differ depending on the area where the raw material comes from.

Table 1 shows the properties of turpentine and diesel. It is noticeable that the values for turpentine are somewhat different from those of diesel fuel. Namely, the cetane number is slightly lower. However, the calorific value is higher. Despite this, turpentine has the potential to be well mixed with diesel to create a new fuel blend. Table 2 shows the typical chemical composition of this substance.

**Table 1.** Typical properties of diesel and turpentine, adapted from [23,27–30].

| Properties | Turpentine Oil | Diesel |
|---|---|---|
| Formula | $C_{10}$-$H_{16}$ | $C_{12}$-$H_{23}$ |
| Molecular weight (g/mol) | 136 | 167 |
| Boiling point (°C) | 150–180 | 180–340 |
| Specific gravity | 0.86–0.9 | 0.83 |
| Kinematic viscosity @ 40 °C (cSt) | 1.3 | 2.5 |
| Latent heat of vaporisation (kJ/kg) | 305 | 230 |
| Flash Point (°C) | 38 | 74 |
| Auto-ignition temperature (°C) | 220–255 | 254–285 |
| Calorific value (kJ/kg) | 44,400 | 42,500 |
| Cetane index | 20–25 | 40–55 |
| Density (kg/m$^3$) | 860–900 | 830 |

**Table 2.** Chemical composition of turpentine [24].

| N° | Component | RI | % |
|---|---|---|---|
| 1 | Tricyclene | 930 | 0.13 |
| 2 | α-Thujene | 935 | 0.33 |
| 3 | α-Pinene | 940 | 81.41 |
| 4 | α-Fenchene | 949 | 0.01 |
| 5 | Camphene | 952 | 3.1 |
| 6 | β-Pinene | 979 | 6.47 |
| 7 | β-Myrcene | 991 | 0.04 |
| 8 | α-Terpinene | 1019 | 0.11 |
| 9 | Limonene | 1032 | 1.9 |
| 10 | ϒ-Terpinene | 1063 | 0.06 |
| 11 | α-Terpinolene | 1089 | 0.03 |
| 12 | *Cis*-β-Terpinene | 1144 | 0.09 |
| 13 | *Trans*-β-Terpinene | 1163 | 0.1 |
| 14 | α-Terpineol | 1190 | 0.06 |
| 15 | α-Terpinen-7-al | 1282 | 0.38 |
| 16 | α-Terpinyle acetate | 1350 | 0.13 |
| 17 | α-Longipinene | 1351 | 0.01 |
| 18 | Longicyclene | 1372 | 0.11 |
| 19 | β-longipinene | 1398 | 0.09 |
| 20 | Longifolene | 1403 | 2.4 |
| 21 | β-Caryophylene | 1427 | 2.37 |
| 22 | α-Humulene | 1461 | 0.03 |
| 23 | Germacrene-b | 1555 | 0.33 |
| 24 | β-Caryophyllene oxyde | 1588 | 0.28 |
| | **Retention Index** | | **99.99** |

### 2.3. Experimental Setup

The engine (Figure 3) in which the fuels were tested was a 1.6 L PSA HDI four-cylinder, four-stroke, water-cooled, turbocharged diesel engine with an intercooler in the intake manifold. The only changes made to the engine were related to the turbocharger pressure and the EGR, which was deactivated. These changes were carried out to avoid an excessive number of parameters being varied at the same time, which would have made it impossible to suitably analyse the effect of fuel blends on the behaviour of the engine. Namely, the turbocharger pressure was set at a fixed value throughout the speed range in order to have the same boost pressure (0.5 bar) for all tests independently of the speeds and loads tested. A similar statement may be made regarding the EGR levels. Although these alterations somewhat limited the scope of the results obtained, they allowed a more objective assessment of the impact of turpentine blending for the different test conditions. The original ECU was used, but it was reprogrammed to eliminate errors associated with using the engine outside of the vehicle and also to deal with the changes made to the engine, as mentioned above. A rotary encoder measuring both 1 and 2500 pulses per revolution was used on the electromechanical brake shaft to monitor its speed. The brake controller varied the braking torque so that it kept the engine speed within ±5 rpm of the setpoint. The engine was equipped with a high-pressure fuel pump and a common rail fuel system. The injection command was carried out electronically through the electronic control unit (ECU). A high-pressure sensor inserted into the common rail was responsible for monitoring the fuel pressure. To facilitate the analysis of the engine behaviour, the EGR was not active.

**Figure 3.** Experimental engine test stand.

There was no particle filter, which allowed a direct assessment of the particle emissions through opacity measurements. The technical information of the engine is highlighted in Table 3. The monitoring of the temperatures of the coolant, intake manifold, lubricant and exhaust flow was performed using type-K thermocouples connected to a National Instruments (NI 9214,) data acquisition system connected to a graphical interface. The engine load setting was monitored in the same interface with the help of the data acquisition board from National Instruments (NI 9201). The data collection had the same time reference because all the programs were in the same interface.

**Table 3.** Test engine specifications.

| | |
|---:|:---|
| Engine: | in-line 4 |
| Fuel: | diesel |
| Fuel injection: | common rail |
| Displacement: | 1560 cm$^3$ |
| Bore/Stroke: | 75/88.3 mm |
| Valves: | 16 valves |
| Turbo-charging: | turbo + intercooler |
| Compression ratio: | 18:1 |
| Power: | 75 HP/56 kW @ 4000 RPM |
| Torque: | 170 Nm @ 1700 RPM |

The engine was connected to a Telma AD6-55 electromagnetic brake with eddy currents to dissipate the power produced [31]. The torque was measured with an S-type 500 kgf Zemic B3G load cell with a combined full-scale error of ±0.02%. This translated into a maximum error of ±0.1 Nm in the engine torque for considering the load cell installation and the 3.86:1 transmission relation. Regarding power, it was obtained from multiplying the torque by the angular speed. The error in power calculation for the worst combination of speed and torque values was 0.93%. The calibration curve of the cell can be seen in Figure 4.

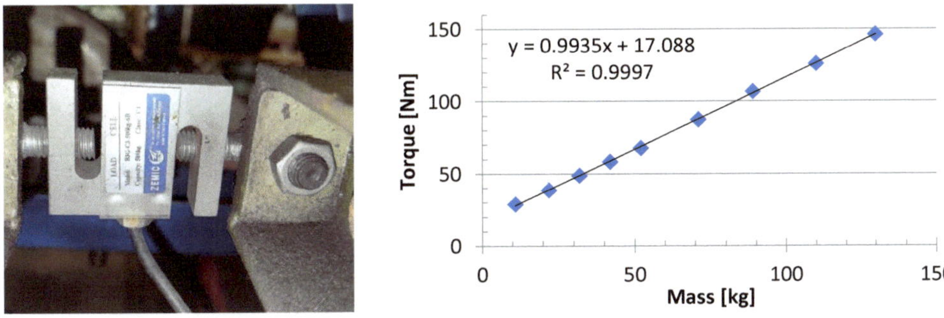


(a) (b)

**Figure 4.** (**a**) S-type torque cell and (**b**) calibration curve, with calibration points in blue.

A model AVL DIGAZ 4000 Light analyser from the company AVL for measuring HC, $CO_2$, CO and $NO_x$ emissions was used to determine the concentrations of pollutant emissions from the exhaust gas. The smoke was measured with an AVL DISMOKE 4000 device. The details of the measuring devices and the measurement precision are presented in Tables 4 and 5.

**Table 4.** Technical specifications of the AVL gas analyser.

| Measurement Data | Measurement Range | Resolution |
|---|---|---|
| CO: | 0–10% Vol. | 0.01% Vol. |
| $CO_2$: | 0–20% Vol | 0.1% Vol. |
| HC: | 0–20,000 ppm Vol | 1 ppm |
| $NO_x$: | 0–5000 ppm Vol. | 1 ppm |
| $O_2$: | 0–25% Vol. | 0.01% Vol. |
| Engine speed: | 250–9990 rpm | 10 rpm |
| Oil temperature: | 0–150 °C | 1 °C |

**Table 5.** Technical specifications of the AVL smoke analyser.

| AVL DiSmoke 4000 | Measurement Range | Resolution |
|---|---|---|
| Opacity | 0–100% | 0.10% |
| Absorption (K-value): | 0–99.99 $m^{-1}$ | 0.01 $m^{-1}$ |
| Acceleration time: | 0–5 s | 0.05 s |
| Engine speed: | 250–9990 rpm | 10 rpm |
| Oil temperature: | 0–150 °C | 1 °C |
| Ignition angle TDC sensor: | −60–100 °c.a. | 0.1 °c.a |
| Ignition angle stroboscope: | 0–60 °c.a. | 0.1 °c.a. |

Fuel consumption was measured using a Kern PCB3500-2 (Figure 5) precision weighing scale with a ±0.01 g precision [32]. The scale was connected to the data acquisition system for the continuous and real-time recording of the experimental data. The consumption was calculated by extracting the slope of the mass vs time curve through least-squares analysis fit to the stabilised portion of the curve. Average sample rates higher than 3 Hz were acquired. Measurements spanning several minutes were performed so that the number of samples used to extract the consumption value (slope of the curve) were close to or even surpassed a thousand samples. The oscillation of the consumption was exceptionally low, as seen in Figure 6 The $R^2$ fit was mostly higher than 0.999, as seen in the example displayed in Figure 6. Thus, the random error related to the fit was negligible. The external temperature, air humidity and atmospheric pressure at which the tests were conducted were also recorded and saved with a thermo-hygro-barometer [33].

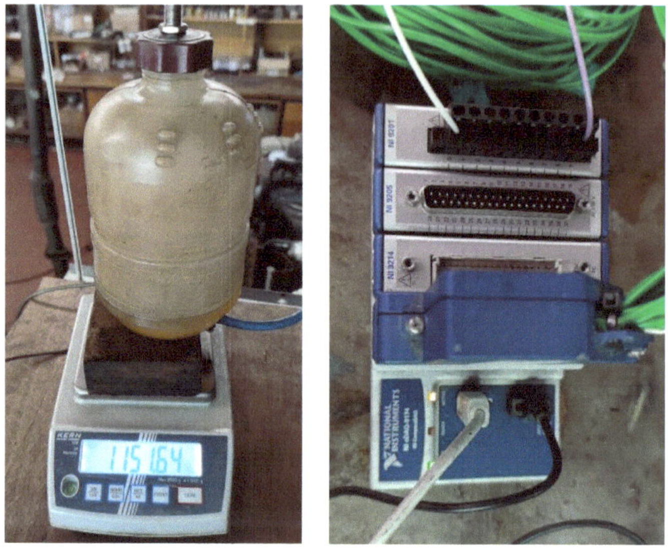

**Figure 5.** Fuel consumption measurement and data acquisition board.

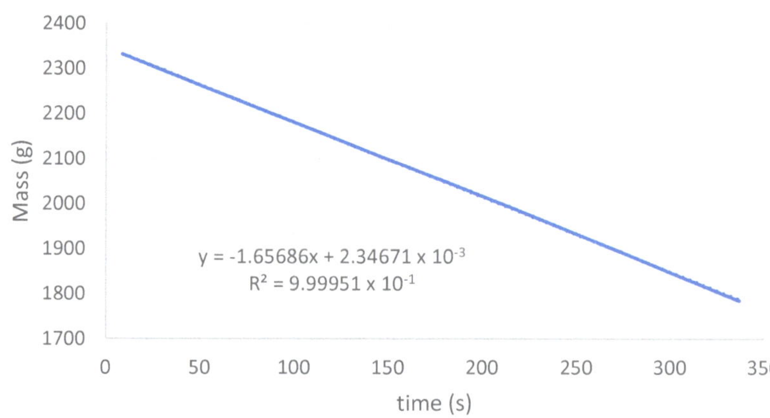


**Figure 6.** Example of the evolution of fuel mass in the tank over time to enable calculation of consumption including linear regression fit, in which the slope is the resulting mass flow rate of fuel.

Each test was performed at least three times. The results were recorded only once the engine operation was stabilised. This occurred once the coolant and lubricant temperatures, as well as the pollutant levels, were stabilised. Measurement of the average fuel consumption was conducted for at least 2 min, but it was often conducted a lot more. The results represent the average values of several repetitions of each test. Differences between repetitions were generally quite low. For instance, deviations in the fuel consumption errors from the average were in the range of 1%, while those of pollutants were generally below 4%, except for opacity, which sometimes reached slightly above 10% from the average.

### 2.4. Performance Parameter Calculation

A measure of the overall efficiency of an engine is given by the brake thermal efficiency. The brake thermal efficiency is the ratio of the energy produced by the engine to the fuel energy. The brake thermal efficiency was calculated using Equation (1). For a variation in the calorific value of 1% the maximum combined error in efficiency would be lower than 3%.

$$BTE = \frac{BrakePower\ (\mathrm{kW}) \cdot 3600}{FuelFlow\left(\frac{\mathrm{kg}}{\mathrm{h}}\right) \cdot CalorificValue\left(\frac{\mathrm{kJ}}{\mathrm{kg}}\right)} \cdot 100\ [\%] \quad (1)$$

### 2.5. Test Parameters

Table 6 displays the parameters tested. In particular, two different speeds and engine loads were tested for all the fuel blends. Two baseline torques of 58 Nm and 87 Nm were imposed for the baseline fuel. These torques corresponded to the engine loads represented in the table for each torque/speed combination. The engine loads were chosen to simulate a vehicle in motion in the highest gear, at 90 km/h and 120 km/h along flat and mildly sloped roads. These speeds are typical of the national roads and highway roads of the European Union.

**Table 6.** Test parameters.

| Engine Speed [RPM] | 1700 | | 2250 | |
|---|---|---|---|---|
| Equivalent driving conditions | National road 90 km/h | | Highway 120 km/h | |
| | horizontal | mild slope | horizontal | mild slope |
| Engine torque (baseline fuel—diesel) [N·m] | 58 | 87 | 58 | 87 |
| Power (baseline fuel—diesel) [kW] | 10.4 | 15.3 | 13.6 | 20.6 |
| Engine load (all fuels) [%] | 27.1 | 34.0 | 31.6 | 37.3 |
| Turpentine incorporations tested [%]<br>Example: 5T95D = 5% turpentine + 95% diesel | 5%, 10%, 15%, 20%, 30% | | | |

Then, these same engine loads were imposed for all the fuel blends, resulting in slightly different torques depending on the fuel performance.

## 3. Results and Discussions

The present section describes the results of the experimental tests, such as the performance characteristics and the pollutant emissions of the engine that used diesel as a reference fuel and mixtures of diesel with turpentine at the same engine load with the same operating parameters.

### 3.1. Performance Characteristics

The performance parameters described in this section are the brake thermal efficiency BTE (in %), torque (N·m) and power (kW).

#### 3.1.1. Brake Thermal Efficiency

The BTE (in %) of the engine for the different concentrations of mixtures of biofuel with diesel is presented in Figure 7. As can be seen, the turpentine addition always had a positive effect relative to diesel except for at the high-speed, low-load setting, at which the improvement was only visible for the highest level of turpentine incorporation.

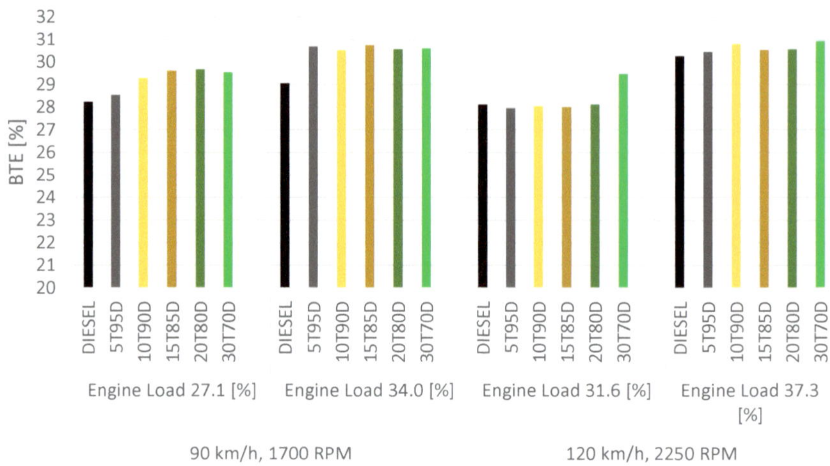


**Figure 7.** Brake thermal efficiency (%) of the engine for different fuel blends.

Regarding the general trend, the differences in efficiency were fairly small, with a maximum difference of 1.5 percentage points in terms of efficiency being obtained, meaning that there was a relative difference of 5.2% between the pure diesel and any of the turpentine blends. Strictly speaking, there was no set of loads/speeds in which the efficiency always increased with increasing the turpentine incorporation. However, when accounting for oscillations, the trend can be considered to have roughly increased when the turpentine incorporation increased in two of the four conditions.

Despite the small differences between the blends, the oscillations around the mean value obtained between the tests carried out with the same conditions were also small, generally between 0.1% and 3%. Therefore, although the specific differences between the different blends were too close to be assessed, the general trends seemed to be sufficiently visible to allow a qualitative comparison and to evaluate the addition of turpentine as mostly positive. This was similar to what some authors have found and might be attributed to a better atomisation of the fuel, better vaporisation and a reduction in the viscosity of the mixture. In addition, the presence of oxygen in the fuel helped it to burn more efficiently, increasing the BTE of the engine [34,35]. Some other authors have found the efficiency to decrease, attributing this to the negative effect of the lower cetane number of turpentine [36]. However, this tended not to be the case for the present study. In fact, for a high load and at low speed (corresponding to a vehicle speed of around 90 km/h), the mixture with the highest turpentine incorporation registered a 1.5% increased efficiency. The highest efficiency was recorded by the mixture 30T70D for a high speed and at both loads, with values of 1.3% and 0.7%, respectively, being obtained. These results seem interesting regarding the potential of turpentine. The fact that the negative effect of a lower cetane number was not observed might be related to the fact that low boost pressures were used. Under these conditions, the improvement in combustion due to the higher oxygen content seems to have played a larger net positive effect than the challenges to combustion due to the higher ignition delay of lower-cetane-number fuels.

#### 3.1.2. Fuel Consumption

The variations in the fuel consumption can be seen in Figure 8. It should be noted that a higher fuel consumption does not necessarily mean a lower efficiency because of the different calorific values of the two components and the different powers produced. Also, a higher speed also tends to induce a higher consumption because of the higher number of cycles per unit time. Curiously, the high-load/low-speed tests recorded a higher absolute consumption than the low-load/high-speed tests. The lowest fuel consumption values were recorded for the 20T70D mixture. A slight trend seemed to be that the lowest consumption tended to occur for the extremes of pure diesel and the highest turpentine incorporation. Nevertheless, these results did not allow the extraction of a lot of conclusions given that too many parameters affected them.

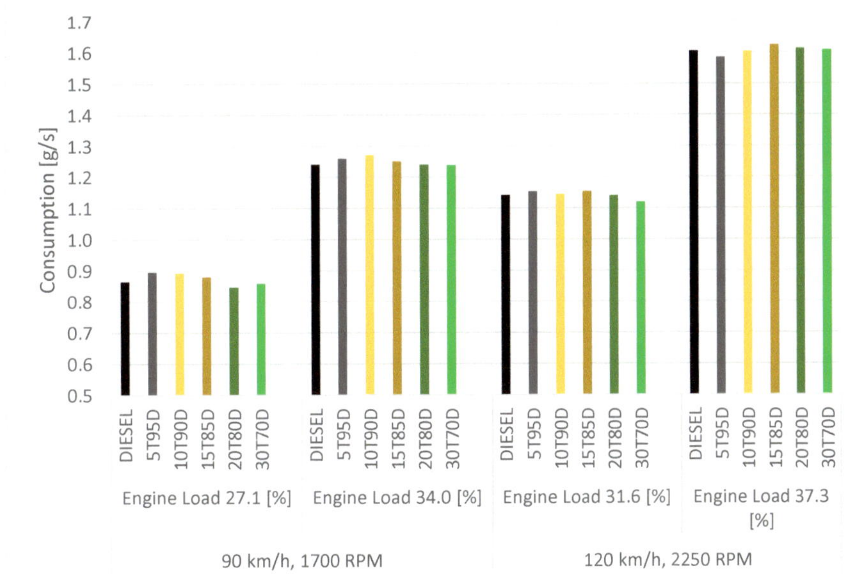


**Figure 8.** Fuel consumption (g/s) of the engine for different fuel blends.

### 3.1.3. Torque and Power

In Figure 9, the variation in the torque with turpentine incorporation can be seen for the four different conditions tested.

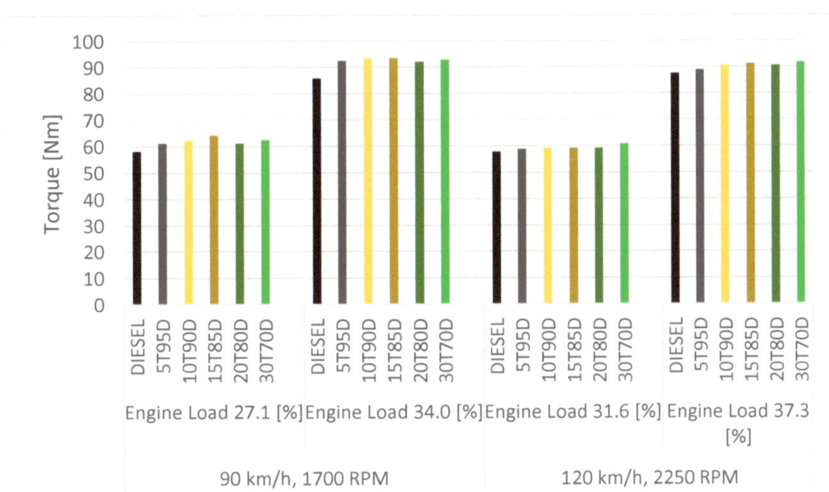


**Figure 9.** Torque (N·m) of the engine with different blends.

For the low load and with an engine speed of 1700 RPM, the 15T85D mixture registered the highest increase in torque, which was 7.9% higher than that of diesel. For the engine speed of 2250 RPM at low load, the 30T70D mixture recorded the highest performance, with a torque that was 5.2% higher than that of diesel. For the high engine load at an engine speed of 1700 RPM, the 15T85D mixture registered the highest increase in torque relative to the baseline fuel, specifically with an increase of 6.8%. The 30T70D mixture had an increase of 4.7% at a high load and high speed of the engine.

Figure 10 displays the brake power of the engine obtained with several fuel blends. At low speed, the increases were around 9% for the fuel with a 15% incorporation of turpentine for both the low-speed and low-engine-load settings. For the high-speed setting, the fuel with a 30% incorporation of turpentine displayed the highest increase in power, which was close to 5% for both engine load settings. These improvements were related to the improvement in torque, as already discussed above. These improvements seem interesting for the potential of turpentine mixtures in engines.

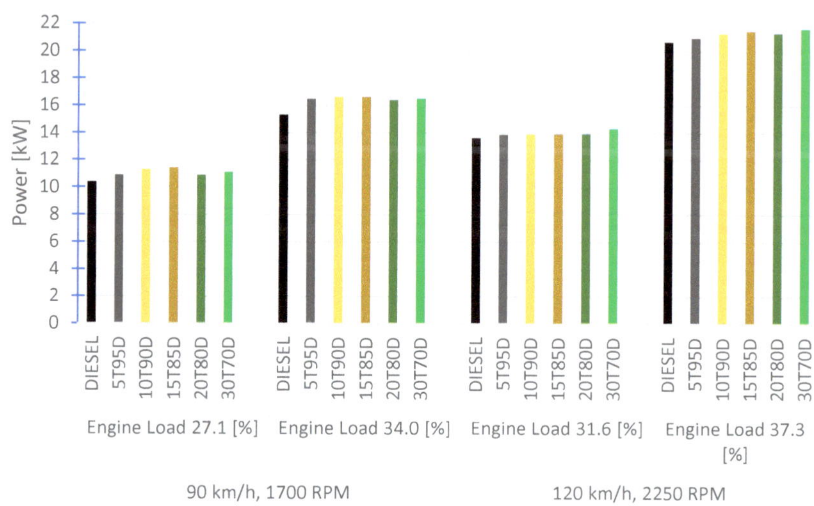


**Figure 10.** Brake power (kW) of the engine with different blends.

### 3.2. Emission Characteristics

The emissions of unburned hydrocarbons (HC) can be seen in Figure 11. The highest HC emissions for all engine loads were recorded for the mixture with the highest percentage of turpentine in diesel (30T70D). The HC emissions for this mixture at 1700 RPM were 24.6% and 37.1% higher than those of the reference fuel for the two different engine loads. For 2250 RPM, the HC emissions for 30T70D were 17.3% and 25.3% higher than those of diesel. However, it should be noted that the incorporation of turpentine also improved the torque. For this reason, the specific emissions did not increase as much.

Figure 12 displays the emission of smoke for the several fuel blends. The most apparent fact was that the test conditions with high load and low speed provided much higher levels of smoke than all the other test conditions, and this happened to all the fuel blends. This signalled a sub-optimal combustion, which was probably due to the fact the turbocharger did not deliver a large amount of air at 1700 RPM and high load. However, for these operating conditions, the incorporation of turpentine provided a very substantial reduction in smoke emissions for all the blends, with a maximum reduction of 24.8% relative to diesel. The formation of smoke is intimately associated with the air–fuel ratio and the combustion effectiveness [16]. The oxygen that was present in the composition of the bio-additive tended to cause the smoke emissions to decrease considerably because a more complete combustion could take place. Similar conclusions were found in works such as that of [37], where the authors mixed oxygenated fuel and reported a decrease in smoke emissions. The trends were difficult to assess, although the highest incorporation of turpentine provided good results for all but one of the test conditions, i.e., the one with a lot of smoke.

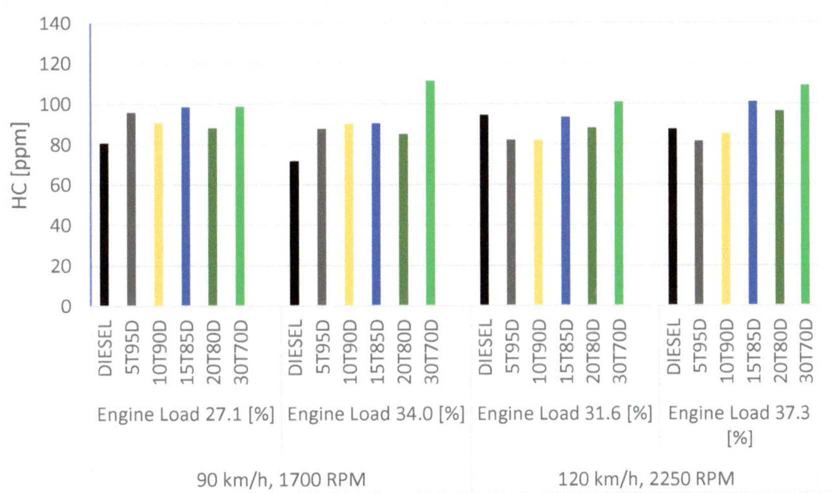


**Figure 11.** HC (ppm) emission of various blends.

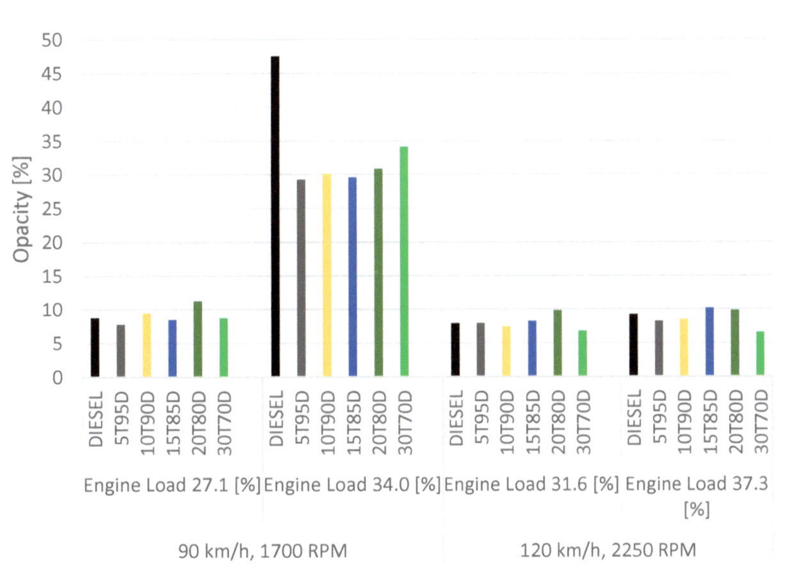


**Figure 12.** Smoke emission (opacity %) of various blends.

For the engine rotation of 2250 RPM, the 30T70D mixture obtained the lowest value compared to the reference fuel.

Figure 13 shows the level of $NO_x$ emissions. It was observed that as the concentration of turpentine increased, the $NO_x$ emissions tended to be higher. This was likely due to the presence of oxygen in the additive, which promoted $NO_x$ production. Also, it is known that an increase in the combustion performance tends to increase the temperature and indirectly increase the emission of $NO_x$ due to the favourable environment for the appearance of nitrogen oxides. The highest recorded values were for the 30T70D mixture. Again, the increase in specific emissions was not as high because the incorporation of turpentine had a positive effect on the torque.

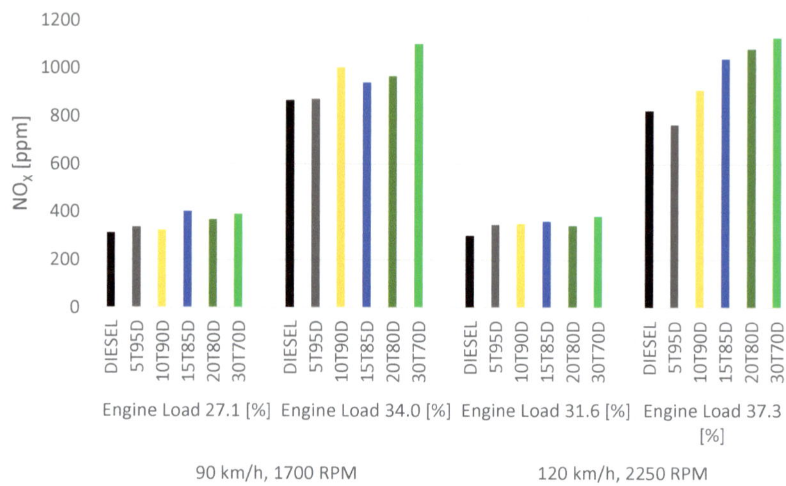


**Figure 13.** $NO_x$ (ppm) emissions of the various blends.

## 4. Conclusions

In the present work, incorporations of turpentine into regular diesel of between 5 and 30% were assessed in a direct-injection diesel engine used in light-duty vehicles. The objective was to assess whether it would be possible to maintain or even improve the performance and emissions parameters of the engine. One of the main advantages of this incorporation being feasible would be a direct cut in fossil greenhouse gas emissions.

The comparisons were made in a slightly different way compared to the existing literature. Namely, comparisons for conditions typical of road vehicle driving were made, with the comparison of the different fuel blends being carried out for fixed values of the engine load.

For the conditions tested, the following conclusions may be drawn:

- The differences obtained in the performance parameters such as torque, power and efficiency were not big, so the discussion of these results is necessarily limited. Nevertheless, a qualitative assessment may be conducted.
- The incorporation of turpentine into regular diesel fuel was beneficial for the brake thermal efficiency for most of the turpentine incorporations tested. Namely, improvements of up to 5.5% were recorded at high loads and low speeds for the fuel with highest turpentine incorporation (30%). The brake torque always increased when incorporating turpentine into diesel fuel. Increases of nearly 8% and 7% were obtained for the fuels with 15% and 30% incorporations of turpentine. The improvement in the brake power was even better, with 9% being obtained with a 15% turpentine incorporation at the low-speed setting, while a 5% improvement was achieved for the high-speed setting with a 30% incorporation of turpentine.
- Regarding pollutant emissions, the results were mixed. On the one hand, the level of unburned hydrocarbons (HC) was almost always higher with the turpentine–diesel mixtures than with pure diesel, although the specific emissions (per unit torque or power produced) were not as high because the torque was also higher. On the other hand, this study found that the particulate-matter-induced opacity, commonly referred

to as smoke, displayed either higher or lower values with the turpentine incorporation when compared to diesel. However, a sharp drop in the opacity occurred in the case of low speeds and high loads for all the turpentine blends. This might be related to the positive effect of the slight oxygen content of turpentine on the combustion. Also, the highest incorporation of turpentine (30%) had a positive effect on the smoke reduction in almost all cases.

- Unfortunately, the level of nitrogen oxides ($NO_X$) produced registered a considerable increase in relation to the concentration of turpentine, which was probably due to not only the extra oxygen in the composition but also due to the improved performance of the engine. In fact, it is known that an increase in combustion performance is generally accompanied by a rise in $NO_X$ emissions.

As an overall evaluation of the results obtained, it seems reasonable to assume that incorporations of turpentine of up to 30% could represent a good opportunity for a proportional reduction in fossil fuel consumption and the associated $CO_2$ emissions due to the neutral or even negative contribution of turpentine coming from sustainable pine forest management (pines are not cut down for turpentine extraction). This substitution could be achieved with minimal modifications to engines, and even a slight increase in torque, power and efficiency can be obtained in most driving conditions, with potentially lower smoke emissions at high engine loads and low speeds.

The higher emissions of HC and $NO_X$ seem to be an objective disadvantage of turpentine incorporation. However, these results were obtained without implementing advanced aftertreatment. A full analysis of the impact on the viability of turpentine due to these emissions would need further assessment, which was out of the scope of the present study.

Turpentine as a fuel might come closer to becoming economically competitive with the advent of the latest-generation biotechnology-enhanced pines. And in the case of Portugal, Pinus Pinaster is among the most-explored tree species, and it could be explored for turpentine production without losing its economic value for the wood and paper paste markets in which it is currently explored commercially [38].

**Author Contributions:** Conceptualization, R.M.C., J.M., M.G. and F.P.B.; Methodology, R.M.C. and F.P.B.; Validation, F.P.B.; Formal analysis, J.M. and F.P.B.; Investigation, R.M.C., T.-C.C., E.O. and F.P.B.; Resources, J.M., M.G. and F.P.B.; Data curation, R.M.C., J.M. and F.P.B.; Writing—original draft, R.M.C. and F.P.B.; Writing—review & editing, J.M., M.G. and F.P.B.; Visualization, R.M.C.; Supervision, J.M., F.P., K.U., I.V.I. and F.P.B.; Project administration, J.M. and F.P.B.; Funding acquisition, J.M. and F.P.B. All authors have read and agreed to the published version of the manuscript.

**Funding:** This research was funded by the following projects, institutions and funding agencies: Research Project UIDB/04077/2020 from the Mechanical Engineering and Resource Sustainability Center—MEtRICs—through Fundação para a Ciência e a Tecnologia (FCT), Norte2020, Compete2020, under the PORTUGAL 2020 Partnership Agreement, through Portuguese national funds of FCT/MCTES (PIDDAC) and the European Regional Development Fund; Research project DREAM (Dynamics of the REsources and technological Advance in harvesting Marine renewable energy), supported by the Romanian Executive Agency for Higher Education, Research, Development and Innovation Funding—UEFISCDI—grant number PN-III-P4-ID-PCE-2020-0008.

**Data Availability Statement:** The data presented in this study are available on request from the corresponding author.

**Conflicts of Interest:** The authors declare no conflict of interest.

## References

1. Ramalingam, S.; Thamizhvel, R.; Sudagar, S.; Silambarasan, R. Production of third generation bio-fuel through thermal cracking process by utilizing Covid-19 plastic wastes. *Mater. Today Proc.* **2023**, *72*, 1618–1623. [CrossRef] [PubMed]
2. Brodny, J.; Tutak, M. The analysis of similarities between the European Union countries in terms of the level and structure of the emissions of selected gases and air pollutants into the atmosphere. *J. Clean. Prod.* **2020**, *279*, 123641. [CrossRef] [PubMed]
3. Senra, M.; McCartney, S.N.; Soh, L. The effect of bio-derived additives on fatty acid methyl esters for improved biodiesel cold flow properties. *Fuel* **2019**, *242*, 719–727. [CrossRef]

4. Fathurrahman, N.A.; Wibowo, C.S.; Nasikin, M.; Khalil, M. Optimization of sorbitan monooleate and $\gamma$-$Al_2O_3$ nanoparticles as cold-flow improver in B30 biodiesel blend using response surface methodology (RSM). *J. Ind. Eng. Chem.* **2021**, *99*, 271–281. [CrossRef]
5. de Almeida, V.F.; García-Moreno, P.J.; Guadix, A.; Guadix, E.M. Biodiesel production from mixtures of waste fish oil, palm oil and waste frying oil: Optimization of fuel properties. *Fuel Process. Technol.* **2015**, *133*, 152–160. [CrossRef]
6. Bozbas, K. Biodiesel as an alternative motor fuel: Production and policies in the European Union. *Renew. Sustain. Energy Rev.* **2008**, *12*, 542–552. [CrossRef]
7. Gasol, C.M.; Salvia, J.; Serra, J.; Antón, A.; Sevigne, E.; Rieradevall, J.; Gabarrell, X. A life cycle assessment of biodiesel production from winter rape grown in Southern Europe. *Biomass-Bioenergy* **2012**, *40*, 71–81. [CrossRef]
8. Pande, T.; Pani, S.; Hiran, S.; Rao, V.; Shah, H.; Vishwanathan, K. Turpentine poisoning: A case report. *Forensic Sci. Int.* **1994**, *65*, 47–49. [CrossRef]
9. Yumrutaş, R.; Alma, M.H.; Özcan, H.; Kaşka, Ö. Investigation of purified sulfate turpentine on engine performance and exhaust emission. *Fuel* **2008**, *87*, 252–259. [CrossRef]
10. Arpa, O.; Yumrutas, R.; Alma, M. Effects of turpentine and gasoline-like fuel obtained from waste lubrication oil on engine performance and exhaust emission. *Energy* **2010**, *35*, 3603–3613. [CrossRef]
11. Venkatesan, V.; Nallusamy, N. Pine oil-soapnut oil methyl ester blends: A hybrid biofuel approach to completely eliminate the use of diesel in a twin cylinder off-road tractor diesel engine. *Fuel* **2019**, *262*, 116500. [CrossRef]
12. García, D.; Ramos, Á.; Rodríguez-Fernández, J.; Bustamante, F.; Alarcón, E.; Lapuerta, M. Impact of oxyfunctionalized turpentine on emissions from a Euro 6 diesel engine. *Energy* **2020**, *201*, 117645. [CrossRef]
13. Sahay, S. (Ed.) *Handbook of Biofuels*; Academic Press: Cambridge, MA, USA, 2022. [CrossRef]
14. Van Basshuysen, R.; Schäfer, F. *Internal Combustion Engine Handbook*, 2nd ed.; SAE International: Warrendale, PA, USA, 2016.
15. Anand, B.P.; Saravanan, C.; Srinivasan, C.A. Performance and exhaust emission of turpentine oil powered direct injection diesel engine. *Renew. Energy* **2010**, *35*, 1179–1184. [CrossRef]
16. Agarwal, D.; Sinha, S.; Agarwal, A.K. Experimental investigation of control of $NO_x$ emissions in biodiesel-fueled compression ignition engine. *Renew. Energy* **2006**, *31*, 2356–2369. [CrossRef]
17. Kumaran, D.; Rajendran, M.; Kumaravelan, R.; Gandhi, V.C.S. Testing of three-fuel mixture in a four-stroke single cylinder direct injection diesel engine. *Trans. Famena* **2013**, *37*, 75–86.
18. Karikalan, L.; Chandrasekaran, M. Performance and pollutants analysis on diesel engine using blends of Jatropha Biodiesel and Mineral Turpentine as fuel. *Int. J. Environ. Sci. Technol.* **2016**, *14*, 323–330. [CrossRef]
19. Panneerselvam, N.; Ramesh, M.; Murugesan, A.; Vijayakumar, C.; Subramaniam, D.; Kumaravel, A. Effect on direct injection naturally aspirated diesel engine characteristics fuelled by pine oil, ceiba pentandra methyl ester compared with diesel. *Transp. Res. Part D Transp. Environ.* **2016**, *48*, 225–234. [CrossRef]
20. Vallinayagam, R.; Vedharaj, S.; Yang, W.; Lee, P.; Chua, K.; Chou, S. Combustion performance and emission characteristics study of pine oil in a diesel engine. *Energy* **2013**, *57*, 344–351. [CrossRef]
21. Huang, H.; Teng, W.; Liu, Q.; Zhou, C.; Wang, Q.; Wang, X. Combustion performance and emission characteristics of a diesel engine under low-temperature combustion of pine oil–diesel blends. *Energy Convers. Manag.* **2016**, *128*, 317–326. [CrossRef]
22. Distillation. Available online: https://www.britannica.com/summary/distillation (accessed on 30 May 2023).
23. Yadav, A.S.; Singh, P.; Sahu, R.; Thangamuthu, P.; Narain, R.S.; Rao, Y.A.; Balu, A.; Panda, B.P.; Sharma, A. Potential utilization of turpentine oil as an alternative fuel. *Mater. Today Proc.* **2022**, *63*, A1–A8. [CrossRef]
24. Merghni, A.; Lassoued, M.A.; Rasoanirina, B.N.V.; Moumni, S.; Mastouri, M. Characterization of Turpentine nanoemulsion and assessment of its antibiofilm potential against methicillin-resistant Staphylococcus aureus. *Microb. Pathog.* **2022**, *166*, 105530. [CrossRef] [PubMed]
25. Panithasan, M.S. *Exploring the Preparation and Usage of Low Viscous Pine Oil and Rice Husk Nano Additives in a CRDI Engine under the Effect of Water Emulsion*; SAE Technical Paper; SAE: Troy, MI, USA, 2020. [CrossRef]
26. Gobalakichenin, D.; Gnanamoorthi, V. *Assessing the Combined Outcome of Rice Husk Nano Additive and Water Injection Method on the Performance, Emission and Combustion Characters of the Low Viscous Pine Oil in a Diesel Engine*; SAE Technical Paper; SAE: Troy, MI, USA, 2019. [CrossRef]
27. Nanditta, R.V.; Deenadayalan, S.R.; Manikandaraja, G.; Udayakumar, P. Performance and emission characteristics of CI engine fuelled with turpentine oil-diesel blend with diethyl ether as additives. *IOP Conf. Ser. Mater. Sci. Eng.* **2020**, *912*, 042075. [CrossRef]
28. Ballesteros, R.; García, D.; Bustamante, F.; Alarcón, E.; Lapuerta, M. Oxyfunctionalized turpentine: Evaluation of properties as automotive fuel. *Renew. Energy* **2020**, *162*, 2210–2219. [CrossRef]
29. ICSC 1561-DIESEL FUEL No. 2. Available online: https://www.ilo.org/dyn/icsc/showcard.display?p_lang=en&p_card_id=1561&p_version=2 (accessed on 22 June 2023).
30. ICSC 1063-TURPENTINE. Available online: https://www.ilo.org/dyn/icsc/showcard.display?p_lang=en&p_card_id=1063&p_version=2 (accessed on 22 June 2023).
31. Telma. Available online: https://pt.telma.com/applications/industriel (accessed on 30 May 2023).
32. Kern. Available online: https://docs.rs-online.com/a64e/0900766b81683823.pdf (accessed on 30 May 2023).
33. Index, H. Datasheet Thermo/Hygro/Barometer. 2017, Volume 127, pp. 1160–1161. Available online: https://docs.rs-online.com/1164/0900766b81599e70.pdf (accessed on 22 June 2023).

34. Panithasan, M.S.; Venkadesan, G. Evaluating the outcomes of a single-cylinder CRDI engine operated by lemon peel oil under the influence of DTBP, rice husk nano additive and water injection. *Int. J. Engine Res.* **2023**, *24*, 308–323. [CrossRef]
35. Dubey, P.; Gupta, R. Influences of dual bio-fuel (Jatropha biodiesel and turpentine oil) on single cylinder variable compression ratio diesel engine. *Renew. Energy* **2018**, *115*, 1294–1302. [CrossRef]
36. Panithasan, M.S.; Malairajan, M.; Balusamy, M.; Venkadesan, G. *Evaluating the Engine Emission Outcomes from a CRDI Engine Operated with Low Viscous Bio-Oil Blends under the Influence of Diethyl Ether*; Springer: Singapore, 2021; pp. 265–276. [CrossRef]
37. Yoshimoto, Y.; Kinoshita, E.; Shanbu, L.; Ohmura, T. Influence of 1-butanol addition on diesel combustion with palm oil methyl ester/gas oil blends. *Energy* **2013**, *61*, 44–51. [CrossRef]
38. Martins, J.; Brito, F.P. Alternative Fuels for Internal Combustion Engines. *Energies* **2020**, *13*, 4086. [CrossRef]

**Disclaimer/Publisher's Note:** The statements, opinions and data contained in all publications are solely those of the individual author(s) and contributor(s) and not of MDPI and/or the editor(s). MDPI and/or the editor(s) disclaim responsibility for any injury to people or property resulting from any ideas, methods, instructions or products referred to in the content.

energies

MDPI

Article

# Influence of Performance Packages on Fuel Consumption and Exhaust Emissions of Passenger Cars and Commercial Vehicles under WLTP

**Luca Marchitto [1], Maria Antonietta Costagliola [1,*] and Alessandra Berra [2]**

[1] Istituto di Scienze e Tecnologie per l'Energia e la Mobilità Sostenibili, National Research Council of Italy, Via G. Marconi, 4, 80125 Naples, Italy; luca.marchitto@stems.cnr.it
[2] Dorf Ketal B.V., Schimmelt 2-16, 5611 ZX Eindhoven, The Netherlands; alessandra.berra@dorfketal.com
* Correspondence: mariaantonietta.costagliola@stems.cnr.it

**Abstract:** The transportation sector is responsible for about 16% of worldwide greenhouse gas emissions. Despite efforts for a sensible reduction by means of new technologies' development, the average age of a vehicle fleet is 12.3 years in the European Union. In light of this, actions aiming at improving the efficiency of circulating vehicles can prove effective in the short to mid-term. Introducing performance packages in standard fuels could allow a reduction in the $CO_2$ emissions of whole vehicle fleets without any modification to powertrain. Such a kind of additive is generally used in premium fuels; deposit control additives can reduce or control the deposits at intake valves and at nozzle holes with benefits for the fuel efficiency and exhaust emissions. Further improvements in combustion phasing can be achieved with cetane/octane improver. This paper aims to assess the influence of two performance packages on the exhaust emissions and fuel consumption of five vehicles set to be as representative as possible of circulating Italian passenger cars and light commercial fleet vehicles (LCVs). Based on the literature datasets, three Euro 4 vehicles were selected with a mileage representative of each single vehicle class: two passenger cars (one spark ignition and one diesel) and an LCV. Further, two diesel Euro 6 vehicles, a passenger car and an LCV, were tested to investigate the effect of fuel additives on the combustion of vehicles compliant with current homologation regulation. Exhaust emissions and fuel consumption were experimentally estimated on a chassis dynamometer over a worldwide harmonized light vehicles test cycle (WLTC) in a climate-controlled laboratory. Each vehicle was preliminarily tested when running with base fuel, then a 3000 km clean-up stage was performed using the additive package. Finally, WLTC tests were repeated. Results demonstrated the efficiency of the performance packages with a reduction between 1.2% (diesel Euro 6 passenger car) and 8.1% (diesel Euro 4 passenger car) in fuel consumption. Similar trends were found for $CO_2$ emissions. Further, a sensible reduction in THCs, CO and PM was found for each vehicle class.

**Keywords:** fuel additive; exhaust emissions; fuel consumption; passenger car; light-duty vehicle

**Citation:** Marchitto, L.; Costagliola, M.A.; Berra, A. Influence of Performance Packages on Fuel Consumption and Exhaust Emissions of Passenger Cars and Commercial Vehicles under WLTP. *Energies* **2024**, *17*, 3356. https://doi.org/10.3390/en17143356

Academic Editors: Davide Lanni and Enzo Galloni

Received: 5 June 2024
Revised: 28 June 2024
Accepted: 2 July 2024
Published: 9 July 2024

**Copyright:** © 2024 by the authors. Licensee MDPI, Basel, Switzerland. This article is an open access article distributed under the terms and conditions of the Creative Commons Attribution (CC BY) license (https://creativecommons.org/licenses/by/4.0/).

## 1. Introduction

The transport sector causes negative impacts on the environment and human health, being responsible for about 25% of the total greenhouse gas (GHG) emissions in Europe [1]. GHG emissions by transport sector are strictly related to the transport demand; they, in fact, dropped substantially in 2020 because of reduced activity during the COVID-19 pandemic. In 2021, they started to increase by almost 8.6%, followed by further growth of 2.7% in 2022.

In future years, they are expected to increase due to the continuous increase in transport activity volume. In 2022, the number of EU-registered passenger cars reached almost 253 million, corresponding to an increase of 7.0% as compared with 2017. Moreover, passenger cars powered by electric battery and hydrogen (alternative fuels) only made up a small share of the new registration of passenger cars in the EU in 2022 [2]. This is in contrast with the EU's aim to be climate neutral by 2050. The European Green Deal has recently recommended a 90% net greenhouse gas emissions reduction by 2040 compared to 1990 levels [3]. Indeed, a sustainable mobility system based on zero emission vehicles and a more rational use of transportation is strongly needed.

In the short term, the use of fuel with performance additives is a viable way to obtain a sudden reduction in fuel consumption and GHG emissions from in-use vehicles. Benefits coming from these additive fuels are expected to be more effective for old vehicles with significant mileage accumulation and scarce maintenance. According the European Automobile Manufacturers' Association (ACEA) [4], in 2022, European circulating fleets of light-duty vehicles had an average age of 12.3 years; the age of passenger cars had a minimum of 7.9 years in Luxemburg and a maximum of 17.3 years in Greece. This age distribution implies a great share of pre-Euro 6 circulating vehicles. For example, in Italy where in 2022 cars were on average 12.5 years old, close to European statistics, two vehicles out of three were pre-Euro 6: Euro 6 vehicles accounted only for 32.6%, Euro 5 for 16.7% and Euro 4 for 23.3% [5].

Many fuel additives have been investigated with a view to reducing fuel consumption and exhaust emissions [6,7].

In general, performance packages are used to improve the function of the engine/vehicle; they differ from finished fuel additives, which are used to make the fuel compliant with legislative standard.

Performance additive packages generally contain DCAs (deposit control additives), corrosion inhibitors and demulsifiers. They might also include additional components such as friction modifiers, antifoams, etc.

DCAs (deposit control additives) allow control over the formation of deposits on the engine components: they are fuel systems for direct and indirect injection internal combustion engines (the most common are the injector-nozzle coking and fouling) and intake valves in the case of port fuel injection engines. The use of DCAs keeps elements of the fuel system clean, reducing emissions of carbon monoxide, unburned hydrocarbons and particulate matter and reducing fuel consumption [8]. Controversial results have been achieved regarding the influence of DCAs on deposits in the combustion chamber. The diffusion of GDI engines has brought attention to this issue: direct injection promotes the formation of deposits in the combustion chamber, partially offsetting the benefits achieved in terms of fuel efficiency and exhaust emissions. The use of DCAs contributes to reduce the formation of carbon deposits on the injector holes, ensuring proper long-term functioning. When the proper dosage is used, deposit control additives should also limit carbon deposits in the combustion chamber. However, an excessive concentration of DCAs can have the opposite effect, promoting deposits' formation [9].

Fuel additive packages can additionally contain friction modifier to control friction and energy losses, hence improving fuel economy [10].

Frausher et al. investigated the effect of the friction modifiers as fuel additives instead of oil additives. In fact, the degradation of oil during the engine run banishes the effect of oil additives. They found that adding a friction modifier directly into the fuel can substitute depleted oil additives and improve friction and wear characteristics, even after engine oil degradation [10].

Some fuel additives can influence the ignition timing. In particular, the use of some additives which improve the autoignition process increases the cetane number of a diesel fuel, involving a reduction in ignition delay in a compression ignition engine. Reducing the ignition delay results in improved start ability, less noise and lower exhaust emissions, especially of NOx. The cetane improver additive category generally includes alkyl nitrates and peroxides [11].

Ashok et al. [12] used a multifunctional additive package including surfactant, de-emulsifier, friction modifier, dispersant, cetane improver, antioxidant and combustion catalyst to improve the fuel economy and exhaust emissions of a single cylinder diesel engine fueled with diesel and biodiesel. For both fuels, the use of additives resulted in a reduction in BSFC, which can be partially attributed to a lower ignition delay. Further, antioxidant and cetane improver limited the formation of NOx.

In an analogous way, the octane quality of a gasoline (resistance to the autoignition) can be improved using high-octane blend components (such as oxygenates or aromatic) or octane-improving additives. I.e., in the past, methylcyclopentadienyl manganese tricarbonyl (MMT) was used, though metalorganic octane boosters have now been phased out in most countries [11]. Components are blended with fuels in low volume percentages; therefore, they alter the properties of the fuel (density, heat value, etc.) in addition to the octane resistance. Additives are added to fuel in a few parts per million without influencing other physical–chemical characteristics of gasoline [13]. A high knock resistance can allow increasing engine compression ratios, with benefits for performance and efficiency. Therefore, considering the need for more efficient engines with the aim of reducing GHG emissions, a growing interest in the use of octane enhancer is expected [14].

Evaluating the efficacy of a performance additive in an internal combustion engine or a vehicle requires long-term tests as its action is mainly focused on the cleaning of deposits and friction reduction and the effect on engine performance and exhaust emissions is not immediate except for the cetane/octane improver. Even though a specific regulation defining a harmonized test procedure does not exist, the European automotive industry, brought together in CEC (Coordinating European Council for the development of performance tests for fuels, lubricants and other fluids), developed a series of protocols to estimate the performance of oils and fuels [15]. Engine and test procedures have been developed for each engine type (PFI spark ignition (SI), DISI and diesel) in order to produce comparable data regarding the influence of additives on defined parameters (deposits on intake valves, nozzle deposits, internal diesel injector deposits (IDID), etc.). These procedures take several engine running hours; i.e., the CEC-F-05-93 and CEC-F-20-98 methods for the evaluation of intake valve deposit formation in SI engines require 60 h tests.

Burke et al. [16] studied in depth the effects of injector-nozzle fouling in diesel engines, demonstrating a relationship with air and fuel pumping losses. Injector fouling tests were performed by means of a modified version of the CEC F-98-08 cycle to avoid issues related to engine power derating due to engine control unit protection algorithms, which could alter results. Further WLTC tests were carried out to analyze the effect of injection fouling during a transient cycle representative of on-road conditions.

CEC F-98-08 test procedure is specifically thought to quantify fouling effect, defining a relationship with the reduction in fuel rate, engine torque and power. Hence, steady-state conditions are established, keeping constant engine speed and accelerator position. In this way, it is easy to quantify the power loss related to fouling and the benefits of additives. On the other hand, data on fuel economy cannot be considered robust as a wider range of operative conditions and transient stages have to be taken into account to be representative of real driving conditions [17].

Few papers report information about the influence of deposit control additives on the performance and exhaust emissions of vehicles under real-world driving conditions.

In this paper, the results of a wide experimental campaign are discussed and carried out to analyze the effect of additive fuel on exhaust emissions and fuel consumption of 5 in-use vehicles: 3 passenger cars and 2 commercial vehicles. Vehicles are homologated with Euro 4 and Euro 6 stages and were chosen to be representative of circulating fleets in Italy. For this purpose, each vehicle was tested in chassis-dynamometer laboratory over WLTC driving cycle.

## 2. Materials and Methods

Performances of fuel additives were evaluated by comparing the exhaust emissions and specific consumption of same vehicle fueled with baseline and treated fuels (baseline fuels + additive). For this purpose, chassis-dynamometer WLTC tests were firstly carried out to measure vehicle emissions with baseline fuel and repeated to characterize vehicle emissions with treated fuels. The concentration of additives in the fuels is a few ppm, and the treated fuels comply with the European standards EN 228 (RON 95 gasoline) and EN 590 (automotive diesel fuel). Therefore, their use does not require any modifications to the powertrain architecture or the after-treatment system, nor is it necessary to optimize the engine control unit calibration. Tests relative to treated fuels were preceded by a mileage accumulation of 3000 km, which is necessary for engine cleaning and conditioning. This section describes the experimental set-up, tested vehicles and fuels.

### *2.1. Experimental Set-Up*

Vehicle emissions were measured during experimental tests on the chassis-dynamometer bench (Schenk, 184 kW, maximum speed 200 km/h) that simulates the dynamic road load resistance during the execution of the driving cycle.

The exhaust gases are collected, diluted by a constant volume dilution system (CVS) and then sent to Horiba-Mexa 7200H gas analyzer for the continuous concentration measurement of carbon monoxide CO and carbon dioxide $CO_2$ by means of a NDIR (non-dispersive infrared) sensor and total hydrocarbons (THCs) with a FID (flame ionization detector), while chemiluminescence method is applied for nitrogen oxides NOx (sum of NO and $NO_2$). At the same time, for each single phase of WLTC cycle, a sample bag is filled during the test to estimate the average values. Fuel consumption is estimated by applying the carbon balance to the exhaust pollutants. Moreover, particulate matter (PM) was sampled on a high-efficiency filter (99.95% aerosol retention 0.3 µm DOP) and quantified throughout gravimetric analysis. For Euro 6 vehicles, particle number (PN) was sampled with a fine particle sampler (FPS, Dekati, Kangasala, Finland) and measured with an electric low pressure impactor (ELPI, Dekati) which counts the number of particles with aerodynamic diameter from 7 nm up to 10 mm. The ELPI analyzer measures the total PN and the particle size distribution into 12 dimensional stages.

In order to verify the effect of fuels in real driving conditions, all the vehicles were tested using worldwide harmonized light vehicles test cycles (WLTC); the certification driving cycle came into force in 2017 with the Euro 6d-temp phase. The WLTC cycle is, in fact, characterized by many transients (Figure 1), and includes four consecutive phases (WLTC low, WLTC medium, WLTC high and WLTC extra-high) with increasing average speed, representative of urban, rural and highway driving mode.

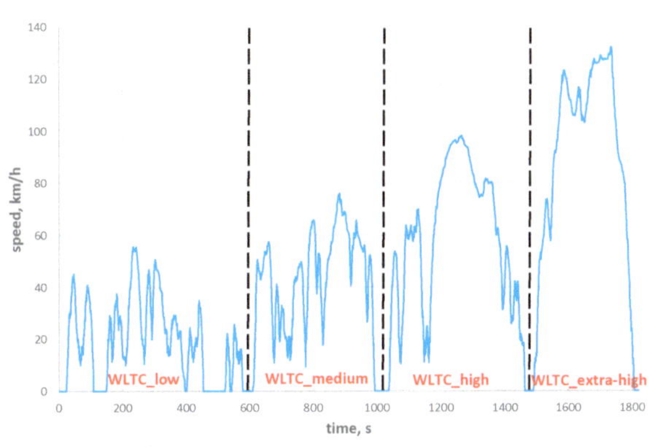


**Figure 1.** WLTC driving cycle.

Cold-start WLTC (vehicle placed at 23 °C with the engine off for at least 6 h before starting the engine) and warm-start WLTC (vehicle with engine fluid temperature higher than 80 °C) were repeated three times with reference and additive fuel, respectively.

A statistical analysis was performed (F-test) for checking the significance of difference in fuel consumption and exhaust emissions between vehicle operations with and without additive, with a confidence threshold level of 95%, over each single stage of WLTC [18].

### 2.2. Vehicles

Vehicle technical characteristics are shown in Table 1.

**Table 1.** Vehicle characteristics.

| | G_PC_EU4 | D_PC_EU4 | D_PC_EU6 | D_LCV_EU4 | D_LCV_EU6 |
|---|---|---|---|---|---|
| Vehicle category | M1 | M1 | M1 | N1 | N1 |
| Fuel | gasoline | Diesel | Diesel | Diesel | Diesel |
| Engine displacement, $cm^3$ | 1242 | 1248 | 1598 | 2287 | 2299 |
| Maximum power, kW | 44 | 55 | 70 | 88 | 81 |
| Curb weight, kg | 1020 | 1235 | 1280 | 1935 | 2046 |
| EU directive | Euro 4 2001/100/CE-B | Euro 4 2003/76/CE-B | Euro 6 2016/646 ZD | Euro 4 2003/76/CE-B | Euro 6 2015/45/CE-B |
| Mileage, km | 175,000 | 176,300 | 220,771 | 145,000 | 258,365 |

Fleet is composed of a Euro 4 gasoline passenger car (G_PC_EU4), a Euro 4 and Euro 6 diesel passenger car (D_PC_EU4 and D_PC_EU6) and a Euro 4 and Euro 6 diesel light commercial vehicle (D_LCV_EU4 and D_LCV_EU6).

The power to mass ratio of all Euro 4 vehicles is comparable (between 43 and 45 W/kg); Euro 6 passenger car and light commercial vehicle have the highest and lowest power to mass ratio (almost 55 and 40 W/kg, respectively).

All the vehicles had a huge mileage, between 145,000 km for EU4 commercial vehicle and over 250,000 km for EU6 commercial vehicle. High mileage was needed to clearly assess the cleaning effect of additive fuels. Tested vehicles were chosen as representative of Italian circulating fleets. They are, in fact, equipped with internal combustion engines, which are, largely speaking, widespread in Italy. It was estimated that almost 10–12% of gasoline and diesel passenger cars and 20% of commercial vehicles circulating in Italy have the same engine technology as the tested vehicles. Moreover, as shown in Figure 2, Euro 4 and Euro 6 vehicles constitute a not-negligible share (almost 53% of passenger cars and 43% of light commercial vehicles) of vehicles circulating in Italy [5].

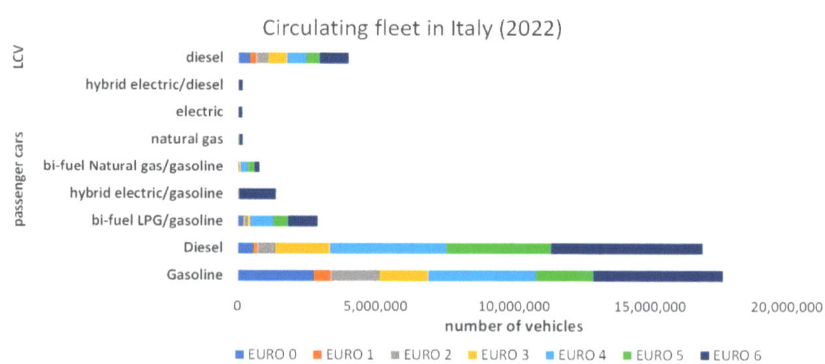


**Figure 2.** Composition of circulating passenger cars and light commercial vehicles (LCVs) in Italy, 2022 [5].

Mileage of vehicles is well above the durability threshold recommended by European legislation (100,000 km for Euro 4 vehicles and 160,000 for Euro 6 vehicles). In order to state the status of vehicle aging, Table 2 summarizes the deterioration factors (i.e., ratio between the aged emissions and the type approval limits) calculated for each vehicle over the homologation driving cycle (NEDC for Euro 4 and WLTC for Euro 6). The most critical ratios are relative to CO emissions for gasoline vehicles since deterioration factor is 2.1, higher than European factor of 1.2, and to the level of THCs + NOx emissions for diesel

Euro 4 vehicles. For these vehicles, in fact, deterioration factor is almost double compared to European threshold of 1.

**Table 2.** Deterioration factors.

| Vehicle | Driving Cycle | CO | HC | NOx | THCs + NOx | PM | PN |
|---|---|---|---|---|---|---|---|
| G_PC_EU4 | NEDC | 2.1 | 1.1 | 0.7 | | | |
| D_PC_EU4 | NEDC | 0.8 | | 2.8 | 2.5 | 1.5 | |
| D_PC_EU6 | WLTC | 0.2 | | 0.6 | 0.3 | 0.4 | 0.0 |
| D_LCV_EU4 | NEDC | 0.2 | | 2.2 | 2.0 | 0.4 | |
| D_LCV_EU6 | WLTC | 0.1 | | 1.3 | 0.8 | 0.3 | 0.0 |

### 2.3. Fuels

The fuels, EN 228 gasoline and EN 590 diesel including 6 to 7% FAME, were supplied by a petroleum and distribution company and sampled from their daily production. An independent inspection company was commissioned to inject the performance package and certify the doping as well as the properties of the fuels before and after the injection of the additives. The fuel drums were sealed by the inspection company and the seal was removed only to refuel the vehicles. Main properties of tested fuels can be found in Tables 3 and 4 for gasoline and diesel fuels, respectively. As aforementioned, additives concentration in the fuels is a few ppm, and treated fuels comply with the European standards EN 228 and EN 590.

**Table 3.** Gasoline fuel's properties.

| Parameter | Method | Units | Lower/Upper Limit | Gasoline | Additive + Gasoline |
|---|---|---|---|---|---|
| Hydrocarbons content | EN ISO 22854 | | | | |
| Aromatics | | % *v*/*v* | -/35.0 | 28.1 | 28.1 |
| Olefins | | % *v*/*v* | -/18.0 | 1.0 | 1.0 |
| Benzene | | % *v*/*v* | -/1.00 | 0.76 | 0.76 |
| Sulphur | EN ISO 20884 | mg/kg | -/10.0 | <3.0 | <3.0 |
| Oxygen compounds | EN ISO 22854 | | | | |
| Methanol | | % *v*/*v* | -/3.0 | <0.1 | <0.1 |
| Ethanol | | % *v*/*v* | -/5.0 | <0.1 | <0.1 |
| Isopropyl alcohol | | % *v*/*v* | | <0.1 | <0.1 |
| Isobutyl alcohol | | % *v*/*v* | | <0.1 | <0.1 |
| Tert-butyl alcohol | | % *v*/*v* | | 0.1 | 0.1 |
| Ethers | | % *v*/*v* | | 13.4 | 13.4 |
| others | | % *v*/*v* | | <0.1 | <0.1 |
| Oxygen content | | % m/m | -/2.7 | 2.5 | 2.5 |
| Corrosiveness to copper | EN ISO 2160 | | Class 1 | Class 1 | Class 1 |
| Distillation | EN ISO 3405 | | | | |
| Final Boiling Point | | °C | -/210.0 | 191.4 | 192.0 |
| Residue | | % *v*/*v* | -/2.0 | 0.9 | 0.9 |
| Evaporated @ 70 °C | | % *v*/*v* | 22.0/50.0 | 49.0 | 49.1 |
| Evaporated @ 100 °C | | % *v*/*v* | 46.0/71.0 | 67.8 | 68.0 |
| Evaporated @ 150 °C | | % *v*/*v* | 75.0/- | 89.8 | 89.9 |
| Doctor test | ASTM D4952 | - | | N | N |
| Gum content | EN ISO 6246 | mg/100 mL | -/5.0 | 1.4 | 1.2 |
| Manganese | EN 16135 | mg/L | -/2.0 | <0.1 | <0.1 |
| Density @ 15 °C | UNI EN ISO 12185 | kg/L | 0.72/0.775 | 0.7338 | 0.7340 |
| Motor Octane Number | EN ISO 5163 | - | 85.0/- | 87.3 | 87.3 |
| Research Octane Number | EN ISO 5164 | - | 95.0/- | 95.3 | 95.3 |
| Pb | UNI EN 237 | mg/L | -/5.0 | <2.5 | <2.5 |
| Flash Point | ASTM D56 | °C | -/21 | <21 | <21 |
| Oxidation stability | EN ISO 7536 | min | 360/- | >400 | >400 |
| Vapour Pressure | EN 13016-1 | kPa | 60.0/90.0 | 63.0 | 62.8 |

**Table 4.** Diesel fuel's properties.

| Parameter | Method | Units | Lower/Upper Limit | Diesel | Additive + Diesel |
|---|---|---|---|---|---|
| Water | EN ISO 12937 | mg/kg | -/200 | 90 | 95 |
| CFPP | EN 116 | °C | -/−10 | −15 | −15 |
| Particulate contamination | EN 12662 | mg/kg | -/24.0 | 17.0 | 15.0 |
| FAME | EN14078 | % v/v | -/7.0 | 6.7 | 6.7 |
| Sulphur | EN ISO 20884 | mg/kg | -/10.0 | 7.5 | 7.5 |
| Copper corrosion (3 h @50 °C) | EN ISO 2160 | | | Class 1A | Class 1A |
| Distillation T95 | EN ISO 3405 | °C | -/360.0 | 354.2 | 354.5 |
| @250 °C | | % v/v | -/65 | 38.4 | 38.4 |
| @350 °C | | % v/v | 85.0/- | 93.9 | 94.0 |
| I.B.P. | | °C | | 174.4 | 180.5 |
| F.B.P. | | °C | | 367.8 | 368.5 |
| PAH | EN 12916 | % m/m | -/8.0 | 4.4 | 4.4 |
| Manganese | EN 16576 | mg/L | -/2.0 | <0.02 | <0.02 |
| Density@ 15 °C | EN ISO 12185 | kg/m$^3$ | 820.0/845.0 | 834.5 | 834.5 |
| Cetane number | ISO 5165 | - | 51.0/- | 52.2 | 56.4 |
| Cetane Index | EN ISO 4264 | - | 46.0/- | 52.8 | 53.0 |
| Lubricity (wsd @ 60 °C) | ISO 12156-1 | µm | -/460 | 336 | 335 |
| Flash Point | EN ISO 2719 | °C | >55 | 68.0 | 67.5 |
| Carbon Residue | ISO 10370 | % m/m | -/0.300 | <0.001 | <0.001 |
| Ash | ISO 6245 | % m/m | -/0.001 | 0.001 | 0.002 |
| Oxidation stability | ISO 12205 | g/m$^3$ | -/25.0 | 2.5 | 1.4 |
| Oxidation stability | UNI EN 15751 | h | | >20 | >20 |
| Viscosity @ 40 °C | EN ISO 3104 | mm$^2$/s | 2.000/4.500 | 2.680 | 2.691 |
| Cloud Point | ASTM D7689 | °C | | −5.0 | −5.0 |

## 3. Results

In this section, the experimental results of vehicle fuel consumption and pollutant exhaust emissions will be discussed.

### 3.1. Influence of Additive Package on Fuel Consumption and $CO_2$ Emissions

Fuel consumption, as liters per 100 km, for all tested vehicles is shown in Figure 3. The bar plot presents average fuel consumption, measured over cold and warm WLTC, for all the vehicles fueled with and without additive. Vehicles are grouped by category (passenger cars, PCs and light commercial vehicles, LCVs). Standard deviations of measurements are represented by red lines. Moreover, for each vehicle, the graph represents the percentage difference in fuel consumption due to the use of fuel with additives.

The fuel with additives provides a fuel saving in all the experimental tested configurations. The percentage of fuel consumption reduction ranges between 1.2% for diesel Euro 6 PCs over warm WLTC and 8.1% for diesel Euro 4 PCs over warm WLTC. Since the measurement repeatability is very high (the coefficient of variation is around 1%), the percentage differences are statistically significant, with 95% of confidence.

As expected, among the Euro 4 vehicles the lowest reduction was found in the case of the spark-ignition vehicle (−2.2% for warm WLTC). As mentioned in Section 2.1, the vehicle is equipped with a PFI system, therefore, the main advantage of using a deposit control additive is related to the removal of deposits from intake valves. Larger benefits were achieved with the diesel Euro 4 PC. In this case, the deposit control additive acts on nozzle coking, reducing the carbonaceous deposits inside the holes, which limit the fuel rate and increase combustion variability. As an effect of clean-up, an improvement in fuel economy of up to 8.1% (warm WLTC) is achieved.

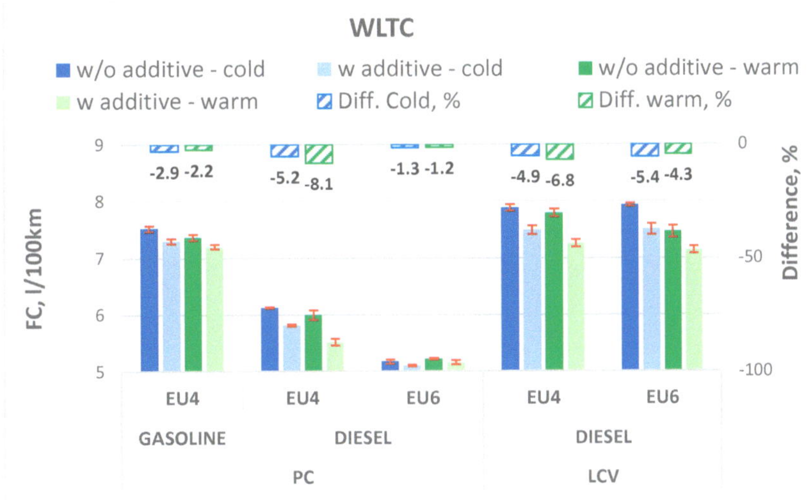


**Figure 3.** Fuel consumption of vehicle fleet.

The lowest reduction in fuel consumption was recorded for the Euro 6 passenger car. This result can be explained by the different ages of the two vehicles (Euro 4 PC registration 2003 and EU6 PC 2016) and the different maintenance records, despite the larger mileage accumulated by the Euro 6 PC. This hypothesis is confirmed by lower exhaust emissions and measurement variability, as apparent in the following section.

Further, the Euro 6 PC vehicle was used for covering long distance journeys, therefore, the mileage was accumulated mainly on highways; its driving, indeed, was characterized by high and constant driving speed and less time was spent in urban ambient areas, where the speed is lower and transient stages are frequent.

For both homologation classes of light commercial vehicles, a strong reduction in fuel consumption was found, ranging between 4.3% and 6.8%.

$CO_2$ emissions follow the same trend as fuel consumption (Figure 4), being the main product of fuel combustion. The use of additive packages, indeed, involves a reduction in $CO_2$ exhaust emissions, which is higher than 4% for all diesel vehicles with the exception of the Euro 6 passenger car. Its $CO_2$ emission reductions, in fact, are much more similar to gasoline vehicle ones (lower than 2%).

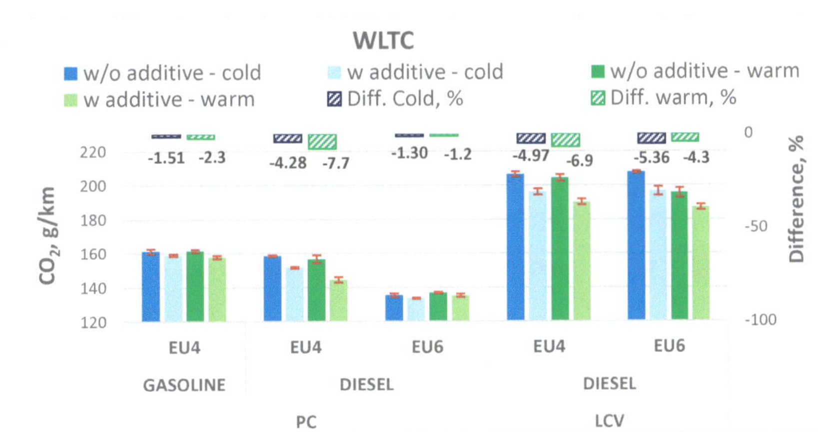


**Figure 4.** $CO_2$ emissions of vehicle fleets.

For a deeper understanding of additive effect on $CO_2$ emissions, it is interesting to analyze each single stage of the WLTC. With this aim, Figure 5 reports $CO_2$ emissions as a function of the average speed. The points on the curves correspond to the four driving cycle phases.

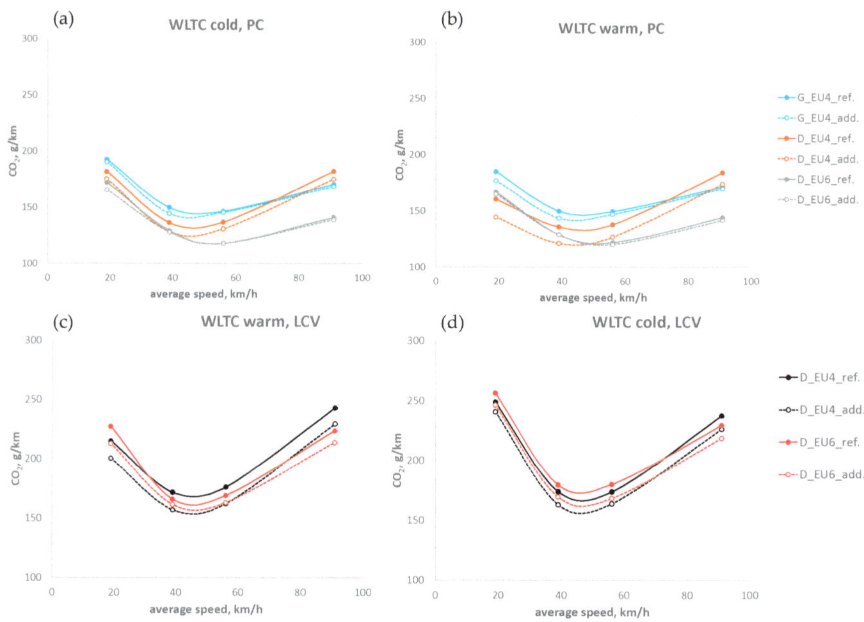


**Figure 5.** $CO_2$ exhaust emissions for different vehicle speeds, (**a**) PC over cold WLTC; (**b**) PC over warm WLTC; (**c**) LCV over cold WLTC and (**d**) LCV over warm WLTC.

$CO_2$ emission curves are well fitted by a 2nd order polynomial function for all the tested vehicles. Minimum $CO_2$ emissions are measured between 40 and 50 km/h. Figure 5a reports the trend for PCs during the cold WLTC. The action of additive fuels in terms of $CO_2$ emissions and fuel consumption reduction is dependent on the homologation class and engine type. For the Euro 4 SI vehicle, the reduction mainly occurs during the low-medium speed driving cycle ($CO_2$: −1.1% urban low speed phase and −3.5% rural medium speed phase and fuel consumption: −8.5% urban phase and −3.5% rural phase), in which the clean-up action carried out by the fuel with additives reduces the combustion instability linked to the high number of gear changes and low rotation speeds, and promotes a more complete combustion. Negligible differences were found in the main road high speed phase, where the engine operative conditions are close to its maximum efficiency (medium-high load and low-medium engine speed), while a reduction was recorded in the highway extra-high-speed phase, where consumption is higher due to an increase in the load and engine speed. As expected, the $CO_2$ emissions follow the same trend as the fuel consumption, with the exception of the urban phase, where the CO emissions were one magnitude larger than in the other phases. Using additives resulted in a CO reduction of about 61% (−9 g/km) in this phase, which explains the difference between CO2 percentage difference and fuel consumption.

Similar to the Euro 4 SI PC, for the Euro 6 diesel PC, additives provided a reduction in $CO_2$ mainly during the low-medium speed driving cycle, with a maximum improvement of 3.7%, corresponding to a reduction in fuel consumption of 3.8% in the urban phase.

A different behavior was found for the Euro 4 diesel PC: in this case, $CO_2$ emissions and fuel consumption were reduced over the whole speed range by using additive fuels. However, it can be noted that the urban phase is often characterized by the strongest fuel saving. This behavior can be attributed to the removal of deposits on nozzle-holes since injector fouling impacts on vehicle drivability over the whole range of operative conditions, with rough idling, stalling and reduced power output [19].

### *3.2. Influence of Additive Package on Exhaust Emissions*

The CO emissions of all the tested vehicles are presented in Figure 6. The bar plot refers to average emissions, measured over cold and warm WLTCs, for all the vehicles fueled with reference and additive fuels.

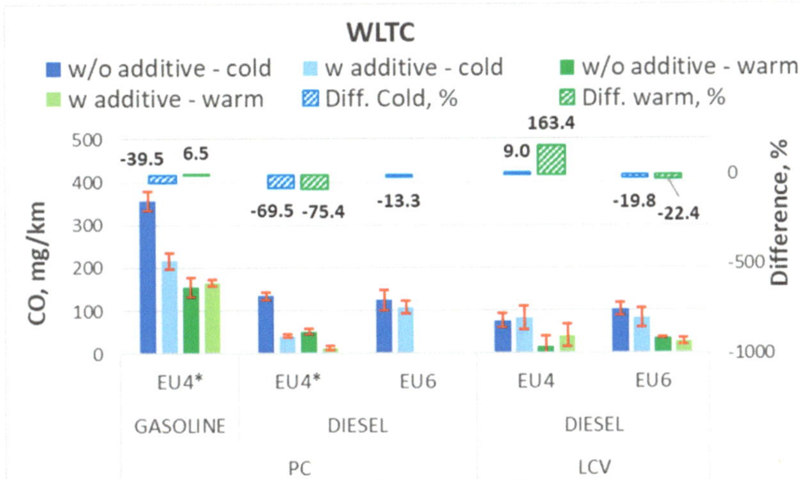


**Figure 6.** CO exhaust emissions (* data divided by 10).

In order to represent the average data of the whole fleet in the same plot, CO emissions of gasoline and diesel EU4 passenger cars were divided by 10.

As expected, the SI vehicle, fired with a rich-stoichiometric gasoline–air charge, produces carbon monoxide levels that are greater by an order of magnitude compared to diesel engines running with a diluted diesel–air charge. Switching from Euro 4 to Euro 6 PCs results in a further reduction of an order of magnitude. In the case of the warm WLTCs, Euro 6 PC emissions were close to zero for both fuels.

In general, additives provide a reduction in CO emissions, with a maximum benefit of −75.4% in the case of the Euro 4 diesel PC for the warm WLTC. An increase in CO was found in the case of the Euro 4 LCV even though the percentage differences were not statistically significant due to the low absolute CO emissions (lower than 83 mg/km), which were abundantly below the emission standards (500 mg/km for both Euro 4 and Euro 6).

The maximum CO emission was recorded in the case of the Euro 4 SI vehicle during the cold WLTC due to a different combustion mode compared to diesel. To better understand the effect of additives on CO emissions and, more generally, on combustion, in SI vehicles, it is useful to analyze the time trend air–fuel equivalence ratio $\lambda$. This trend is shown with the vehicle speed profile in Figure 7a. Figure 7b represents the instantaneous (solid line) and cumulative (dotted line) emissions of CO during the cold WLTC with and without additives.

As expected, in correspondence with the CO peaks, the air–gasoline mixture is enriched with values of $\lambda$ up to approximately 0.7 in the case of fueling without additives. It is worth noting that the engine control strategy provides a stoichiometric mixture content ($\lambda = 1$) throughout the whole engine operating map with the exception of the maximum load conditions, corresponding to the highest speed WLTC section. Rich mixture strategies are typically adopted also in the urban phase during cold-start conditions. Furthermore, during this cold phase the three-way catalyst is at temperatures lower than the activation temperature; therefore, its efficiency is low for the oxidation of CO and THCs and the reduction in NOx.

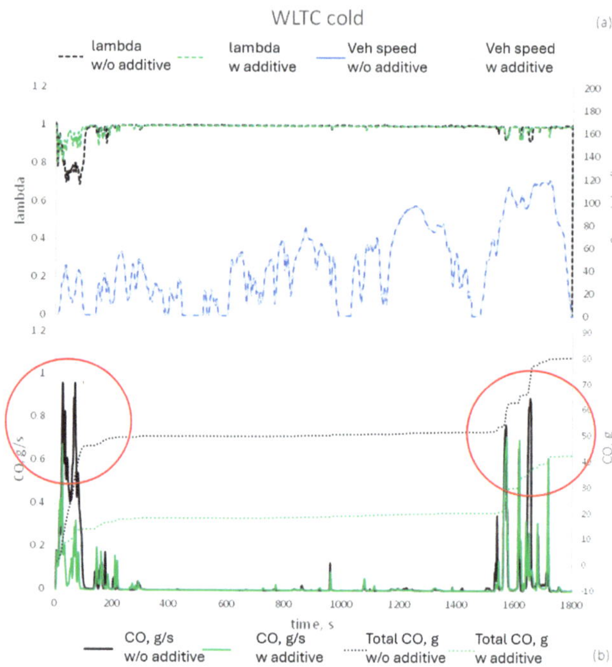


**Figure 7.** Instantaneous air–fuel equivalence ratio λ and vehicle speed (**a**) and instantaneous and cumulative CO emissions on the WLTC cold cycle with and without additives (**b**) for Euro 4 SI PC.

As is evident in Figure 7a, while in the maximum speed phase, the trend of the air–fuel equivalence ratio is substantially independent of the fuel, while in the urban phase a strong enrichment of the mixture was achieved in the case without additives. This difference can be explained by the cleaning effect of the additives, which allow a better dynamic response of the engine to acceleration and gear changes and, consequently, a more complete combustion with a reduction in CO emissions (Figure 7b) and specific consumption.

Figure 8 represents the instantaneous emissions of CO and vehicle speed profiles over the cold WLTC with and without additives for the Euro 4 LCV.

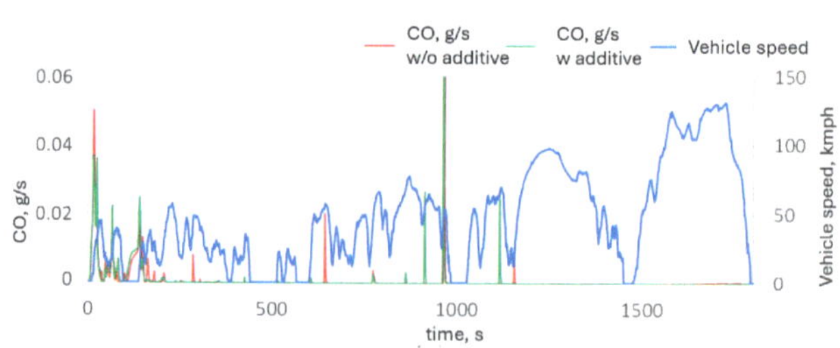


**Figure 8.** Instantaneous CO emissions and vehicle speed on the WLTC cold cycle with and without additives for Euro LCV.

As aforementioned, CO emissions are too low to appreciate statistically significant variations with the fuel switching. The main contribution to CO emissions comes from the urban phase of the WLTC, including the cold-start operation. In this stage, the oxidation catalyst has not yet reached the activation temperature (light-off).

The continuous profile of CO emissions shown in Figure 8 highlights that the light-off phase lasts for approximately 200 s from the start of the cycle. Once the catalyst has reached the activation temperature, CO emissions become zero except for the emission peaks observed during some rapid transient stages of the driving cycle, probably due to an enrichment of the air/fuel mixture in the internal combustion engine. These peaks are non-repeatable and independent of the type of fuel. It is worth noting a zero-emission profile during the highway phase.

The THC emissions of all tested vehicles are presented in Figure 9. The bar plot refers to average emissions, measured over cold and warm WLTCs for all the vehicles fueled with reference and additive fuels.

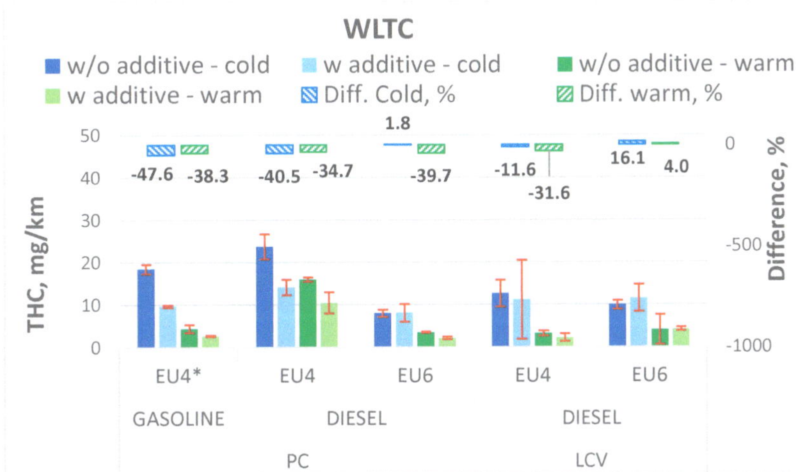


**Figure 9.** THC exhaust emissions (* data divided by 10).

In order to represent the average data of the whole fleet in the same plot, THC emissions of gasoline EU4 passenger cars were divided by 10.

Using additives results in THC emission reductions for all the tested vehicles with the exception of the Euro 6 LVC. Nevertheless, the low emission level (less than 11 mg/km) combined with the high emission variability do not to allow us to appreciate statistically significant differences. As expected, the maximum level of THC emissions was achieved by the SI Euro 4 PC during the cold WLTC (187 mg/km). The maximum efficiency of the additives, with a reduction of 47.6% in THCs, was found under the same operative conditions: the THC emissions dropped to 98 mg/km, falling in the Euro 4 limit of 100 mg/km on the NEDC cycle (homologation standard for this vehicle), and in the Euro 5 limit of 100 mg/km on the WLTC cycle. The warm WLTC provided lower THC emissions compared to the cold WLTC: during the cold start in the first phase of the driving cycle, when exhaust temperature is too low to activate the three-way catalyst, a high number of transients and lower combustion stability occur, producing the maximum level of emissions. The activation of the catalyst in the warm WLTC cuts down the amount of THCs at the exhaust.

The NOx emissions of all tested vehicles are shown in Figure 10. In agreement with the increase in fuel efficiency and the reduction in CO and THCs, a slight increase in NOx was found when vehicles were fueled with fuel and additives. The bar plot refers to average emissions, measured over cold and warm WLTCs, for all the vehicles fueled with reference and additive fuels. The maximum increase was found for the SI Euro 4 vehicle due to the improvement in combustion efficiency, which, in turn, resulted in an increase in combustion temperature combined with a reduction in air–fuel mixture enrichment, contributing to the combustion temperature being raised. On the other hand, it is worth point out that NOx emissions remained below Euro 4 emissions limits.

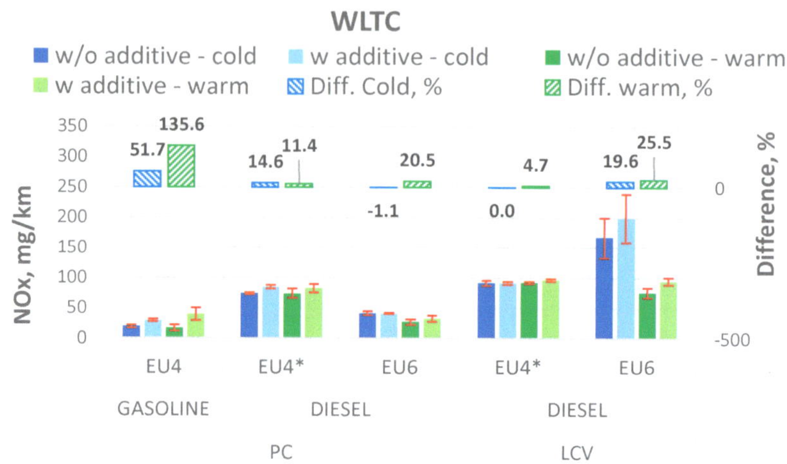


**Figure 10.** NOx exhaust emissions (* data divided by 10).

Diesel vehicles provided larger NOx emissions compared to gasoline vehicles except for the Euro 6 diesel PC, which produces similar emission levels, abundantly below the Euro 6 diesel standard limit, thanks to the action of the selective catalytic reduction (SCR) system. In this case, using additives resulted in a slight decrease in NOx emissions for the cold WLTC and an increase of 20.5% for the warm WLTC. However, both differences were out of the 95% confidence level of statistical analysis.

The Euro 6 LCV provided larger NOx emissions compared to the Euro 6 PC, especially during the cold WLTC, double that recorded in the case of the warm WLTC, indicating the poor efficiency of SCR under cold-start conditions. However, these values are well below those of the Euro 4 LCV, which emitted NOx levels greater by an order of magnitude compared to the Euro 6 LCV.

Figure 11 reports PM emissions of all tested vehicles that are presented in Figure 6. The bar plot refers to average emissions, measured over cold and warm WLTCs, for all the vehicles fueled with reference and additive fuels. In agreement with the increase in NOx emissions, a reduction in PM was achieved when an additive was used. A maximum reduction in PM emissions of 68.8% (from 60.7 mg/km w/o additives to 18.9 mg/km with additive) was recorded for the Euro 4 diesel PC during the warm WLTC.

The Euro 4 diesel vehicles' PM results were one magnitude larger than those for other vehicles: the difference, compared with other diesel vehicles, is due to the introduction of the diesel particulate filter (DPF) to meet Euro 5 and Euro 6 limits.

Euro 6 vehicles emitted PM levels strongly below the emission standard limits of 5 mg/km, with values ranging between 1.0 mg/km for LCVs, warm WLTC, w/o additives to 2.9 mg/km for PCs, warm WLTC, w/o additives. So, low values also implied high variability measurements and, hence, the variations induced by using additives were not significant.

As required by exhaust emissions standards, PN emissions were estimated for Euro 6 vehicles (Figure 12). In agreement with the PM results, PN emissions were abundantly below the Euro 6 limit of $6 \times 10^{11}$ particles/km: Euro 6 PC values were lower by two orders of magnitude compared to the limit, while LCV levels were lower by an order of magnitude. The average PN emissions were affected by a notable variability due to the low concentrations. Therefore, the differences induced by using additives were not statistically significant.

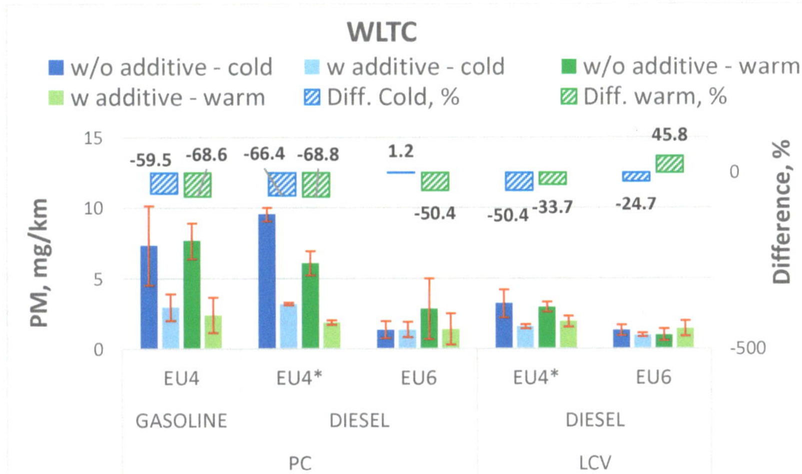


Figure 11. PM exhaust emissions (* data divided by 10).

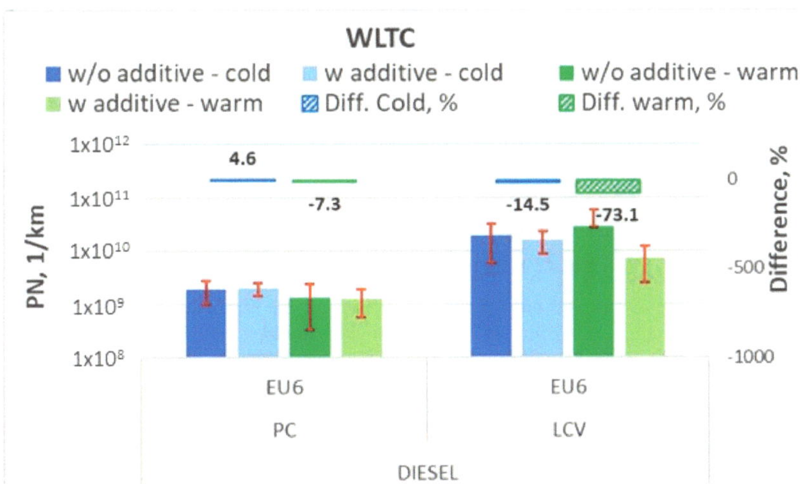


Figure 12. PN exhaust emissions.

## 4. Conclusions

An experimental activity was conducted to assess the influence of two performance packages (diesel and gasoline) on the exhaust emissions and fuel consumption of five vehicles set to be as representative as possible of circulating Italian passenger cars and light commercial vehicles fleets. Based on the literature datasets, three Euro 4 vehicles were selected with a mileage representative of each single vehicle class: two passenger cars (one spark ignition and one diesel) and a light commercial vehicle. Further, two diesel Euro 6 vehicles, a passenger car and an LCV were tested to investigate the effect of fuel additives on the combustion of vehicles compliant with the current Euro 6 homologation regulation.

Exhaust emissions and fuel consumption were experimentally estimated on a chassis dynamometer over WLTCs in a climate-controlled laboratory. Each vehicle was preliminarily tested running with base fuel, then a 3000 km clean-up stage was performed using the additive package; finally, WLTC tests were repeated.

The main findings can be summarized as follows:

- Performance packages provide a reduction in fuel consumption for all tested vehicles between 1.2% (diesel Euro 6 passenger car) and 8.1% (diesel Euro 4 passenger car). Diesel light commercial vehicles and Euro 4 passenger cars were more sensitive to the influence of additives. This behavior is attributed to the deposit removal effect of additives reducing injector-nozzle fouling. The benefits over Euro 6 diesel passenger cars were limited, indicating a good level of maintenance of engine parts, as confirmed by close-to-zero exhaust emissions of CO, THCs and PM.
- A gasoline vehicle was equipped with a PFI system, therefore, the main advantage of using a performance additive is related to the removal of deposits from intake valves and a reduction in friction loss.
- Using additives in Euro 4 passenger cars provided a strong reduction in CO, THCs and PM. In agreement with PM reduction, an increase in NOx was found.
- When used in spark-ignition engines, additives promoted a more stable combustion during the transient stages, especially in cold-start conditions, as demonstrated by the reduction in fuel charge enrichment in the urban phase compared to the case without additives.
- For Euro 6 vehicles, the exhaust emissions measurement showed a relatively high variability due to extremely low values recorded for all monitored species. CO and THC emissions are typically low for this class of engine, while NOx and PM/PN emissions were very low due to after-treatment systems installed on Euro 6 vehicles. An exception was found for Euro 6 LCVs under cold-start conditions due to the late activation of the SCR system, resulting in high levels of NOx emissions.

**Author Contributions:** Conceptualization, M.A.C., A.B. and L.M.; methodology, M.A.C., A.B. and L.M.; formal analysis, M.A.C. and L.M.; investigation, M.A.C. and L.M.; writing—original draft preparation, M.A.C., A.B. and L.M.; writing—review and editing, M.A.C., A.B. and L.M. All authors have read and agreed to the published version of the manuscript.

**Funding:** This research received no external funding.

**Data Availability Statement:** Data is contained within the article.

**Acknowledgments:** Authors would like to thank Rocco Giuzio and Alfredo Mazzei for their valuable support in the experimental activity.

**Conflicts of Interest:** Author Alessandra Berra, who provided technical advisor on the performance packages benefits and application, was employed by the company Dorf Ketal B.V. The remaining authors declare that the research was conducted in the absence of any commercial or financial relationships that could be construed as a potential conflict of interest.

## References

1. EEA—European Environment Agency. *Transport and Environment Report 2022*; EEA Report No. 07/2022; European Environment Agency: Copenhagen, Denmark, 2022.
2. Eurostat. *Passenger Cars in the EU*; Eurostat: Luxembourg, 2023; ISSN 2443-8219.
3. EC—European Commission, 2019. The European Green Deal. Brussels, 11.12.2019. COM(2019) 640 Final. Available online: https://eur-lex.europa.eu/legal-content/EN/TXT/PDF/?uri=CELEX:52019DC0640 (accessed on 1 March 2024).
4. ACEA, 2024. Vehicles on European Roads. Available online: https://www.acea.auto/files/ACEA-Report-Vehicles-on-European-roads-.pdf (accessed on 1 March 2024).
5. ACI, 2023. Autoritratto 2022. Available online: https://www.aci.it/laci/studi-e-ricerche/dati-e-statistiche/autoritratto/autoritratto-2022.html (accessed on 1 March 2024).
6. Siddartha, G.; Ramakrishna, S.; Kujur, P.; Rao, A.; Dalela, N.; Yadav, A.; Sharma, A. Effect of fuel additives on internal combustion engine performance and emissions. *Mater. Today* **2022**, *63*, A9–A14.
7. Zhu, Q.; Zong, Y.; Tan, Y.R.; Lyu, J.Y.; Pan, J.; Zhou, X.; Liu, H.; He, S.; Chen, W.; Yu, W.; et al. Comparative analysis of PODE3 and PODE4 fuel additives for emission reduction and soot characteristics in compression ignition engines. *Energy* **2024**, *286*, 129498. [CrossRef]
8. Ershov, M.A.; Savelenko, V.D.; Makhova, U.A.; Kapustin, V.M.; Potanin, D.A.; Habibullin, I.F.; Lazarev, V.E.; Abdellatief, T.M.M.; Abdelkareem, M.A.; Olabi, A.G. New insights on introducing modern multifunctional additives into motor gasoline. *Sci. Total Environ.* **2022**, *808*, 152034. [CrossRef] [PubMed]

9. Awad, O.; Xiao, M.; Kamil, M.; Zhou, B.; Ali, O.M.; Shuai, S. A Review of the E ects of Gasoline Detergent Additives on the Formation of Combustion Chamber Deposits of Gasoline Direct Injection Engines. *SAE Int. J. Fuels Lubr.* **2021**, *14*, 13–25. [CrossRef] [PubMed]
10. Frauscher, M.; Agocs, A.; Wopelka, T.; Ristic, A.; Ronai, B.; Holub, F.; Payer, W. Improving sustainability by enhanced engine component lifetime through friction modifier additives in fuels. *Fuel* **2024**, *358*, 130102. [CrossRef]
11. Bennett, J. Advanced fuel additives for modern internal combustion engines. In *Alternative Fuels and Advanced Vehicle Technologies for Improved Environmental Performance: Towards Zero Carbon Transportation*; Folkson, R., Ed.; Woodhead Publishing: Sawston, UK, 2014; pp. 165–194.
12. Ashok, B.; Nanthagopal, K.; Chyuan, O.H.; Le, P.T.K.; Khanolkar, K.; Raje, N.; Raj, A.; Karthickeyan, V.; Tamilvanan, A. Multi-functional fuel additive as a combustion catalyst for diesel and biodiesel in CI engine characteristics. *Fuel* **2020**, *278*, 118250. [CrossRef]
13. Di Girolamo, M.; Brianti, M.; Marchionna, M. Octane enhancers. In *Handbook of Fuels: Energy Sources for Transportation*; WILEY-VCH Verlag GmbH & Co. KgaA: Weinheim, Germany, 2021; pp. 403–430.
14. Herreros, J.M.; Oliva, F.; Zeraati-Rezaei, S.; Tsolakis, A.; Delgado, J. Effects of high octane additivated gasoline fuel on Three Way Catalysts performance under an accelerated catalyst ageing procedure. *Fuel* **2022**, *312*, 122970. [CrossRef]
15. Bennett, J. Additives for spark ignition and compression ignition engine fuels. *Proc. Inst. Mech. Eng. Part D J. Automob. Eng.* **2018**, *232*, 148–158. [CrossRef]
16. Burke, R.D.; Madamedon, M.; Williams, R. Newly identified effects of injector nozzle fouling in diesel engines. *Fuel* **2020**, *278*, 118336. [CrossRef]
17. Zare, A.; Bodisco, T.A.; Nabi, M.N.; Hossain, F.M.; Ristovski, Z.D.; Brown, R.J. A comparative investigation into cold-start and hot-start operation of diesel engine performance with oxygenated fuels during transient and steady-state operation. *Fuel* **2018**, *228*, 390–404. [CrossRef]
18. Johnson, L.A. Analysis of variance of parameter estimates: *F* tests and *t* tests. *Anal. Biochem.* **1992**, *206*, 195–201. [CrossRef] [PubMed]
19. Edney, M.K.; Barker, J.; Reid, J.; Scurr, D.J.; Snape, C.E. Recent advances in the analysis of GDI and diesel fuel injector deposits. *Fuel* **2020**, *272*, 117682. [CrossRef]

**Disclaimer/Publisher's Note:** The statements, opinions and data contained in all publications are solely those of the individual author(s) and contributor(s) and not of MDPI and/or the editor(s). MDPI and/or the editor(s) disclaim responsibility for any injury to people or property resulting from any ideas, methods, instructions or products referred to in the content.

*energies*

MDPI

*Article*

# A CFD Modelling Approach for the Operation Analysis of an Exhaust Backpressure Valve Used in a Euro 6 Diesel Engine

**Francisco J. Martos [1], José A. Soriano [2], Andrei Braic [2], Pablo Fernández-Yáñez [2] and Octavio Armas [2,*]**

[1] Escuela de Ingenierías Industriales, Universidad de Málaga, c/Doctor Ortiz Ramos, s/n, 29071 Málaga, Spain

[2] Escuela de Ingeniería Industrial y Aeroespacial, Campus de Excelencia Internacional en Energía y Medioambiente, Universidad de Castilla-La Mancha, Av. Carlos III, s/n, 45071 Toledo, Spain

* Correspondence: octavio.armas@uclm.es

**Abstract:** Harvesting residual thermal energy from exhaust gases with thermoelectric generators is one of the paths that are currently being explored to achieve more sustainable and environmentally friendly means of transport. In some cases, thermoelectric generators are installed in a by-pass configuration to regulate the mass flow entering the thermoelectric generator. Some manufacturers are using throttle valves with electromechanical actuators and electronic control in the exhaust pipe to improve techniques for active control of pollutant emissions in reciprocating internal combustion engines, such as the exhaust gas recirculation. The above-mentioned circumstances have motivated the approach of this work: computational fluid dynamics (CFD) modelling of the operation of a throttle valve used for establishing adequate exhaust backpressure conditions to achieve the low pressure exhaust gas recirculation in Euro 6 engines. The aim of this model is to understand the flow control process with these types of valves in order to incorporate them in an exhaust system that will include two thermoelectric generators used to convert residual thermal energy into electrical energy. This work presents a computational model of the flow through the throttle valve under different temperatures and mass flow rates of the exhaust gas with different closing positions. For all cases, the values of the pressure drop were obtained. In all cases studied, the level of agreement between the modelled and experimental results exceeds 90%. The developed model has helped to propose a correlation to estimate the mass flow rate of exhaust gas from easily measurable quantities.

**Keywords:** compression ignition engines; simulation; back pressure valve; performance

**Citation:** Martos, F.J.; Soriano, J.A.; Braic, A.; Fernández-Yáñez, P.; Armas, O. A CFD Modelling Approach for the Operation Analysis of an Exhaust Backpressure Valve Used in a Euro 6 Diesel Engine. *Energies* **2023**, *16*, 4112. https://doi.org/10.3390/en16104112

Academic Editors: Constantine D. Rakopoulos, Davide Lanni and Enzo Galloni

Received: 29 March 2023
Revised: 3 May 2023
Accepted: 11 May 2023
Published: 16 May 2023

**Copyright:** © 2023 by the authors. Licensee MDPI, Basel, Switzerland. This article is an open access article distributed under the terms and conditions of the Creative Commons Attribution (CC BY) license (https://creativecommons.org/licenses/by/4.0/).

## 1. Introduction

The use of valves in reciprocating internal combustion engines, both spark and compression ignition engines, was a very common solution for mass flow regulation. Most typical examples are the use of throttle valves in carburetors [1–3] or in exhaust gas recirculation (EGR), for modifying the exhaust gas vorticity at the entry to the inlet manifold [3] or, during the last years, for regulating the back pressure in the exhaust pipe for controlling the high- and low-pressure EGR [4]. In this case, the authors proposed to change between high- and low-pressure EGR configuration depending on the engine operation for increasing the potential of use of both systems. A similar work is presented in [5], but in this case, the authors concluded that the angle of electronic-variable valves has impact on the back pressure in exhaust systems.

Other types of valves such as spool valves [6] or angle-seat valves [7] have also been studied and characterized for gas flow control. X. Pan et al. [6] studied the discharge coefficients of a servo-spool valve under laminar and turbulent flows. The authors established a mathematical model for the orifice flow by means of computational fluid dynamics (CFD) for solving the discharge coefficient and Reynolds number. In [7], the authors proposed a method for characterizing discharge and resistance coefficients of air discharged from a reservoir. In this work, the authors used as a test valve an angle-seat valve with different opening sections. Gülmez and Özmen [8] presented a sensitivity analysis of the effect of a backpressure valve installed in the exhaust pipe of a diesel engine. They determined the level of the negative effect of the backpressure valve on volumetric efficiency of the engine among other parameters.

Harvesting residual thermal energy from exhaust gases is one of the paths that are currently being explored to achieve more sustainable and environmentally friendly means of transport [8,9]. In a previous work [10], the authors of this paper experimentally studied the effect of the installation in the exhaust pipe of a thermoelectric generator on the pumping losses of a Euro 3 diesel engine.

These facts have motivated the approach of this work: the thermo-fluid dynamic modelling of the operation of a current throttle valve used for establishing adequate exhaust backpressure conditions to achieve low-pressure exhaust gas recirculation in Euro 6 engines. The aim of this model is to understand the flow control process with these types of valves to incorporate them in an exhaust system that will include two thermoelectric generators to convert residual thermal energy into electrical energy as energy-harvesting technology. This work presents a computational fluid dynamic model that calculates the exhaust gas flow around of the back pressure throttle valve under different temperatures and mass flow rates of the exhaust gas with different closing positions. For all cases, the values of the pressure drop were obtained. In all cases studied, the level of agreement between the modelled and experimental results exceeds 90%.

The methodology, experimental facilities and techniques used are presented in Section 2. Section 3 describes the proposed 3D model of the backpressure valve and the fluid-mechanical model used. The experimental results of the tests, the results of the proposed model and its validation are developed in Section 4, while conclusions are presented in Section 5.

## 2. Experimental Installation and Methodology

Figure 1 shows different views of the experimental installation used in this work. As a gas generator, an early Euro 6 Diesel automotive engine with R9M code, mounted in NISSAN Xtrail vehicles, was used.

(a) (b)

**Figure 1.** (**a**) View of the instrumented engine exhaust system. (**b**) Front view of the engine test bench.

The main characteristics of the engine tested are: 1.6 L, 4-cylinder, 4-stroke, with both hot flow & high-pressure and cooled flow and low-pressure exhaust gas recirculation

(EGR). The engine is also equipped with a variable geometry turbocharger with an intercooler and a common-rail injection system with fuel multi-injection. The exhaust system includes a diesel oxidation catalyst and a diesel particle filter as aftertreatment devices. The main values of the engine output are maximum-rated power 96 kW (at 4.000 $min^{-1}$) and maximum-rated torque 320 Nm (at 1750 $min^{-1}$). The engine was coupled to an AVL asynchronous electric machine working as a dynamometer. All the tests were carried out by means of the $\alpha / n$ control mode, through the AVL Emcon 400 control system. Fuel consumption was conditioned and measured by means of an AVL FuelExact 740. Thermodynamic diagnosis was carried out using an AVL IndiMicro 602 (which includes an in-cylinder pressure sensor GH14P and its glow plug adaptor AG03) and an AVL IndiCOM2013(2.5) system for combustion analysis. For registering both the engine speed and the piston position, both trigger and crankshaft angle pulses were provided by an AVL model 365C optical angle encoder coupled to the engine crankshaft.

Figure 2 shows a view of the exhaust back pressure valve (Pierburg, model PBT GF-30) under study in this work. This valve was originally used for the operation of the low-pressure and cooled EGR of the engine.

**Figure 2.** View of the exhaust back pressure valve.

Although the valve is normally electronically controlled by the electronic control unit (ECU) of the engine, for this work, the valve was independently controlled. Valve position was controlled by means of an electronic module designed for that purpose. It consists of a L298N DC motor driver, a MCP3208 high-precision analog-to-digital converter and an Atmega328p microcontroller. A block diagram of the designed electronic module is presented in Figure 3.

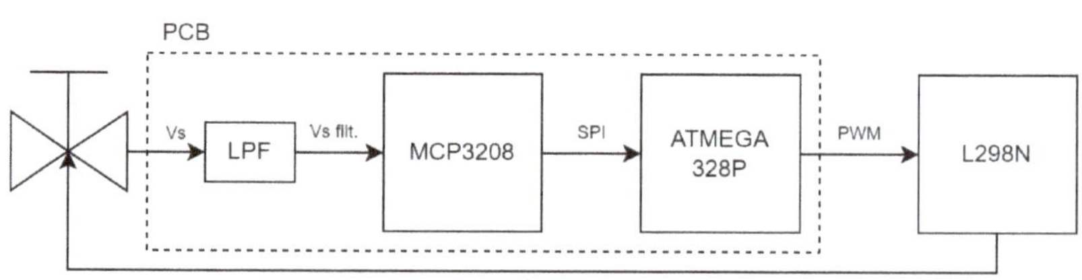


**Figure 3.** Scheme of the connection of the control module.

The regulator designed for the valve control is presented in Figure 4. This regulator was digitally implemented in the microcontroller as the sum of two different control algorithms: a sliding control to compensate for the force generated by the spring that has the valve on the shaft and a PID (proportional integral derivative) modified to the needs of the plant.

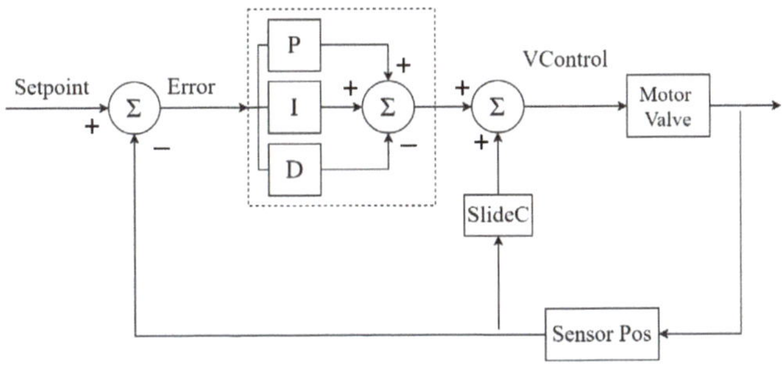


**Figure 4.** Block diagram of the regulator used for the valve control.

By complementing these algorithms with analog and digital filters, it is possible to position the butterfly with an error of only 0.7 degrees. The control is capable of effectively compensating for the disturbances generated by the thrust of the gas as well as eliminating electromagnetic interference from the environment.

Experimental tests have helped to understand the dependence between the back-pressure valve closing angle, engine behaviour and exhaust gas flow. However, the main objective of the experimental data was to validate the proposed model. From the results of all simulations, a correlation was proposed with which the exhaust gas mass flow can be determined as a function of easily measurable engine variables. Being able to control the mass flow rate through the valve is essential in order to optimise the operating conditions of a thermoelectric generator installed in the exhaust pipe of an engine.

## 3. Model Description

### *3.1. Back Pressure Valve*

In Figure 5a,b, it is shown how the backpressure valve is not symmetrical. It is convex on the upstream flow side and concave on the downstream flow side. Additionally, it is not completely circular. The diameter of the valve at its central part is greater than the inner diameter of the pipe, so that when the valve is completely closed, touching the inside of the pipe, the angle between the plane of the valve and the axial axis of the pipe is approximately 70 degrees. The 3D model of the valve is shown in Figure 5c,d for the upstream flow side and downstream flow side, respectively.

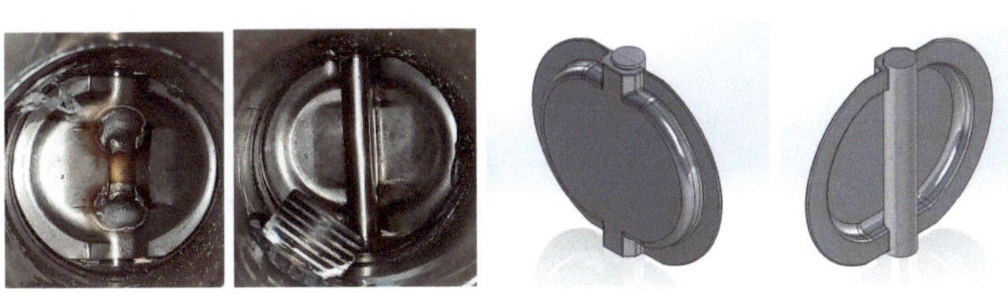

(a) (b) (c) (d)

**Figure 5.** Physical model of the valve: (**a**) upstream side of the valve, (**b**) downstream side of the valve, (**c**) upstream side 3D model, and (**d**) downstream side 3D model.

Six models were created, one for each valve opening angle. As shown in Figure 6, the simulated valve openings range from 0 to 50 degrees in increments of 10 degrees. Figure 6 shows the isometric view of the six developed models.

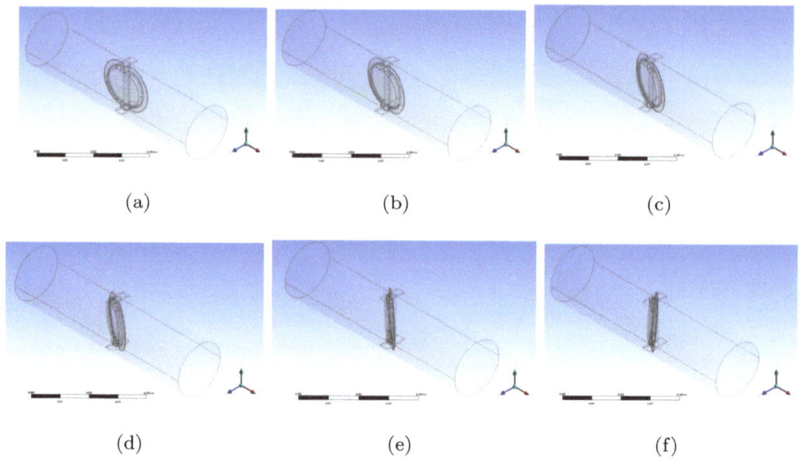


**Figure 6.** Three-dimensional model for each angle: (**a**) 0, (**b**) 10, (**c**) 20, (**d**) 30, (**e**) 40, and (**f**) 50 degrees.

The meshing of any of the models in Figure 6 was carried out taking into account the interaction of the fluid with the walls of the pipe and the valve using both hydrodynamic and thermal boundary layers. The inflation technique was used so that the maximum $y^+$ throughout the domain is less than 5, as shown in Figure 7. In this way, the shear stresses due to viscosity within the boundary layer are as close to reality as possible [11]. Four-node linear tetrahedrons and six-node linear wedges were used in the meshing. A mesh convergence study determined that the number of three-dimensional elements in the resulting mesh was between 3,400,000 and 4,000,000, depending on the valve opening. The quality of the mesh measured by the orthogonal quality parameter was around 0.15 in all cases, as a minimum value. As an example, Figure 7 shows the meshing of the model shown in Figure 6d.

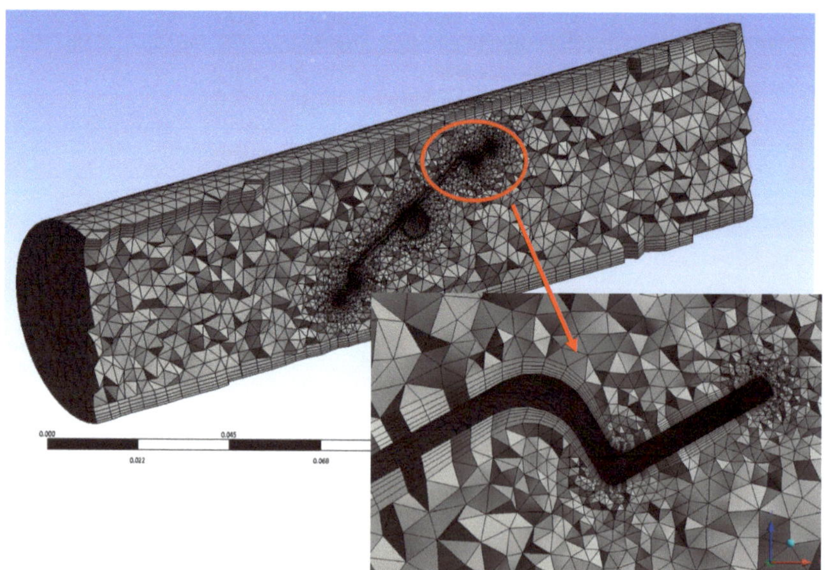


**Figure 7.** Axial–vertical cross-section of a 3D meshed model of a valve rotated 30 degrees from fully open.

It was considered that the radial temperature difference in the tube is negligible, and since the modelled valve is installed in a stationary engine test bench, heat dissipation outside the tube by natural convection, using the Churchill–Chu correlation, and radiation were considered. The inlet and outlet boundary conditions are of known mass flow rate and pressure, respectively. The turbulence boundary conditions are turbulent intensity equal to 5% and turbulent viscosity ratio equal to 10. The pressure–velocity coupling with the SIMPLE scheme was used as the solution method. Second-order spatial discretisation was used for pressure, density, momentum and energy, while first-order discretization was used for turbulent kinetic energy and turbulent dissipation rate. A convergence criterion of $10^{-3}$ was used for all variables, except for energy, which was set to $10^{-6}$.

For each valve opening angle, seven gas inlet temperatures to the valve were simulated, ranging from 200 to 500 °C in steps of 50 °C, and nine mass flow rates, ranging from 20 to 100 g/s in steps of 10 g/s. Therefore, a total of 6 × 7 × 9 = 378 simulations were carried out. The geometries were created in SolidWorks®, the meshings were conducted in Ansys® Meshing® 2021 R2, and the simulations in Ansys® Fluent® 2021 R2.

The computer used to tune the models and perform sensitivity studies for each mesh was a Workstation Dell Precision 5820 with 28-core Intel Xeon W-2175, 2.5 GHz. The supercomputer used to perform the 378 simulations was Picasso, belonging to the University of Malaga, with 30,616 cores and 156 TB of RAM.

### 3.2. Turbulence Model

Mass flow rate was between 20 and 100 g/s. The average gas temperature was between 200 and 500 °C; thus, the dynamic viscosity was between $2.60 \times 10^{-5}$ and $3.62 \times 10^{-5}$ Pa·s.

Therefore, the average Reynolds number was between 13,500 and 94,200, indicating that the flow was predominantly turbulent. Turbulent flows can be simulated through computational fluid dynamic modelling using three different approaches: DNS (direct numerical simulation) [12], LES (large eddy simulation) [13,14], and RANS (Reynolds-averaged Navier–Stokes) [15]. The citation order corresponds to the decreasing calculation precision and increasing computational time cost.

As the bibliography review showed, most used methods for modelling the internal flow inside the exhaust pipe were Reynolds-averaged Navier–Stokes. Among the works which used the RANS method, most of the consulted works modelled the turbulence by means of $k - \varepsilon$ equations [16–19]. This is justified by the reasonable balance between the computational time cost and the precision of the calculation method.

The present work is based on renormalisation group (RNG) $k - \varepsilon$ turbulence modelling. This mathematical technique is used for solving the Navier–Stokes equations (Equations (1) and (2)) applied to each fluid. An exhaustive description of the RNG theory and its application to turbulent flows can be found in [20].

$$\frac{\partial}{\partial t}(\rho k) + \frac{\partial}{\partial x_i}(\rho k u_i) = \frac{\partial}{\partial x_j}\left(\alpha_k \mu_{eff} \frac{\partial k}{\partial x_j}\right) + G_k + G_{bu} - \rho\varepsilon - Y_M \tag{1}$$

$$\begin{gathered}\frac{\partial}{\partial t}(\rho \varepsilon) + \frac{\partial}{\partial x_i}(\rho \varepsilon u_i) = \frac{\partial}{\partial x_j}\left(\alpha_\varepsilon \mu_{eff} \frac{\partial \varepsilon}{\partial x_j}\right) + \\ C_{1\varepsilon}\frac{\varepsilon}{k}(G_k + C_{3\varepsilon}G_{bu}) - C_{2\varepsilon}\rho\frac{\varepsilon^2}{k} - R_\varepsilon\end{gathered} \tag{2}$$

In Equations (1) and (2), $G_k$ is the turbulence kinetic energy generation caused by the mean velocity gradients, $G_{bu}$ is the generation of turbulence kinetic energy due to the floatability, while $Y_M$ is the contribution of the dilatation variability in compressible turbulence to the overall dissipation rate. The terms $\alpha_k$ and $\alpha_\varepsilon$ are the inverse of the Prandtl numbers for $k$ and $\varepsilon$, respectively.

For modelling thermal and hydrodynamic boundary layers, an enhanced wall treatment was used in Equation (2). Both pressure gradients and thermal effects were taken into account inside the boundary layers.

Enhanced wall treatment for the ε-equation is a near-wall modelling method that combines a two-layer model with so-called enhanced wall functions. If the near-wall mesh is fine enough to be able to resolve the viscous sublayer (typically with the first near-wall node placed at $y^+ \approx 1$), then the enhanced wall treatment will be identical to the traditional two-layer zonal model. However, the restriction that the near-wall mesh must be sufficiently fine everywhere might impose too large a computational requirement. Ideally, one would like to have a near-wall formulation that can be used with coarse meshes (usually referred to as wall-function meshes) as well as fine meshes (low-Reynolds number meshes). In addition, excessive error should not be incurred for the intermediate meshes where the first near-wall node is placed neither in the fully turbulent region, where the wall functions are suitable, nor in the direct vicinity of the wall, where the low-Reynolds number approach is adequate.

According to reference [21], the thermophysical properties of the exhaust gas are similar to those of air as a real gas at the same temperature and pressure. The majority of the composition of both gases is nitrogen. Both the density and viscosity of air vary with local pressure and temperature. The density of air was modelled using the Redlich–Kwong equation of state and dynamic viscosity using the Sutherland equation.

## 4. Results and Discussion

### 4.1. Test Results

#### 4.1.1. Engine Operating Conditions

In Table 1, the measured average values for the four engine operating modes with the backpressure valve fully open are shown. $\dot{m}$ is the mass flow rate of exhaust gases. $T_{up}$ and $p_{up}$ are the temperature and pressure of the exhaust gases before the backpressure valve. As shown in Figures 8 and 9, for each of the four engine operating modes, the backpressure valve was progressively closed in steps of 10% up to 50%, and from that point onwards, in steps of 5%. For operating mode A, the backpressure valve was closed up to 80%, for mode B up to 75%, for mode C up to 60%, and for mode D up to 50%. This sequence was chosen to avoid affecting the normal behaviour of the engine, so that the extra pressure drop caused by the valve closure was always less than 160 mbar.

**Table 1.** Engine operating conditions.

| **Operating Mode:** | **A** | **B** | **C** | **D** |
|---|---|---|---|---|
| Symbol in Figures | ○ | △ | □ | ◊ |
| Engine speed (min$^{-1}$) | 1500 | 1500 | 2000 | 2500 |
| Torque (Nm) | 50 | 100 | 150 | 200 |
| BMEP (bar) | 3.98 | 8.07 | 11.77 | 16.50 |
| $\dot{m}$ (g/s) | 19.47 | 26.88 | 47.75 | 73.78 |
| $T_{up}$ (°C) | 241.12 | 309.83 | 353.46 | 432.42 |
| $p_{up}$ (mbar) | 3.38 | 6.15 | 22.50 | 52.81 |

#### 4.1.2. Valve Closing

As shown in Figure 8, the mean instantaneous pressure inside the combustion chamber was experimentally measured for each engine operating mode and backpressure valve closure. A total of 1440 points per thermodynamic cycle of in-cylinder pressure were recorded (each pressure value was recorded every 0.5 degree of crank angle). For all operating modes, 20 thermodynamic cycles were recorded and used to calculate the mean in-cylinder pressure.

For all engine operating modes, except for the high-load mode (mode D), a slightly decreasing trend is observed between the measured instantaneous pressure inside the combustion chamber and the degree of closure of the backpressure valve, as shown in

Figure 8a–c. This trend is consistent with that reported in [22]. This behaviour is consistent with the BMEP obtained for each mode (A, B, and C), as shown in Figure 9a. It can be concluded that, since the engine is turbocharged, a lower mass flow rate of exhaust gases through the turbine, as shown in Figure 9b, implies lower power production in the turbine and therefore lower power in the cylinder-boosting compressor, which slightly decreases the mean instantaneous pressure inside the combustion chamber.

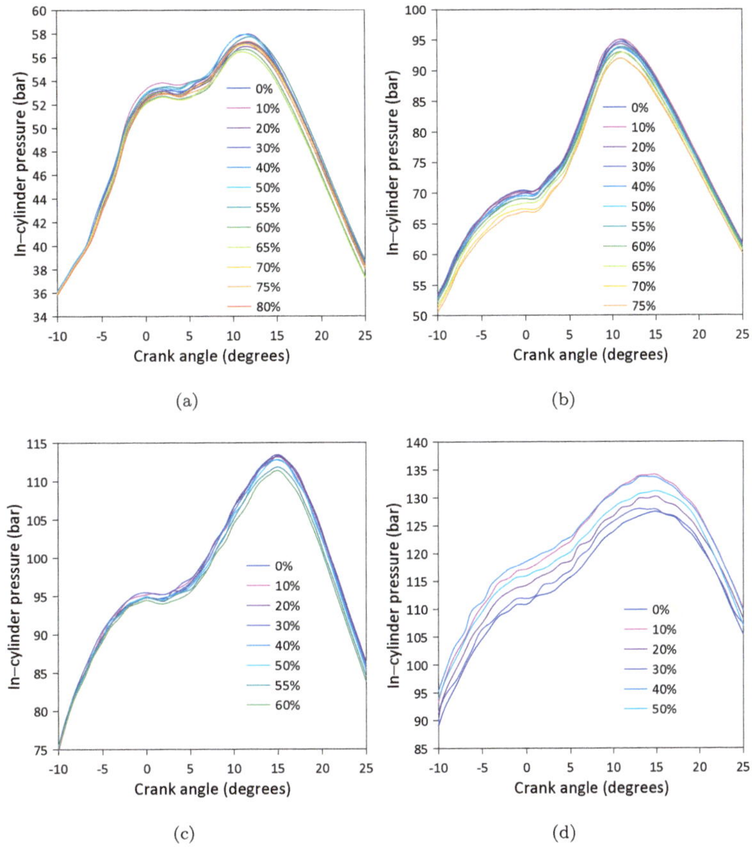


**Figure 8.** In−cylinder pressure vs. crank angle. (**a**) Mode A, (**b**) Mode B, (**c**) Mode C and (**d**) Mode D.

From Figure 8d, no clear trend can be observed between the instantaneous mean pressure inside the cylinder and the closure of the back pressure valve for the high-load operating mode (Mode D). This may be because, regardless of the decrease in the mass flow rate of exhaust gases circulating through the turbine, as shown in Figure 9b, the energy of the exhaust gases is more than enough to power the turbine and, consequently, the compressor that supercharges the engine.

From Figure 9d, it can be observed that with the increase in the degree of closure of the backpressure valve and therefore the decrease in the mass flow rate, ash shown in Figure 9b, the average temperatures upstream of the backpressure valve increase. This may be due to the decrease in the circulation velocity of the exhaust gas, which leads to a lower convective heat transfer coefficient within the exhaust pipe and consequently to a lower rate of dissipation of heat power.

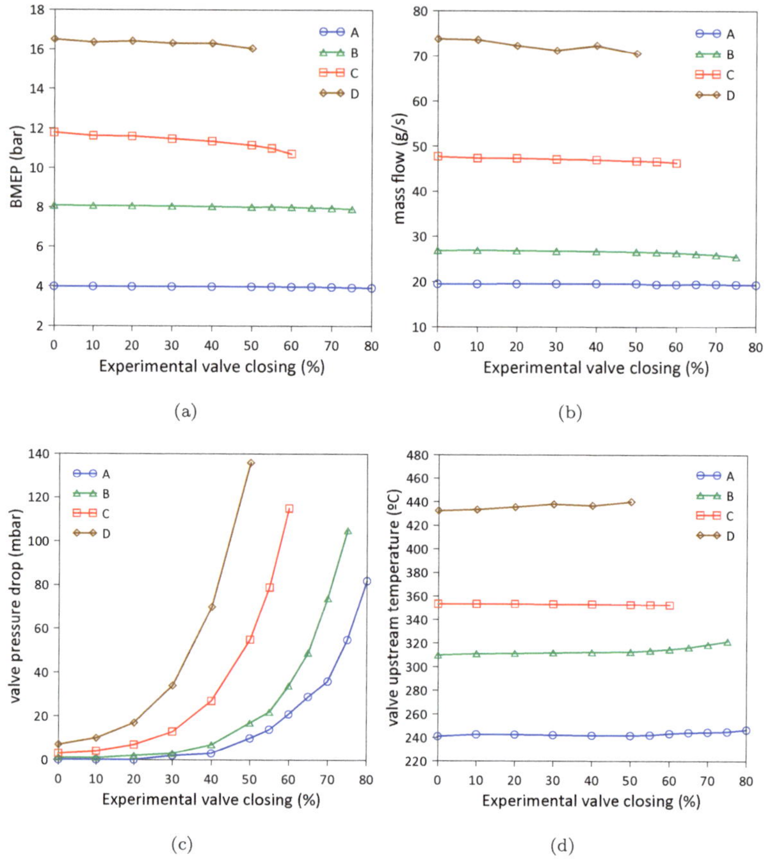


**Figure 9.** Average experimental data (**a**) BMEP, (**b**) mass flow, (**c**) valve pressure drop and (**d**) valve upstream temperature vs. valve closing.

Figure 9c shows how the increase in the degree of closure of the back pressure valve leads to a potential increase in the pressure drop of the exhaust gas as it passes through the back pressure valve [7].

### *4.2. Model Results*

As an example, Figure 10 shows the velocity, temperature, and pressure results in the axial plane, and Figure 11 shows the streamlines of a simulation with the valve rotated 30 degrees from fully open, a mass flow rate of 60 g/s, and an upstream exhaust gas temperature of 400 °C. The results obtained are consistent with [22]. It can be observed that the acceleration of the gas as it passes through the valve implies a significant decrease in temperature, as a result of the transfer of thermal energy to kinetic energy, and therefore an increase in pressure drop.

As discussed in Section 3.1, a total of 378 simulations were carried out, with the following combinations: mass flow rate equal to {20, 30, 40, 50, 60, 70, 80, 90, 100} (g/s), valve rotation angle with respect to the fully open position equal to {0, 10, 20, 30, 40, 50} degrees, and gas temperature upstream of the valve equal to {200, 250, 300, 350, 400, 450, 500} (°C). The results of all these simulations are represented in Figure 12, classified by the valve rotation angle.

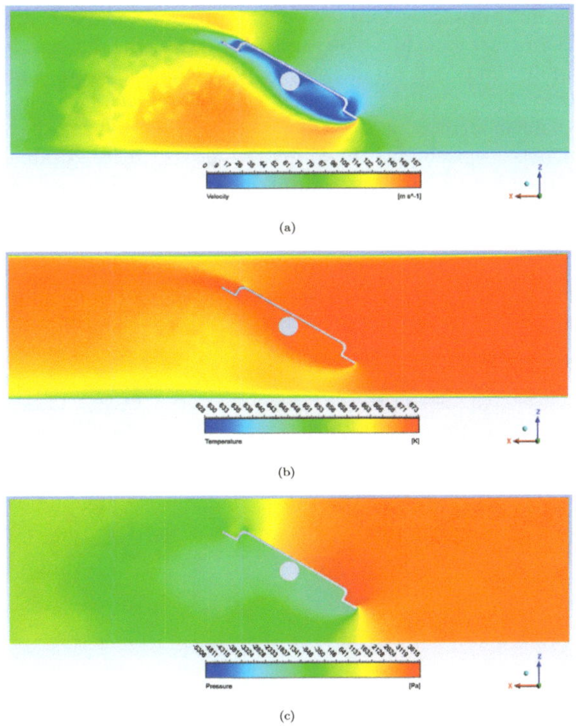


**Figure 10.** Valve closing 30 degrees. Velocity (**a**), temperature (**b**) and pressure (**c**) field in axial plane.

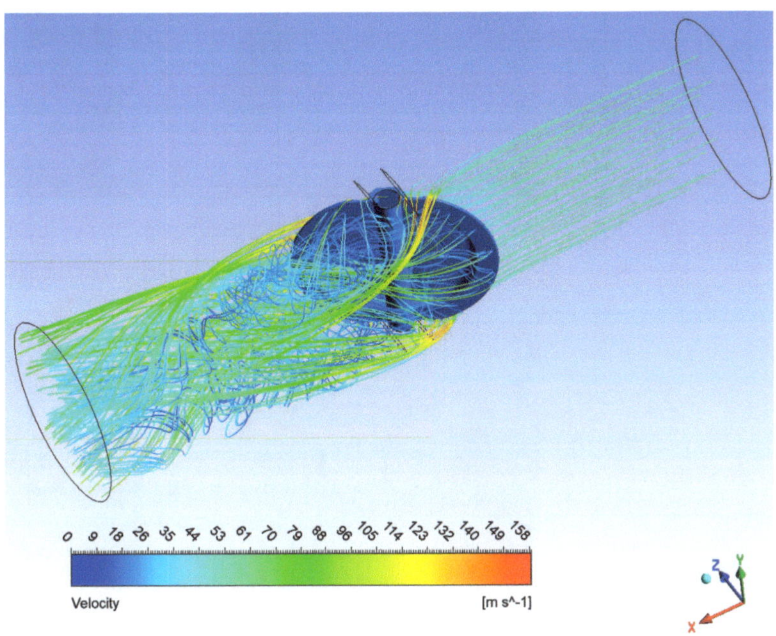


**Figure 11.** Streamlines in the valve closing 30 degrees.

In Figure 12, the colour map represents the drop in exhaust gas pressure in the valve under different conditions. Contour lines of constant exhaust gas pressure drop are shown.

It can be observed that the valve angle and the exhaust gas mass flow rate are the two factors that most affect pressure drop, while the temperature of the exhaust gas has a less significant effect. An increase in the valve closure angle results in an increase in pressure drop [5], due to the increased flow velocity. An increase in the exhaust gas mass flow rate results in an increase in pressure drop, as the flow velocity increases and the pressure drop increases with higher flow rates. A slight increase in exhaust gas temperature results in a minor increase in pressure drop due to a decrease in density, resulting in a minor increase in gas velocity, while the mass flow rate remains constant.

In Figure 12, the experimental measurements of the four modes tested experimentally are shown. The percentage of valve closure in which the experimental pressure drop in the valve is of the same order as the pressure drop resulting from the modelling is shown in parentheses.

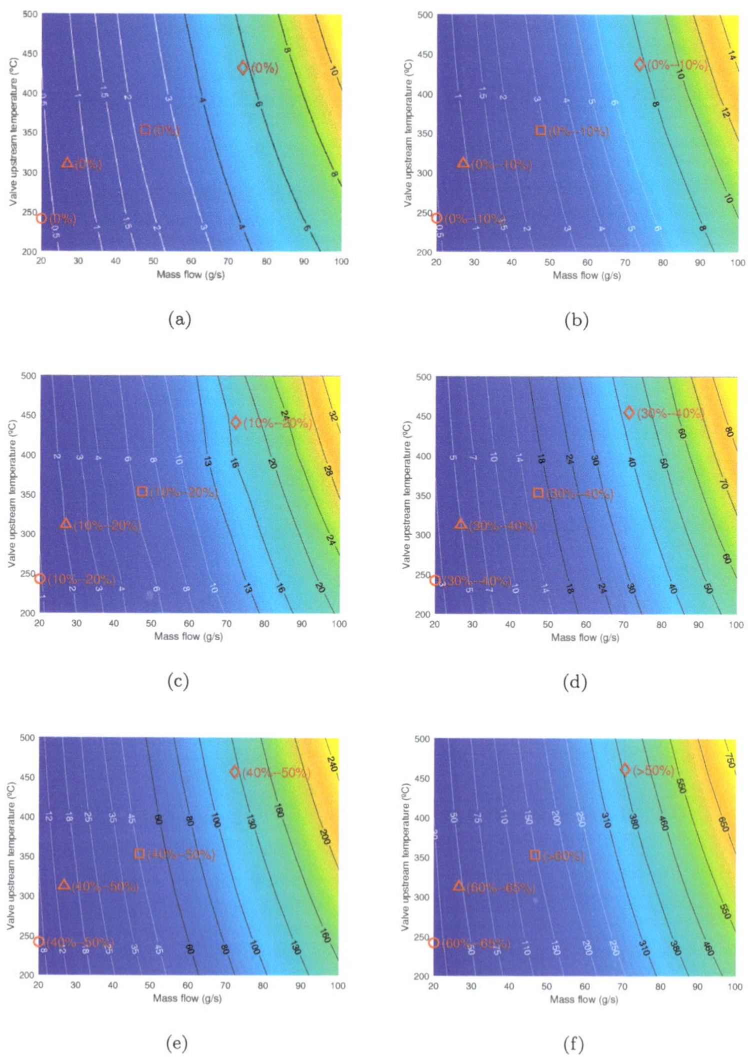


**Figure 12.** Pressure drop (mbar) vs. valve upstream temperature and mass flow. Closing angle: (**a**) 0, (**b**) 10, (**c**) 20, (**d**) 30, (**e**) 40 and (**f**) 50 degrees.

In experimental tests of flow in a back pressure valve, it is more cost-effective to measure the pressure drop using a differential pressure gauge than to measure the mass flow rate. Therefore, a correlation is proposed based on the results of the simulations that relate the four variables, Equation (3). The result of this correlation is the mass flow rate $\dot{m}$ in (g/s). A correlation coefficient of $R^2 = 0.984$ and a mean squared error of 0.32 were obtained.

$$\dot{m} = 181.8(1 - \sin(\theta))^{1.51} T^{-0.28} \Delta p^{0.50} \quad (3)$$

In Equation (3), $\theta$ is the angle of rotation of the valve with respect to the fully open position, $T$ is the temperature of the gas upstream of the valve in °C, and $\Delta p$ is the pressure drop of the gas through the valve in mbar.

The experimental valve closing angle is difficult to measure, and even if it is measured, the actual value may differ from the assumed one. To quantify this deviation, the angle that the valve should have ($y$-axis of Figure 13) was compared with the experimental valve closing angle measured ($x$-axis of Figure 13).

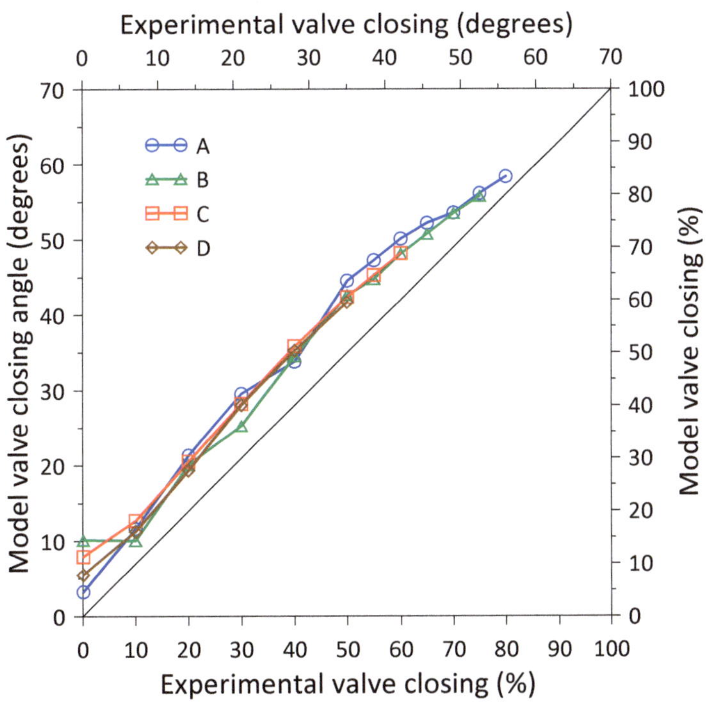


**Figure 13.** Model valve closing angle vs. experimental valve closing angle.

In Figure 13, it is shown that there is a linear correlation between the experimentally measured valve closing angle and the valve angle obtained with the correlation of Equation (3). For all test modes with closing angles less than 50 degrees, there is a deviation of about 5 degrees between the two angles, with the experimentally measured angle being lower than the one obtained by the correlation.

## 5. Conclusions

The fluid-dynamic behaviour of an electronically controlled exhaust backpressure valve was analysed, operating at different opening degrees under different speed regimes and load degrees of a Euro 6 internal combustion engine. The following conclusions can be drawn from this research:

- The negative effect that the closure of the backpressure valve has on the instantaneous in-cylinder pressure was observed. Assuming no effect on the average pressure inside the cylinder with full opening, a reduction in the in-cylinder pressure was measured between 1.3% and 3.9% for valve closures between 50% and 80%, respectively, for medium–high load levels.
- It was observed that with an increase in the closure of the pressure relief valve, the effective average pressure of the engine decreases between 0.07 and 1.06 bar for valve closures between 50% and 80%, respectively, for medium–high load levels. However, for that range of valve closures, increases in the temperature of the exhaust gases upstream of the valve are observed between 5.7 and 11.7 °C.
- The development of this work has made possible to propose a CFD model. With this model, the flow around the valve was simulated under different temperatures and the mass flows of exhaust gas and different closing positions of the backpressure valve. For all cases, pressure drop values were obtained. The level of agreement between the results obtained by the model compared to experimental measurements is around 91%.
- The developed model helped to propose a correlation to estimate the mass flow rate of exhaust gas from easily measurable quantities on a test bench, such as gas temperature, valve closing angle, and pressure drop.
- A positive deviation of around 5 degrees of backpressure valve closing was obtained between the modelled angle and the experimental angle for all engine operating conditions and all valve closing angles.

**Author Contributions:** Conceptualization, O.A. and F.J.M.; methodology, O.A., J.A.S., A.B., F.J.M., and P.F.-Y.; formal analysis, F.J.M., P.F.-Y. and O.A.; investigation, J. A.S., A.B.; resources, O.A.; data curation, A.B. and F.J.M.; writing—review and editing, O.A., F.J.M. and P.F.-Y.; supervision, project administration, funding acquisition, O.A. All authors have read and agreed to the published version of the manuscript.

**Funding:** The authors wish to thank the financial support provided by: (i) the Spanish Ministry of Science, Innovation and Universities to the project RECUPERA, Ref.: RTI2018-095923-B-C21 and (ii) the government of Castilla-La Mancha community to the project ASUAV, Ref. SBPLY/19/180501/000116.

**Data Availability Statement:** Not applicable.

**Acknowledgments:** The authors wish to thank the technical support to Nissan Europe Technology Center for the engine supply and different auxiliary systems and the Repsol Technology Lab for the supplied fuel.

**Conflicts of Interest:** The authors declare no conflict of interest.

## Abbreviations

| | |
|---|---|
| $C$ | constant |
| $G$ | generation |
| $k$ | turbulence kinetic energy |
| $\dot{m}$ | mass flow rate |
| $p$ | pressure |
| $t$ | time |
| $T$ | temperature |
| $u$ | velocity |
| $x$ | direction |
| $Y_M$ | fluctuating dilatation |
| $\alpha$ | inverse Prandtl number |
| $\varepsilon$ | rate of dissipation |
| $\mu$ | dynamic viscosity |
| $\rho$ | density |

Subscripts

| | |
|---|---|
| *bu* | buoyancy |
| $\varepsilon$ | rate of dissipation |
| *i* | index |
| *j* | index |
| *k* | turbulence kinetic energy |
| *up* | upstream |

Acronyms

| | |
|---|---|
| BMEP | Brake Mean Effective Pressure |
| CFD | Computational Fluid Dynamic |
| DNS | Direct Numerical Simulation |
| ECU | Electronic Control Unit |
| EGR | Exhaust Gas Recirculation |
| LES | Large Eddy Simulation |
| PID | Proportional Integral Derivative |
| RANS | Reynolds Averaged Navier–Stokes |
| RNG | Re-Normalization Group |

## References


1. Jovaj, M.S. *Motores de Automóvil*; Editorial MIR: Moscow, Russia, 1982.
2. Heywood, J.B. *Internal Combustion Engine Fundamentals*; McGraw-Hill Education: New York, NY, USA, 2018.
3. Payri González, F.; Desantes Fernández, J.M. *Motores de Combustión Interna Alternativos*; Editorial Universitat Politécnica de Valencia: Valencia, Spain, 2011.
4. Luján, J.M.; Guardiola, C.; Pla, B.; Reig, A. Switching strategy between HP (high pressure)-and LPEGR (low pressure exhaust gas recirculation) systems for reduced fuel consumption and emissions. *Energy* **2015**, *90*, 1790–1798. [CrossRef]
5. Choe, M.; Park, C.; Choi, D. Flow analysis on formation of back-pressure in exhaust system applying electronic-variable valve. *Indian J. Sci. Technol.* **2015**, *8*, 21. [CrossRef]
6. Pan, X.; Wang, G.; Lu, Z. Flow field simulation and a flow model of servo-valve spool valve orifice. *Energy Convers. Manag.* **2011**, *52*, 3249–3256. [CrossRef]
7. Long, C.; Guan, J. A method for determining valve coefficient and resistance coefficient for predicting gas flowrate. *Exp. Therm. Fluid Sci.* **2011**, *35*, 1162–1168. [CrossRef]
8. Gülmez, Y.; Özmen, G. Effect of exhaust backpressure on performance of a diesel engine: neural network based sensitivity analysis. *Int. J. Automot. Technol.* **2022**, *23*, 215–223. [CrossRef]
9. Fernández-Yáñez, P.; Romero, V.; Armas, O.; Cerretti, G. Thermal management of thermoelectric generators for waste energy recovery. *Appl. Therm. Eng.* **2021**, *196*, 117291. [CrossRef]
10. Ezzitouni, S.; Fernández-Yáñez, P.; Sánchez, L.; Armas, O.; Soto, F. Effect of the use of a thermoelectric generator on the pumping work of a diesel engine. *Int. J. Engine Res.* **2021**, *22*, 1016–1027. [CrossRef]
11. Tuncer, C. *Analysis of Turbulent Flows*; Elsevier: Amsterdam, The Netherlands, 2004.
12. Reveillon, J.; Péra, C.; Bouali, Z. Examples of the potential of DNS for the understanding of reactive multiphase flows. *Int. J. Spray Combust. Dyn.* **2011**, *3*, 63–92. [CrossRef]
13. Piomelli, U. Large-eddy simulation: Achievements and challenges. *Prog. Aerosp. Sci.* **1999**, *35*, 335–362. [CrossRef]
14. De Villiers, E. *The Potential of Large Eddy Simulation for the Modeling of Wall Bounded Flows*; Imperial College of Science, Technology and Medicine: London, UK, 2006.
15. Versteeg, H.K.; Malalasekera, W. *An Introduction to Computational Fluid Dynamics: The Finite Volume Method*; Pearson Education: London, UK, 2007.
16. Kang, C.W.; Yi, C.S.; Jang, S.M.; Lee, C.W. A Study of the Measurement of the Flow Coefficient $C_v$ of a Ball Valve for Instrumentation. *J. Korean Soc. Manuf. Process. Eng.* **2019**, *18*, 103–108. [CrossRef]
17. Laouar, S.; Sakib, M.N.; Al, S.M.; Navasardyan, M.V.; Kutsenko, K.V. Pressure drop in valve for different open flow areas. *Proc. J. Phys. Conf. Ser. Iop Publ.* **2020**, *1439*, 012009. [CrossRef]
18. Valdés, J.R.; Rodríguez, J.M.; Saumell, J.; Pütz, T. A methodology for the parametric modelling of the flow coefficients and flow rate in hydraulic valves. *Energy Convers. Manag.* **2014**, *88*, 598–611. [CrossRef]
19. Lisowski, E.; Filo, G. Analysis of a proportional control valve flow coefficient with the usage of a CFD method. *Flow Meas. Instrum.* **2017**, *53*, 269–278. [CrossRef]
20. Orszag, S.A.; Yakhot, V.; Flannery, W.S.; Boysan, F.; Choudhury, D.; Maruzewski, J.; Patel, B. Renormalization Group Modeling and Turbulence Simulations. In Proceedings of the International Conference on Near-Wall Turbulent Flows, Tempe, AZ, USA, 15–17 March 1993.

21. Lapuerta, M.; Armas, O.; Hernández, J. Diagnosis of DI Diesel combustion from in-cylinder pressure signal by estimation of mean thermodynamic properties of the gas. *Appl. Therm. Eng.* **1999**, *19*, 513–529. [CrossRef]
22. Ma, Z.; Zhang, K.; Xiang, H.; Gu, J.; Yang, M.; Deng, K. Experimental study on influence of high exhaust backpressure on diesel engine performance via energy and exergy analysis. *Energy* **2023**, *263*, 125788. [CrossRef]


**Disclaimer/Publisher's Note:** The statements, opinions and data contained in all publications are solely those of the individual author(s) and contributor(s) and not of MDPI and/or the editor(s). MDPI and/or the editor(s) disclaim responsibility for any injury to people or property resulting from any ideas, methods, instructions or products referred to in the content.

*energies*

MDPI

*Review*

# CFD Simulation of Pre-Chamber Spark-Ignition Engines—A Perspective Review

**Soo-Jin Jeong**

Alternative Fuel Power System R&D Department, Korea Automotive Technology Institute, Cheonan-si 31214, Republic of Korea; sjjeong@katech.re.kr; Tel.: +82-41-559-3059

**Abstract:** The growing demand to reduce emissions of pollutants and $CO_2$ from internal combustion engines has led to a critical need for the development of ultra-lean burn engines that can maintain combustion stability while mitigating the risk of knock. One of the most effective techniques is the pre-chamber spark-ignition (PCSI) system, where the primary combustion within the cylinder is initiated by high-energy reactive gas jets generated by pilot combustion in the pre-chamber. Due to the complex physical and chemical processes involved in PCSI systems, performing 3D CFD simulations is crucial for in-depth analysis and achieving optimal design parameters. Moreover, combining a detailed CFDs model with a calibrated 0D/1D model is expected to provide a wealth of new insights that are difficult to gather through experimental methods alone, making it an indispensable tool for improving the understanding and optimization of these advanced engine systems. In this context, numerous previous studies have utilized CFD models to optimize key design parameters, including the geometric configuration of the pre-chamber, and to study combustion characteristics under various operating conditions in PCSI engines. Recent studies indicate that several advanced models designed for conventional spark-ignition (SI) engines may not accurately predict performance under the demanding conditions of Turbulent Jet Ignition (TJI) systems, particularly when operating in lean mixtures and environments with strong turbulence–chemistry interactions. This review highlights the pivotal role of Computational Fluid Dynamics (CFDs) in optimizing the design of pre-chamber spark-ignition (PCSI) engines. It explores key case studies and examines both the advantages and challenges of utilizing CFDs, not only as a predictive tool but also as a critical component in the design process for improving PCSI engine performance.

**Keywords:** pre-chamber ignition engine; turbulent jet ignition; computational fluid dynamics; turbulence–chemistry interaction; wall heat transfer; combustion model; turbulence model

**Citation:** Jeong, S.-J. CFD Simulation of Pre-Chamber Spark-Ignition Engines—A Perspective Review. *Energies* **2024**, *17*, 4696. https://doi.org/10.3390/en17184696

Academic Editors: Davide Lanni and Enzo Galloni

Received: 3 July 2024
Revised: 2 September 2024
Accepted: 19 September 2024
Published: 20 September 2024

**Copyright:** © 2024 by the author. Licensee MDPI, Basel, Switzerland. This article is an open access article distributed under the terms and conditions of the Creative Commons Attribution (CC BY) license (https://creativecommons.org/licenses/by/4.0/).

## 1. Introduction

Given the current global energy demand and shortages, developing technological solutions that enhance fuel conversion efficiency in internal combustion engines is imperative. Two prominent strategies are lean burn combustion and stoichiometric combustion with exhaust gas recirculation (EGR), each offering distinct advantages and challenges.

Leaner combustion strategies are recognized for their potential to boost thermal efficiency and reduce pumping losses [1,2]. Studies have demonstrated that lean fuel–air mixtures significantly reduce fuel consumption and NOx emissions while improving thermal efficiency [1,2]. However, applying lean burn strategies to gasoline engines presents several challenges. One major limitation is the incompatibility with three-way catalytic converters, which are essential for reducing emissions in gasoline engines. Additionally, to meet current NOx emission regulations, $\lambda$ levels must be increased to 1.8–2.0 [2], surpassing the lean stability limit ($\lambda \sim 1.4$). Moreover, leaner air–fuel mixtures can lead to serious issues such as combustion stability degradation, increased cycle-by-cycle variation, reduced thermal efficiency, and elevated unburned hydrocarbon (UHC) emissions.

In contrast, maintaining a stoichiometric air-to-fuel ratio ($\lambda = 1$) while reducing oxygen concentration through EGR offers another viable approach to improving SI engine efficiency [3,4]. This method allows using three-way catalytic converters, which are highly effective in reducing NOx emissions [4]. Additionally, as a diluent, EGR reduces the knock tendency at low-speed, high-load conditions and enhances fuel economy by eliminating the need for fuel enrichment [5]. However, this approach is also associated with challenges such as increased cycle-to-cycle variability, reduced flame propagation speed, and the risk of misfires, challenges typically seen in lean combustion without EGR.

Previous literature [6,7] has identified that the primary obstacles in lean burn operations, whether using air or EGR as diluents, are the increased energy required to initiate combustion and the low flame propagation velocity. Therefore, the most significant challenge in implementing lean combustion technology is developing an ignition system capable of reliably igniting a lean fuel–air mixture. Lean-burn internal combustion engines, which aim to reduce nitrogen oxide (NOx) emissions, require substantial ignition power and widespread ignition sources charge to ignite and burn the lean premixed charge effectively. This strategy is crucial for enhancing efficiency and reducing residual hydrocarbon emissions.

To address this limitation, researchers have proposed and thoroughly examined alternative ignition approaches such as laser-induced plasma ignition [8,9], diesel pilot injection [10], and pre-chamber ignition systems. Despite extensive research efforts, only a few of these methods have transitioned into commercial applications. Laser-induced plasma ignition technology offers precise control over ignition timing and high-energy ignition potential. However, challenges such as high costs, complexity in implementation, and sensitivity to environmental conditions have limited its widespread commercial adoption, keeping the technology mainly in the research phase. Diesel pilot injection also has drawbacks, including increased engine complexity due to the additional fuel source and the reliance on auto-ignition of diesel, which requires high in-cylinder temperatures and offers less precise control. Consequently, the advantages of pilot injection may be limited under specific operating conditions.

All Turbulent Jet Ignition (TJI) systems are equipped with a spark plug, a compact pre-chamber (<3% of clearance volume), and one or more orifices through which the pre-chamber's reacting contents are injected into the main chamber. The spark plug ignites the mixture in the pre-chamber, creating a pressure differential that drives the flame through the nozzle. This results in a jet of intermediate combustion products containing active radicals and high-temperature burned materials, which rapidly ignite the charge in the main chamber, initiating fast and turbulent combustion.

Recently, pre-chamber spark-ignition (PCSI) technology has been actively implemented as an ignition method not only in highly downsized TGI engines [11,12] and heavy-duty gas engines [13,14] but also in hybrid vehicle engines [15]. The PCSI systems have recently been applied to carbon-neutral fueled passenger car engines for automotive applications.

Summarizing the research results of the PCSI system above, it can be observed that by supplying a higher amount of energy in the main chamber at the start of combustion, stable combustion can be achieved under very lean conditions ($\lambda > 2.5$) for gasoline-fueled engines [16,17] and $\lambda$ of up to 2.6 for heavy-duty natural gas engines with near zero NOx emission [18]. Recently, a pre-chamber was applied to a gasoline-fueled PFI engine, achieving 52.5% indicated thermal efficiency and stable ultra-lean burn combustion at a lambda of 2.4 [19]. These make PCSI a more practical and economical lean combustion technique than lean combustion methods.

### *1.1. Pre-Chamber Ignition Concept*

The first pre-chamber combustion engine is often attributed to the Ricardo Dolphin engine, developed in the early 20th century [20]. Sir Harry Ricardo, a pioneering British engineer, designed this engine to improve combustion efficiency and reduce knock in gasoline engines.

Pre-chamber combustion technology later advanced into the jet igniter system [2,4,21,22], characterized by a significantly smaller orifice that connects the main combustion chamber to the pre-chamber cavities. The reduced orifice size accelerates the flow of the burning mixture, effectively quenching the flame while allowing reactive radical species to ignite farther from the pre-chamber. Jet ignition technology was first conceptualized in the late 1950s by Nikolai Semenov, renowned for his contributions to the general theory of chemical chain reactions. In 1981, the LAG system was integrated into the Volga vehicle's powertrain, using a cam-driven injector to introduce a rich mixture ($\lambda = 0.5$) into the pre-chamber, which then ignited an ultra-lean mixture ($\lambda = 2$) in the main cylinder [22,23].

The detailed development process and characteristics of the PCSI system are thoroughly described in previously published review papers [22,23]; therefore, this paper will omit those details for brevity. The most successful representative example of a modern PCSI engine with a divided chamber (pre-chamber) is Honda's Compound Vortex Controlled Combustion (CVCC) system, developed between 1968 and 1972 [24] and first introduced in the 1975 Honda Civic. This technology was innovative at the time as it met the suddenly stringent emission regulations of the California Air Resources Board without the need for a catalytic converter.

However, the CVCC system was not continuously applied to Honda cars for several reasons, such as advancements in emission control technology, cost and complexity, and the shift to direct fuel injection. As emission regulations became stricter, newer and more effective technologies were developed. Catalytic converters and advanced electronic fuel injection systems became standard, offering better performance, efficiency, and lower emissions compared to the CVCC system. Direct fuel injection systems provided better control over the combustion process, improving fuel efficiency and lower emissions. These systems became more prevalent in the automotive industry, replacing older technologies like CVCC. Additionally, emission standards continued to evolve, requiring more sophisticated and integrated approaches to meet the new requirements. The CVCC system, while effective in its time, could not keep up with the increasingly stringent regulations without significant modifications.

Pre-chamber combustion techniques have recently gained renewed attention as a critical technology for environmentally friendly, next-generation engine-based vehicles. This can be attributed to advancements in machining, production technologies, air/fuel ratio control techniques, and improvements in turbulence flow control both inside and outside the pre-chamber. These improvements have enhanced combustion stability under lean burn conditions and advanced knock control techniques. Furthermore, the development of combustion visualization techniques and chemiluminescence using Rapid Compression Expansion Machines (RCEMs), along with advancements in CFD analysis techniques, has significantly contributed to more detailed analysis and understanding of turbulent jet flame development and combustion characteristics.

This section presents a concise overview of the combustion strategies and principles employed in the PCSI engine, which has recently attracted significant interest.

This ignition strategy can be implemented in two main ways: active [25,26] (or scavenged) and passive (or unscavenged) systems [27–29]. Figure 1 schematically compares the operating principles of active and passive pre-chambers with those of a conventional SI engine.

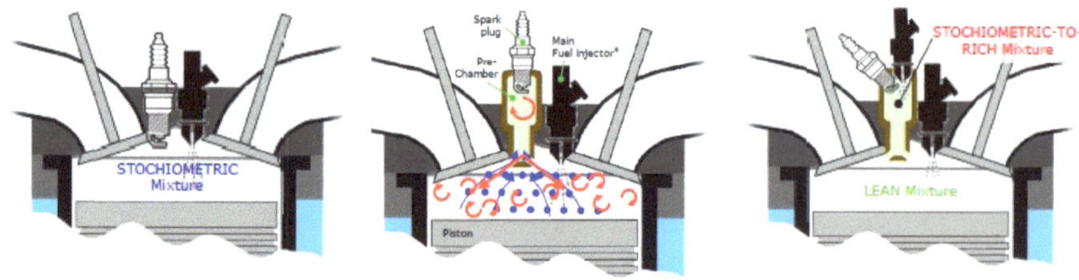


(**a**) Conventional S.I. (**b**) Passive Pre-Chamber (**c**) Active Pre-Chamber

**Figure 1.** The basics of pre-chamber combustion engine [30].

The active system features a dedicated fuel injector or a miniature check valve within the pre-chamber that does not allow the backflow of pressurized gases and combustion products from the pre-chamber to the fuel line, ensuring the air-to-fuel mixture remains optimal for combustion near stoichiometric conditions [31]. Hence, in an active, fueled PCSI system, an auxiliary fueling event occurs in the pre-chamber via an injector or check valve, forming a stoichiometric or fuel-rich mixture near the spark plug. In contrast, the passive system simplifies the design by excluding this secondary fuel injector, which lowers assembly and packaging costs, making it a viable option for passenger vehicles. Thus, fuel injection is performed solely in the main chamber, utilizing either port fuel injection (PFI) or direct injection (DI), ensuring an identical air/fuel ratio in both combustion chambers. The air–fuel mixture then flows into the pre-chamber through interconnection orifices. Consequently, the geometry of the pre-chamber and connecting pipelines plays a crucial role in the combustion process. Numerous studies have confirmed the importance of this ignition system over the past decade. Several studies examined the fundamental aspects of jet ignition using both experimental and simulation methods [32] and the effects of pre-chamber configuration on jet dynamics and combustion physics [33,34].

Recently, Benajes et al. [35,36] have assessed the synergies between pre-chamber ignition and other strategies aimed at enhancing engine thermal efficiency, such as lean burn and CO2-free fuel use, as well as hydrogen [37–39] and ammonia [36].

Furthermore, numerous studies have been conducted to optimize the geometry of the pre-chamber to maximize the effects of the turbulence–chemistry interaction caused by the shear stress of the flow through the holes. Numerous previous studies [14,39–41] investigated how nozzle diameter and pre-chamber volume influence combustion performance for this goal.

### 1.2. *Review Objectives*

Review papers on the performance enhancement and emission reduction effects of PCSI combustion engines, as well as EGR and lean burn limits, have been consistently published from 2010 to 2023 [2,22,42–44]. This indicates ongoing research and technological advancements in this field. Numerous studies have focused on improving the accuracy of CFD analysis techniques to numerically model the complex physical phenomena of PCSI engines more accurately.

Pre-chamber combustion techniques have recently gained renewed attention as a key technology for environmentally friendly, next-generation engine-based vehicles [42,44]. However, PCSI engines are strongly influenced by turbulence–chemistry interactions between the turbulent flame jet generated in the pre-chamber and the turbulent characteristics such as swirl and tumble within the main chamber [2,44]. Therefore, the optimal design of PCSI engines requires a thorough analysis and prediction of complex physical phenomena, including multi-mode combustion, optimization of the nozzle diameter and number, orifice length, shape, and volume of the pre-chamber, the stretching and quenching of the turbulent jet torch, the rapid changes in turbulence length scales, and the thermal mixing between the turbulent jet and lean mixture in the main chamber [2,43]. For this purpose, the utilization of three-dimensional computational fluid dynamics (CFDs) capable of simulating physicochemical processes in PCSI systems is essential. This study aims to review and analyze the technical achievements and limitations of CFD analysis techniques in PCSI engine research over the past 20 years and to discuss the prospects for future advancements in CFD technology.

This paper reviews the evolution and application of CFDs in PCSI engines, with a detailed look at the numerical modeling of the complex physicochemical processes involved in the PCSI ignition system and the contemporary issues regarding the limitations of commercial CFD codes. Finally, the prospects for future advancements in CFD technology are discussed. We explored the relevant literature in-depth using the Web of Science Core Collection, SAE Mobilus, and Google Scholar.

This paper is organized as follows: Section 2 presents the major achievements of CFD applications on PCSI engine design, focusing on optimizing geometric configurations and combustion characteristics within the pre-chamber. Section 3 provides an in-depth analysis of the CFD modeling techniques employed in PCSI engines, including the turbulence and combustion models used in the simulations. Section 4 discusses the interaction between turbulence and chemistry, emphasizing its impact on ignition, flame propagation, and overall combustion performance. Finally, Section 5 concludes this paper by summarizing key findings and suggesting prospects for future advancements in CFD technology for PCSI engines. Each section is thoroughly analyzed to provide a comprehensive understanding of the topics covered.

## 2. Major Achievements of CFD Applications on PCSI Engine Design

Since the early 2000s, research on PCSI using CFD simulation has primarily focused on optimizing the geometric configuration of the pre-chamber, including the nozzle diameter and number, orifice length, shape, and volume. Recently, these research themes have continued with the application of upgraded and more accurate numerical models. Figure 2 schematically illustrates these research topics utilizing CFDs. The research topics using CFDs can be summarized into two categories.

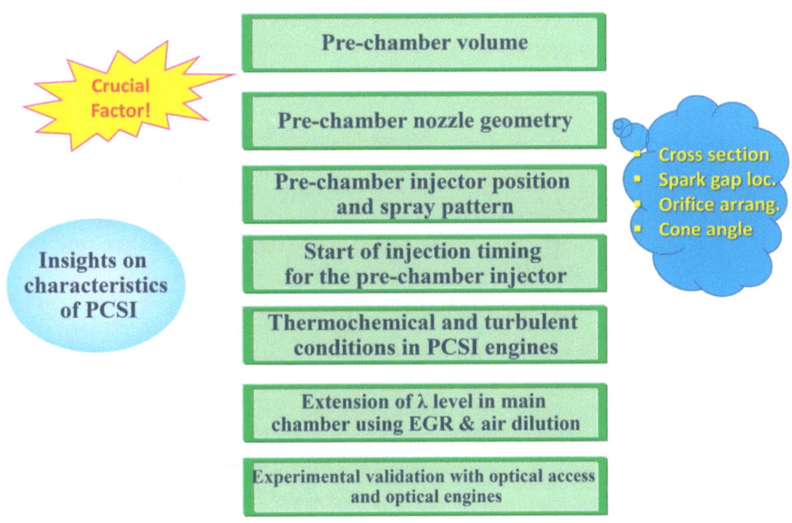


**Figure 2.** Major topics previous and current literature for optimizing PCSI engines.

First, some articles [45,46] study the influence of pre-chamber volume and nozzle diameter on the resultant ignition characteristics using CFD simulation. The impact of the pre-chamber's nozzle orifice diameter and, thereby, the total cross-sectional area of the nozzle orifices has been investigated several times. The effect of the orientation and number of nozzle orifices connecting the pre-chamber and the main combustion chamber, as well as the pre-chamber's internal volume and shape, is evaluated.

Silva et al. [47] developed a CFDs model based on the Reynolds-Averaged Navier–Stokes (RANS) approach for a methane-fueled engine, incorporating a well-stirred reactor combustion model [30,31] along with a detailed methane oxidation mechanism. They investigated the effects of various geometric parameters of the passive pre-chamber, such as the throat diameter, nozzle length, and nozzle diameter, on the overall combustion characteristics. Model validation was performed by comparing pressure profiles in both the main chamber and pre-chamber under motoring conditions at 1200 rpm. The findings revealed that the throat diameter plays a crucial role in influencing pressure accumulation and residence time within the pre-chamber, while the nozzle diameter impacts both the peak pressure and residence time.

Distaso et al. [48] performed 3D CFD simulations on an active pre-chamber ignition system in a lean-burn methane engine, analyzing six phases of scavenging and combustion: filling & scavenging, mixing, flame propagation, ejection, reburning, and expulsion & extraction. Thelen et al. [49–51] conducted extensive CFD modeling of the TJI process. Their 3D CFD simulations, incorporating detailed combustion chemistry, investigated the effects of various orifice diameters (1.0 mm, 1.5 mm, 2.0 mm, and 3.0 mm). The findings indicated that a 1.5 mm orifice diameter provides the quickest ignition and overall combustion, based on pressure data, while a 1.0 mm orifice diameter results in higher jet velocity but a longer burn duration compared to larger diameters.

Secondly, using 3D CFD simulation, research has been conducted on the combustion characteristics within the pre-chamber, which cannot be measured experimentally, and the effects of the turbulent jet generated from the nozzle on the combustion characteristics and NOx emissions in the main chamber. Through these studies, various parametric studies have reported on the optimal geometry of the pre-chamber that can extend the lean limit of the main chamber. Recently, research has been published analyzing the turbulence–flame interaction caused by the hot turbulent jet and the turbulence (swirl and tumble) within the main chamber, along with the predictive limitations of related combustion models used in commercial CFD softwares. In addition, research utilizing optically accessible engines and engine-like geometries is focused on capturing the turbulence and mixture characteristics of hot turbulent jets from pre-chambers while replicating engine-relevant thermodynamic conditions [33,51]. Experiments conducted with rapid compression machines [52–54] equipped with generic pre-chamber geometries have provided valuable insights into combustion dynamics and the influence of nozzle size. The benefit of optical data compared to pressure-only data from traditional metal engines is that it offers additional validation opportunities for model development. By comparing key metrics such as jet exit timing, jet penetration velocity, and cyclic variations, simulations can be validated, enhancing our understanding of the phenomena occurring in the engine under similar conditions [32,55]. From a simulation standpoint, this involves using 0D models [11,56,57] to represent turbulence generation/dissipation, heat transfer, and combustion; 3D RANS calculations with level-set combustion models for industrial CFDs and design optimization [12,13,58–66]; Large Eddy Simulation (LES) for detailed analysis of mixing and combustion behavior [54,67]; and 2D/3D Direct Numerical Simulation (DNS) tools [32,54], which provide highly accurate data for studying fundamental phenomena and refining models. However, a physically accurate analytical model for PCSI combustion has not yet been developed [42,44,68]. So, uncertainties remain in understanding the combustion phenomena, as well as in the accuracy of turbulence and combustion models, especially concerning turbulence–flame interactions [44,68]. These aspects will be discussed in detail in the following sections. Recently, an experimental study has been conducted to optimize all previously researched pre-chamber shapes using Taguchi's method, followed by redesigning to enhance performance [69].

## 3. CFD Modeling PCSI Engines

### 3.1. CFD Software

Developing user-friendly and feature-rich CFD software has greatly expanded its usage across different research fields. This section provides an overview of the significant achievements of CFDs in the PCSI engine over the past 30 years to assess the reasons for the increased usage of CFDs in recent years.

I reviewed various CFD software used for modeling the PCSI engines and the RCEM, as shown in Figure 3. Out of 61 studies [1,11–13,16,29,32,33,38,47,48,51–56,58–100], one did not specify the CFD software employed by the researchers. Most of the remaining articles used the commercial software CONVERGE™ [70] for simulations. Other programs like STAR-CD™ [71], FIRE™ [72], and VECTICS [73] were used much less frequently. CFD codes that appeared in only one study were grouped into an 'Others' category, which includes two programs such as KIVA-V3 [74]. Some studies used open-source software and custom-built codes; three used OpenFOAM [75].

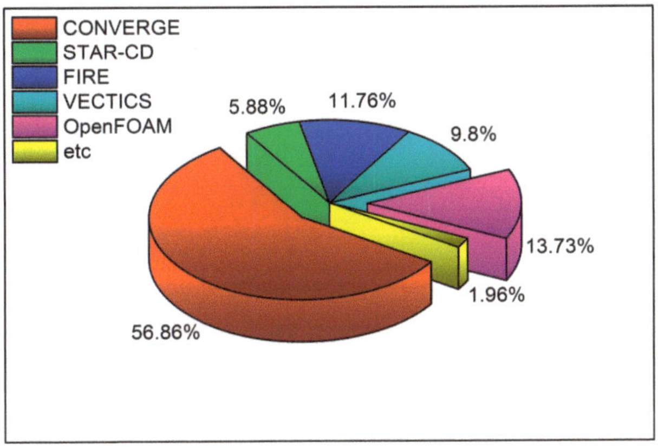


**Figure 3.** Categorical distribution of the type of CFD software for simulating PCSI engines.

The selection of CFD software is primarily influenced by its capabilities, accessibility, and user preferences. Typically updated annually with enhanced features, commercial programs are adept at handling a broad spectrum of simulation tasks, as illustrated in Figure 3. Their user-friendly interfaces further improve their popularity by simplifying the simulation workflow. However, the substantial cost of licenses and limited customization options have driven some users towards open-source alternatives or the development of custom CFD codes. While utilizing these software tools has enhanced the accessibility of CFDs across more comprehensive users, it is imperative to acknowledge that mere accessibility is insufficient in addressing the challenges outlined in PCSI engines, as detailed earlier. The following sections discuss four critical focal points investigated to overcome these challenges. These include the distinctive features and comparative analysis of LES and Reynolds-averaged Navier–Stokes (RANS) models, the establishment of appropriate inflow and outflow boundary conditions, multi-mode combustion models, wall heat transfer, turbulence–chemistry interaction, rigorous verification and validation processes, and development of best practice guidelines in the field.

### 3.2. Turbulence Models

In computational fluid dynamics (CFDs), turbulence modeling can generally be divided into direct numerical simulation (DNS) [101], LES [102,103], and the Reynolds-averaged Navier–Stokes (RANS) model. Due to the challenges of applying DNS to real-world engineering problems, especially in automotive engineering, LES and the RANS model are more frequently utilized.

Turbulence models, especially simpler ones like RANS models, often struggle to accurately predict not only flow separation and reattachment, which are critical in engine intake and exhaust processes, but also the turbulence length scale, which is crucial for determining not only the rate of energy dissipation and mixing efficiency in engine flows but also combustion processes.

Moreover, the flow fields encountered in PCSI engines are highly complex, involving phenomena such as impinging flows, jet-like flow, flow separation, strong swirl, large variable turbulent length scale, and vortex shedding (refer to Section 2). Accurately and universally modeling all these turbulent flow characteristics remains a formidable challenge.

As shown in Figure 4, in the analysis of 52 publications reviewed, 52.8% of studies relied solely on the RNG k-ε turbulence model [103–106], while 11.91% of studies opted for LES alone. Moreover, 7.14% of studies integrated both LES and RANS models in their research. Common RANS turbulence models cited were standard k-ε, renormalization group (RNG) k-ε, and shear stress transport (SST) k-ω [106,107]. Sub-grid scale models for LES included Smagorinsky [107,108] only.

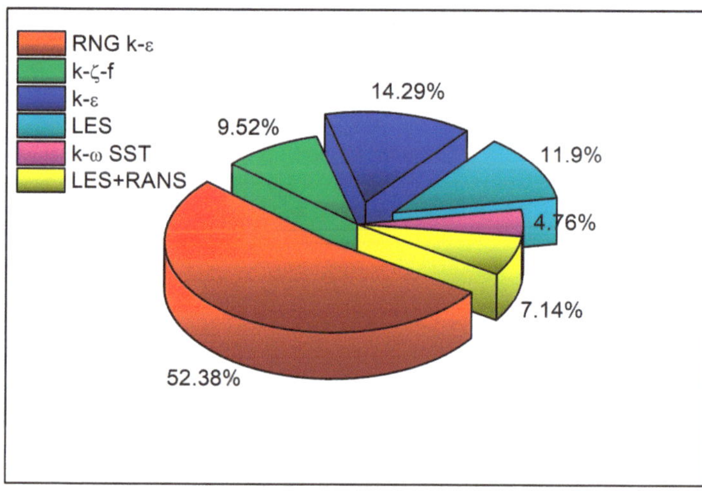


**Figure 4.** Categorical distribution of type of turbulence for simulating PCSI engines.

### 3.2.1. RANS Turbulence Models

The RANS-type turbulence models are simple, computationally inexpensive, and economical [102,107].

Hence, given its simplicity and shorter computation time, the RANS-based model, which represents the average behavior of essential, is typically chosen over the LES model. However, the RANS-type k- ε turbulence models are based on Boussinesq's isotropic eddy viscosity assumptions, and it is well known that they have several problems with deteriorated prediction performance in cases of swirling flow, separation, and reattached flow and flows with large rapid extra strains. Therefore, numerous k-ε model variants have been researched and reported and are currently used to overcome these challenges [104,105]. Numerous k-ε model variants exist due to the ε equation with 3–4 model coefficients, which are empirically derived or determined through ad-hoc methods based on various turbulent flow patterns [103,105,106,109]. Additionally, turbulent flows can vary significantly in their characteristics depending on the geometry, flow velocity, pressure gradients, and other factors. To accurately capture these diverse turbulent flow patterns, modifications to the standard k-epsilon model, particularly in the ε equation, are necessary [104,105]. Each variant attempts to address specific deficiencies of the standard model in certain flow regimes, such as free shear flows, boundary layers, or rotating flows. Consequently, a universally accurate RANS-type turbulence model has not yet been achieved. Therefore, engineers face the challenge of selecting a turbulence model that is appropriate for the specific turbulent flow characteristics of the geometry of interest.

The RNG (Re-Normalization Group) k-ε model [103], which is the most frequently adopted in previous studies for PCSI systems for various engine types [104–106], is one of the variations of the standard k-ε model used for turbulence modeling in various kinds of CFD simulations. It incorporates additional terms that account for the interaction between turbulence and mean strain rate, which is not present in the standard k-ε model, and modifications to the standard k-ε model to improve accuracy for certain flow conditions, particularly for swirling and highly strained flows. To achieve this goal, this model contains a strain-dependent correction term in the constant $C_{1\varepsilon}$ of the production term in the RNG k-ε model's dissipation rate (ε) equation [103–105]. The RNG k-ε model is beneficial for simulations involving complex flow features such as swirling flows, recirculating flows, and flows with high strain rates, making it suitable for jet-like flows and complex industrial applications in which the velocity gradients are significant, causing intense mixing and variations in velocity [70,104,105]. In PCSI internal combustion engines, the scavenging processes in the pre-chamber, along with turbulent flame jets and combustion, create regions of high strain rates. Turbulent jets from pre-chamber nozzles are typical examples of flows with high strain rates. Therefore, for these reasons, applying the RNG k-ε model to PCSI engines extensively is considered practical for engine simulation, as it provides a good balance between accuracy and computational efficiency. However, a noteworthy demerit of the RNG k-ε model is the fact that near-wall treatment of this model can struggle with accurately predicting flows close to walls, particularly in cases involving adverse pressure gradients or separation [70,104,105,109–111]. Hence, one should be careful to use the k-ε model if the turbulent jet interacts significantly with cylinder walls or piston head surface. Additionally, it is clear that the RNG k-ε model is less accurate for detailed structures and does not capture the detailed eddy structures as well as LES [101,103–105].

After the RNG model, the k-ζ-f turbulence model [109–111] is the second most commonly used turbulence model in PCSI engine CFDs simulation. The k-ζ-f model is an extension of the eddy-viscosity concept. It includes three transport equations for turbulence quantities, namely turbulent kinetic energy (k), Turbulence Frequency (ζ), and Dissipation Rate (f). This is the turbulence model in previous studies due to its high accuracy and convergence stability. This model, optimized from Durbin's near-wall turbulence closure model [112,113], is a variant of RSM (Reynolds Stress Model) turbulence models [112–114]. The k-ζ-f model benefits complex flow simulations with significant near-wall effects and flow separation [105,109]. It enhances the standard k-ε turbulence model by introducing the wall-normal velocity fluctuation $v^2$ and its source term f, which incorporate near-wall turbulence anisotropy and non-local pressure-strain effects. The careful introduction of these relaxation terms eliminates the need for damping functions. Additionally, this model improves numerical stability over the original $v^2$-f model by solving a transport equation for the velocity scale ratio $\zeta = v^2/k$ instead of the velocity scale $v^2$. Moreover, this model demonstrates superior predictive accuracy in heat transfer, surface friction at low Reynolds numbers, adverse pressure gradients, and recirculation regions compared to traditional k-ε turbulence models [115,116]. While it shares characteristics with low-Reynolds-number models, it uniquely eliminates the need for wall functions by being applicable near the wall. It introduces a transport mechanism for turbulent energy from the wall, effectively representing near-wall viscous damping effects through an elliptic relaxation equation [109–111]. Hence, this model may offer superior performance in capturing the detailed flow features and interactions in the intake and exhaust processes and the scavenging process inside the pre-chamber, which is crucial for predicting flame propagation, heat transfer, and emissions. Its improved near-wall treatment is advantageous for accurately predicting heat transfer inside orifices of pre-chamber and flow separation around valves and ports.

### 3.2.2. LES Turbulence Model

Due to the substantial turbulent energy and significant influence on momentum transfer and turbulent mixing carried by large eddies, LES methodologies offer superior accuracy compared to RANS turbulence models [117]. LES captures flow structures from the domain scale down to the filter scale, necessitating the significant resolution of high-frequency turbulent fluctuations. This requires using high-order numerical methods or acceptable grid resolution when employing lower-order numerical techniques. Therefore, implementing LES methods in automotive and mechanical engineering necessitates finely resolved grids with grid points positioned close to the boundary layers. This results in significantly heightened computational costs compared to Reynolds-Averaged Navier–Stokes (RANS) methods [118]. Consequently, previous studies have reported only a limited number of simulations utilizing LES, with the majority being conducted on Rapid Compression Machines (RCEMs) [33,54] rather than on full-scale metal engines [67].

Recently, 3D CFD analysis using LES turbulence models was performed to investigate the engine characteristics of active and passive PCSI engines fueled by natural gas for large ships under a single operating condition, both at stoichiometric and lean conditions ($\lambda = 2$). Additionally, the heat release rate curves obtained from the analysis were compared with experimental results [67]. However, to perform accurate CFD simulations of PCSI engines using the LES turbulence model, it is necessary to understand the limitations and characteristics of various sub-models.

The most crucial noteworthy point is that the accuracy of LES heavily relies on the Sub-grid Scale (SGS) models used to represent the unresolved scales. This is because the computational grid limits the size of eddies that can be physically expressed. Despite their presence in the flow field, these eddies cannot be resolved because the CFD mesh lacks the resolution to capture and depict them accurately in CFD simulations. We are mainly concerned with the eddies larger than the mesh size in LES. These eddies are too large to be broken down by molecular viscosity; therefore, we need to find a way to model and remove these eddies from the grid. The turbulent kinetic energy predicted in our LES will be too high if we do not. These eddies are removed by applying an additional stress term to the Navier–Stokes equations, known as the sub-grid scale (SGS) stress. While various SGS models such as the Smagorinsky model [107], dynamic Smagorinsky model [71,72], WALE model [119], and others are available, none are universally applicable to all types of flows, leading to potential inaccuracies in specific scenarios. Different SGS models provide various methods for calculating the sub-grid kinematic viscosity, $\nu_{sgs}$. The sub-grid kinematic viscosity is calculated using Equation (1) provided below:

$$\nu_{sgs} = \left(C_s\Delta\right)^2\sqrt{2S_{ij}S_{ij}} \quad (1)$$

where, $S_{ij}$ is the strain rate of the resolved eddies on the CFD mesh and $\Delta$ is the length scale or grid filter width.

One of the shortcomings of the Smagorinsky Sub-grid Scale model, which is the most frequently adopted in CFD simulations of PCSI combustion, is that it includes model constant Cs called the Smagorinsky coefficient that is not universal and depends on the local flow conditions and the fraction of the cell size that gives the sub-grid length scale. Cs for homogeneous isotropic turbulence is around 0.17. This value is crucial for accurate simulation and comes from an analytical mathematical derivation based on homogeneous isotropic turbulence. It is noted that modern CFD codes use different values of Cs. In STAR-CD [71], Fire [72], Fluent [120], Cs = 0.1, whereas in PHOENICS [121], Cs = 0.17. However, no matter what Cs' values are, this is not true in the case of rotation or near the wall because of too much dissipation near the wall. Additionally, in the Smagorinsky model, sub-grid stress is not damped close to the wall. Hence, some modification to the model is needed. Therefore, various types of sub-grid scale models have been developed. The most frequently used near-wall treatment of the Smagorinsky model is the Van Driest damping function, which is a damping function for proper results in wall-bounded flows [71,72,120].

Figure 5 illustrates the necessary adjustments to the Smagorinsky model for accurate near-wall treatment. To address this, the sub-grid scale stress should ideally approach zero as the wall is approached. This can be achieved through various methods. One option is to implement a different model, such as one based on sub-grid scale kinetic energy, replacing the traditional Smagorinsky model. Alternatively, we can modify the length scale near the wall, reducing it to zero to account for sub-grid scale eddies. Another approach is to adjust the velocity scale, making it less dependent on strain rate. The primary goal of these adjustments is to decrease the sub-grid kinematic viscosity near the wall, reducing the sub-grid scale stress and effectively simulating the damping effects on the eddies.

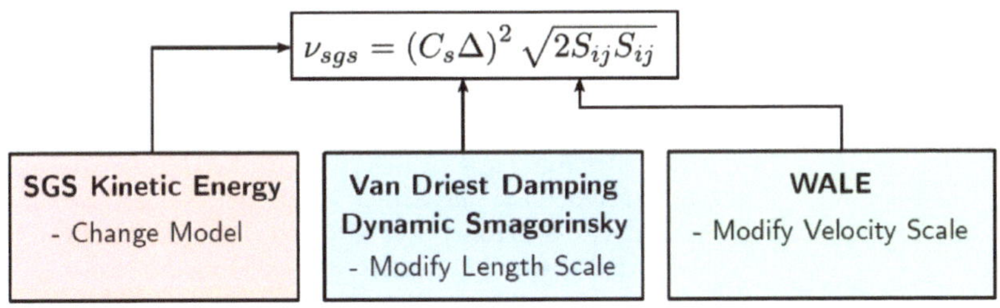


**Figure 5.** Schematic diagram of different options for correcting $\nu_{sgs}$ near the wall [108].

Instead of single user-defined constant Cs, modern CFD codes like Fire and STAR-CD implemented the Dynamic Smagorinsky Sub-grid Scale model [122,123]. This model computes a local time-varying Cs value by test-filtering the flow field at a length scale greater than the grid length scale, which allows it to compute the correct result for wall-bounded flows without the use of damping functions.

The Wall-Adapting Local-Eddy Viscosity (WALE) Sub-grid Scale model [124] is a more advanced sub-grid-scale approach that incorporates an innovative form of the velocity gradient tensor in its formulation and is widely utilized in CFD platforms such as STAR-CCM+ and OpenFOAM [75,125]. Like the Smagorinsky Sub-grid Scale model, it faces the limitation that the model coefficient $C_W$ is not a universal constant. However, it has been observed that the WALE model is generally less sensitive to variations in this coefficient compared to the Smagorinsky model. Another key advantage of the WALE model is that it does not require additional near-wall damping, as it inherently provides accurate scaling near wall boundaries [124].

Figure 5 schematically shows various options for correcting sub-grid scale kinematic viscosity near the wall. Here, Sij is the strain rate of the resolved eddies on the CFD mesh.

Additionally, a Coherent Structure Model (CSM) as a sub-grid-scale model is applied [59,67] and adopted in the Fire Code [72]. It is reported that the CSM gives good predictions and performs almost the same as the dynamic Smagorinsky model for various complex geometries [119].

Additionally, applying appropriate boundary conditions for LES is complex. Inflow boundary conditions, in particular, must accurately represent turbulent fluctuations, which is challenging to achieve in practice. Due to the factors above, achieving dependable LES simulations necessitates more expertise and experience than the requirements for utilizing the RANS model [125].

Bolla et al. [52,54] executed numerical studies of RANS-LES comparison using an RCEM to analyze an automotive-sized scavenged pre-chamber, aiming to compare the two turbulence models' ability to predict turbulence and fuel-air mixing. This study used a Smagorinsky-type sub-grid scale model [126] to compute the unresolved turbulent scales for the LES turbulence model. In contrast, the RANS turbulence model employed the time scale Bounded k-ε Turbulence Model in VECTIS, an enhanced version of the standard k-ε model for high strain rates or strong adverse pressure gradients [73]. The outcomes indicated that the RANS-based model could effectively reproduce the key ensemble-averaged flow patterns seen in LES for two pre-chamber setups. However, RANS often falls short in accurately capturing the radial fuel-air mixing compared to LES.

### 3.3. Physical Phenomena and Combustion Models of PCSI Engines

The most commonly used combustion models for engine simulations are based on the flamelet combustion regime, particularly for spark-ignition (S.I.) engines. In this regime, turbulence can distort and increase the surface area of the flame front while maintaining its inner laminar structure and flame speed. This condition is characterized by a Damköhler number ($Da$) greater than 1 and a Karlovitz number ($Ka$) less than 1.

Numerous previous studies [17,25,29,44,124,127–132] have investigated the operating mechanisms of TJI in PCSI engines, which are schematically illustrated in Figure 6. The characteristics of TJI can be categorized into enrichment, thermal effects, and chemical effects. As shown in Figure 6a, fundamental heat transfer phenomena during flame propagation include thermal quenching, which occurs due to rapid heat transfer to solid surfaces as the flame passes through the nozzle, and hydrodynamic quenching, which happens when the flame mixes with the cool, lean mixture as it enters the main chamber. For a detailed explanation of TJI, please refer to the referenced literature [44]. Recent studies [93,124,128–132] have validated that the two-stage combustion process in the main chamber is comprised of a jet-dominant phase and controlled mixing, which is influenced by the combustion intensity within the pre-chamber, and a flame propagation phase, which depends on the reactivity of the mixture in the main chamber driven by the in-cylinder bulk flow and the associated turbulence.

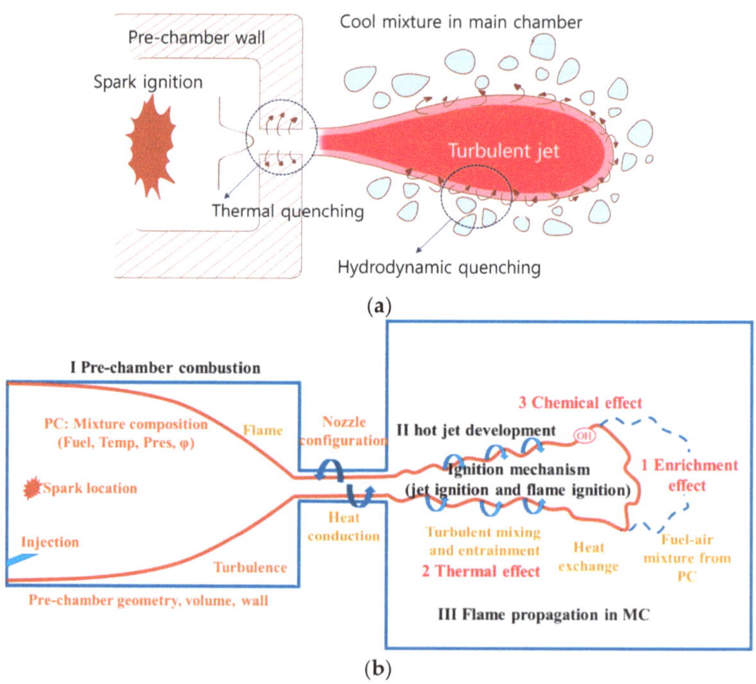


(b)

**Figure 6.** Schematic diagram of (**a**) quenching and (**b**) overview of turbulent jet ignition process [44].

As described above, capturing the flame dynamics across the pre-chamber nozzles is a significant challenge due to the complexity of flame propagation through the orifices [133]. Most previous premixed combustion models were based on the corrugated flame zone; however, recent studies have demonstrated that PCSI engines operate at highly diluted conditions, which approach the thickened flame regime. Consequently, flamelet-based combustion models, conventionally used for SI engine combustion analysis, are unsuitable for application in PCSI engines operating under these ultra-lean conditions. These models fail to capture the combustion behavior of flame quenching and stretching through the orifices, potentially shortening ignition delays in the main chamber [105].

In the studies conducted thus far on 3D CFD combustion analysis of PCSI engines, the combustion models applied have been flamelet-based premixed combustion models originally used for premixed combustion in homogeneously operated spark-ignition engines. However, the PCSI engines operate under very lean conditions ($\lambda > 2$) and in high Karlovitz number ($Ka > 1$) regimes due to the presence of turbulent jets and turbocharging, which result in extremely high turbulence intensities. Therefore, to accurately predict the combustion characteristics of PCSI engines, a combustion model must effectively capture multiple and distributed ignition points within the main chamber, covering both premixed and partially premixed combustion regimes. Furthermore, under the typical operating conditions of TJI systems, characterized by lean mixtures and highly turbulent flow fields, the foundational assumptions of flamelet-based models may no longer be applicable [64]. Therefore, recent studies have raised questions about the predictive performance of flamelet-based combustion models and have made significant efforts to find countermeasures. In this section, we discuss the application cases, comparative predictive performance, and limitations of combustion models that have been applied to the combustion analysis of PCSI engines. Additionally, we review various methods to enhance predictive performance and overcome the limitations of flamelet-based models.

### 3.3.1. Flamelet Assumption

As mentioned above, combustion models for PCSI engines are often based on the so-called flamelet assumption [70–73,132,134], which has frequently been adopted for premixed combustion in homogeneously operated spark-ignition engines [134]. The combustion models based on flamelet assumption simplify the complex interactions between turbulence and chemical reactions by assuming that the flame can be represented as an ensemble of thin, locally laminar flame structures, or "flamelets", embedded within the turbulent flow. They assume a clear separation of scales between the turbulent eddies and the flame thickness. This allows the detailed chemical reactions to be precomputed and stored in a database, which can be accessed during the simulation. The chemical reactions are solved in a laminar flame configuration under temperature, pressure, and mixture composition conditions. The results are stored in flamelet libraries, which provide information about species concentrations, temperatures, and reaction rates as functions of mixture fraction and scalar dissipation rate.

In summary, flamelet models balance accuracy and computational cost, making them suitable for capturing the essential features of premixed combustion in S.I. engines. However, it is necessary to recognize their limitations, especially when dealing with non-premixed or partially premixed combustion regimes. Researchers continue to refine these models and explore more detailed approaches to improve engine combustion simulations.

### 3.3.2. G-Equation Model

A flamelet-based combustion model, the G-equation, is one of the widely adopted combustion models for simulating the combustion processes of PCSI engines in CFD simulations within the engine modeling community. This approach utilizes a level-set method, representing moving interfaces or boundaries on a fixed computational mesh. It is beneficial for problems where the computational domain is divided into two regions separated by an interface. The level-set modeling technique allows the fluid–fluid interface to move within any given velocity field [1,135,136]. Extensive details on the G-equation model are available in the literature [38,52–54,58,60,63,65,82,89,91,95], with only a brief summary provided here. To ensure consistency with the commonly used Favre averaging method in premixed turbulent combustion, the G function and the velocity vector v are decomposed into Favre-averaged components and their corresponding fluctuations. By applying several closure assumptions as outlined by Peters [4], the governing equations are derived, as both the Favre-averaged G function $\widetilde{G}$ and its variance $\widetilde{G''^2}$ are defined [137]. To solve the $\widetilde{G}$ equation, a model for turbulent flame speed is required. The turbulent flame speed is a crucial parameter typically modeled as a function of the laminar flame speed and turbulence characteristics such as turbulent intensity and length scales. This allows the model to incorporate the effects of turbulence on flame propagation without directly solving the detailed turbulence–chemistry interactions.

Within the context of the G-equation, several correlations for turbulent burning velocity from the literature [4,72] have been presented and evaluated in previous studies [138]. Among the most popular is Peters's correlation [139], shown in Equation (2), valid for both large-scale and small-scale turbulence.

$$S_T = S_L + u'\left\{ -\frac{a_4 b_3^2}{2b_1} Da + \left[ \left( \frac{a_4 b_3^2}{2b_1} Da \right)^2 + a_4 b_3^2 Da \right]^{1/2} \right\} \tag{2}$$

where $S_L$ is laminar flame speed, u′ is the fluctuating turbulent component, $\delta_L$ is the laminar flame thickness, $l_t$ is the integral length scale, $b_1$ and $b_3$ are the model constants corresponding to large and small-scale turbulence enhancement, respectively, and Da is the Damkohler number, which is a ratio between the flow time scale $(l_t/u')$ over the chemical time scale $(\delta_L/S_L)$.

The laminar flame speed, $S_L$, depends upon the local pressure, the fresh gas temperature, the local unburned fuel/air equivalence ratio using the Metghalchi and Keck correlation [95,96], and the chemical time arising due to the flame stretching [72,136,138–140]. These common correlations for $S_L$ are equations derived from fitting forms based on combustion experiments conducted over various temperature and pressure ranges. Therefore, outside the range of these correlations, the $S_L$ is calculated using extrapolation methods, which inherently introduce relevant input errors into any combustion model.

Another way to get $S_L$ is to use tabulated values, which were created based on the 30-species skeletal mechanism developed by Lu and Law [91].

The laminar flame thickness is determined by analyzing the temperature gradient along the flame front's normal direction and by using the chemical time. The chemical time is derived from the characteristic time of the laminar flame, which is calculated using the Zeldovich Number, a parameter that depends on the activation energy of the fuel oxidation process.

#### 3.3.3. The Extended Coherent Flame Model (ECFM)

The Extended Coherent Flame Model (ECFM) [72,136] builds upon the basic principles of the Coherent Flame Model (CFM) by incorporating additional features to handle more complex combustion scenarios, which focus on turbulence and flame interaction and includes detailed modeling of how turbulent eddies affect flame stretch and flamelet behavior. This involves correcting for the effects of turbulence on flame stretch, considering the turbulence intensity, and adjusting for the curvature and thermal expansion of flamelets [141]. Thus, the flame stretch is influenced by turbulence and the ratios of turbulent to laminar flame velocities and lengths. It is adjusted for curvature and thermal expansion effects caused by laminar combustion in flamelets based on the assumption of isotropic flame distribution [72].

The model calculates the rate of fuel consumption based on the flame surface density (FSD) and the reaction rate per unit flame surface area.

In the case of the coherent flame model, flame surface area per unit volume is defined as follows:

$$\Sigma = \frac{A_l}{V} \tag{3}$$

Using flamelet assumption, the mean turbulent reaction rate is computed as the product of the flame surface density Σ and the laminar burning velocity $S_L$ as follows:

$$\overline{\rho \dot{r}_{fu}} = -\omega_L \Sigma = \rho_{fu,fr} S_L \Sigma \quad (4)$$

In this, $\omega_L$ is the mean laminar fuel consumption rate per unit surface along the flame front, $\rho_{fu,fr}$ the partial fuel density of the fresh gas, $\rho$ is the density of the fresh gas, and $y_{fu,fr}$ is the fuel mass fraction in the fresh gas.

When combustion starts, several new terms have to be computed. Among them are source terms and two quantities for Equation (2): Σ and $S_L$.

$$\frac{\partial \Sigma}{\partial t} + \nabla \cdot (\Sigma u) = S_{production} - S_{destruction} \quad (5)$$

The first term of the left-hand side is the time-dependent component, and the second term is the convective transport of the FSD. The first term of the right-hand side is the source term, which represents the production of flame surface density that comes essentially from the turbulent net flame stretch; the second term is the sink term, which represents the quenching effect referring to the local extinction or reduction of the flame surface density due to unfavorable conditions, such as excessive strain, heat loss, or insufficient reactants. Hence, the FSD transport equation incorporates these effects into a combined source term, $S_\Sigma$, which includes both the creation and destruction mechanisms:

$$S_\Sigma = S_{production} - S_{destruction} = S_{stretch} + S_{quenching} + S_{other} \quad (6)$$

where $S_{other}$ represents additional source terms that might be relevant depending on the specific combustion scenario.

This approach allows for detailed tracking of how turbulence affects the flame surface and, consequently, the combustion process.

Recently, the ECFM-3Z model has been extensively adopted for 3D CFD analysis of PCSI engines using AVL's Fire CFD code [72]. The ECFM-3Z model is an extension of the ECFM combustion model based on the FSD transport equation and mixing model that can describe premixed inhomogeneous turbulence and diffusion combustion and operates within three distinct zones or regions: fuel, air, and the air–fuel mixture. In this model, the fuel can be represented as a mixture of various components. Both the burnt and unburnt gases are categorized into these three zones.

The extent of mixing among these zones is determined using a characteristic time scale, which is computed based on the k-zeta-f turbulence model [1,58,79,80]. The ECFM-3Z model assumes that the composition of unburnt gases, including air and Exhaust Gas Recirculation (EGR), remains consistent across mixed and unmixed zones. The properties of the burnt gases are calculated based on the reaction progress variable.

In conclusion, this combustion model focuses on flame propagation and the interaction between turbulent flow and flame while utilizing simplified global kinetics for chemical kinetics.

#### 3.3.4. The Multizone Well-Stirred Reactor (MZ-WSR) Model

A homogeneous reactor-type combustion model, MZ-WSR [88], operates on the premise that sharp gradients in temperature and density are unlikely to occur within a cell and models combustion as an ignition-based phenomenon. This model divides the reactor into several well-mixed zones, each assumed to be perfectly mixed with a uniform composition. Additionally, chemical reactions are thought to occur instantaneously within each zone. The governing equation of the MZ-WSR combustion model can be represented as follows:

$$\frac{dY_i}{dt} = \gamma \dot{\omega}_i + \frac{\dot{m}}{V}(Y_{i,in} - Y_i) \quad (7)$$

where $\gamma$ is the multiplier, $Y_i$ is the mass fraction of species i, $\dot{\omega}_i$ the reaction rate of species i, $\dot{m}$ is the mass flow rate, and $Y_{i,in}$ is the mass fraction of species i in the inflow.

The MZ-WSR model, coupled with detailed kinetic calculations, is particularly suitable for modeling the jet ignition process. Therefore, most of the studies [13,38,47,61–64,66,90,91] investigating the PCSI system using WSR models have considered detailed chemical mechanisms derived from GRI Mech 3.0 by Lu and Law [142] by utilizing the integrated chemistry solver known as SAGE [13,88,143], which is included in the commercial code CONVERGE [70]. However, the use of the SAGE model for combustion involves considerable simplification. In the simulations, turbulence models are not applied to the mean chemical production terms in the governing equations. This means the potential impacts of turbulent fluctuations on these terms are not accounted for in the simulations. It should be noted that the influence of turbulence, as modeled using the RANS k-ε turbulence approach discussed earlier, is applied exclusively to the transport equations of mass, energy, and momentum in the averaged equations. Prior numerical investigation into the jet ignition process [143] demonstrated that the WSR assumption could produce cooler temperatures in cells with thin flamelets. This issue can become more pronounced in ultra-lean mixtures. Recently, several papers [63,64] have evaluated the prediction performance of this model by comparing it with the G-equation model for the combustion processes in passive [64] and active PCSI engines [63] fueled by natural gas. This literature showed that the MZ-WSR model predicts faster combustion rates in the main chamber than the G-equation model, which fails to match the pre-chamber combustion phase correctly. However, this model's prediction of the combustion rate in the main chamber matched well with the experimental results. The MZ-WSR model requires that each cell be treated as an individual well-stirred reactor. For this to be accurate, the characteristic time of the turbulence in the cell must be significantly smaller than the characteristic time of the combustion chemistry. In other words, the Damköhler number must be much less than 1 to assume a well-stirred reaction. Therefore, in the case of PCSI combustion, where lean and highly heterogeneous turbulent flows exist, this model can result in significant errors.

Despite its limitations, this model has been extensively used to analyze the combustion processes of PCSI engines within the RANS framework up to the present [26,27]. This is because the MZ-WSR model excels in providing a detailed and chemically accurate representation of combustion processes and allows for more flexibility in adjusting the model to account for different fuels such as natural gas [27,28] and gasoline [26] and dual-fuel combustion conditions [16,31,37] due to its detailed chemical kinetics. This makes it particularly useful for research and development where precise emission predictions and understanding of combustion chemistry are crucial. The experimental validation of this model was conducted using in-cylinder pressure and heat release profiles, and it demonstrated good predictive accuracy for both pressure and the combustion process within the range of $1.6 < \lambda < 2$. [13,88,143].

Recently, when the G-equation has been utilized, the MZ-WSR model is integrated before, during, and after the flame front to calculate the intermediate and post-reaction species instead of using simplified global kinetics [64,65,91]. However, its effects have not been quantitatively proven.

### 3.4. Turbulence–Chemistry Interaction

The interaction between turbulence and chemistry-related quantities is crucial in determining the combustion characteristics of PCSI applications, such as ignition, flame propagation speed in both the pre-chamber and main chamber and burn rate trends. Namely, combustion and turbulent flow are affected by each other within the turbulent reactive flow. Compared to traditional SI engines, the turbulent flow field in PCSI systems is highly inhomogeneous, exhibits large spatial gradients at the jet boundaries, and undergoes rapid temporal evolution. One of the key factors distinguishing the flame evolution in PCSI systems from conventional SI engines is the high level of turbulent fluctuation near the spark plug, driven by jet-to-jet interaction during the scavenging process. This is because this high level of turbulence level makes the flame behavior clearly not isotropic [144]. As the flame front moves through a turbulent flow, it is influenced by both large and small eddies. The larger eddies, which are much bigger than the thickness of a laminar flame, primarily cause the flame to wrinkle and distort. Meanwhile, the smaller eddies are capable of penetrating the flame front, allowing them to directly interact with the combustion chemistry. This interaction enhances turbulent mixing and amplifies diffusive processes within the flame. However, from a macroscopic point of view, the main effect of turbulence on the flame is to increase the propagation speed [145].

Previous literature utilized the source terms of combustion models developed for pre-mixed or non-premixed conventional gasoline S.I. engines to simulate these complex turbulence–chemistry interactions, such as the ECFM-3Z and G-equation models. These models differ in their approaches to representing turbulence–chemistry interactions, leading to variations in calculated laminar flame speeds, turbulence intensities, and spatiotemporal scales. As a result, they produce different turbulent flame speeds, which significantly impact engine output, fuel consumption, and pollutant formation. The predictive accuracy of these models hinges on their ability to accurately calculate turbulent flame speed.

The G-equation model calculates turbulent flame speed using explicit correlations, while the ECFM-3Z model derives it from the Flame Surface Density (FSD) equation. Differences in the velocity scale ratio ($u'/S_L$) and length scale ratio ($l_t/\delta_L$) during combustion, as illustrated in the Borghi–Peters diagram, highlight the distinct approaches these models take in handling the flame brush, particularly in terms of the spatial distribution of the reaction volume. These differences are primarily due to varying turbulence intensities and scales, which affect mean turbulence–flame interaction.

Figure 7 analyzes the behavior of the velocity scale ratio and length scale ratio on the Borghi–Peters diagram for the two combustion models previously discussed under the operating conditions of a GDI engine at 2000 rpm and stoichiometric conditions [132]. As shown in the figure, the ECFM-3Z model indicates that, except for the initial combustion phase, the combustion predominantly falls within the corrugated flamelet regime. In contrast, for the G-equation model using four different turbulent flame speed correlations, most of the combustion is observed to occur within the stirred reaction regime and thin reaction regime. The variation in combustion traces between these models is attributed to the different treatments of the flame brush by each model. This result underscores the significance of turbulence–flame interaction as a critical factor in determining the combustion characteristics in engine analysis. Furthermore, it suggests that the current flamelet-based combustion models may introduce significant errors in the analysis of PCSI engines.

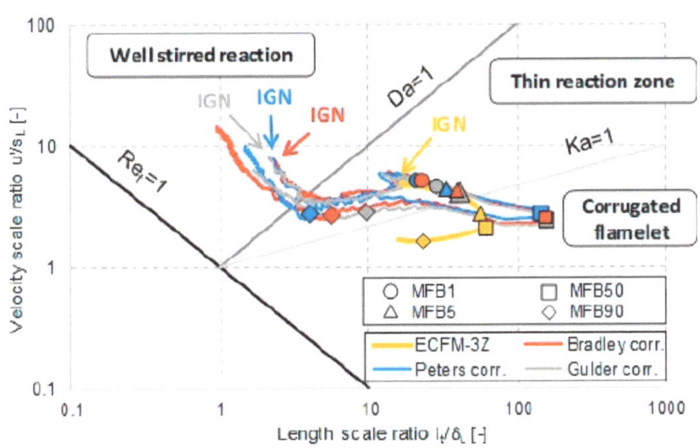


**Figure 7.** Flame evolution on the Borghi–Peters diagram using ECFM-3Z and G-equation with three different turbulent flame speed correlations [132].

As shown in Figure 7, the flamelet-based combustion models exhibit significant differences in turbulence–flame interactions during the combustion process. However, when compared to experimental ensemble-averaged in-cylinder pressure and burn rate traces, these differences are not substantial. Figure 8 illustrates a comparison between pressure traces (left) and heat release rate curves (right) from experiments and various combustion models under the same engine and operating conditions as Figure 7. From an engineering perspective, the results demonstrate overall good agreement with the experimental ensemble averages.

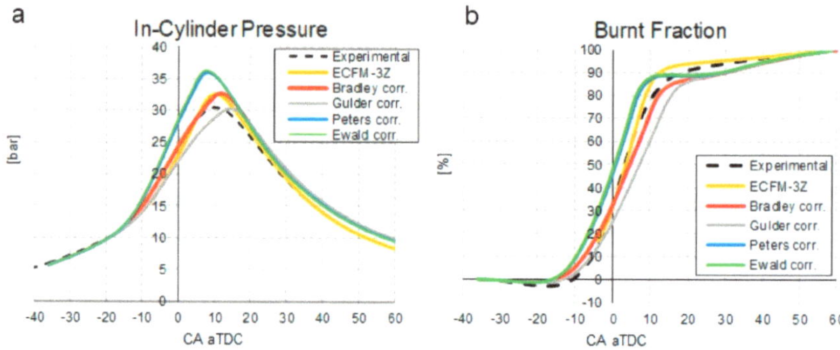


**Figure 8.** Experimental validation of (**a**) averaged pressure trace and (**b**) burn rate trend [132].

The FSD transport equation includes source and sink terms accounting for stretching and quenching effects, which are crucial for accurately modeling the dynamic behavior of the flame front in both laminar and turbulent combustion conditions. By incorporating these effects, the FSD equation comprehensively describes flame surface evolution, which is essential for predicting combustion performance and stability. As illustrated in Equation (6), the stretching effect accounts for the deformation of the flame surface due to the flow field, influenced by both laminar and turbulent strain rates. In turbulent flows, this term can be complex, involving contributions from large-scale and small-scale turbulent eddies stretching.

To simulate stretching and quenching conditions in term $S_\Sigma$ of Equation (7), the intermittent turbulent net flame stretches (ITNFS) model [141] is utilized. The ITNFS model characterizes the interaction between a single vortex and the flame front, extrapolating this interaction to represent the complete turbulent flow. It assumes that each turbulence scale influences the flame independently, with no interaction between different turbulence scales, allowing for the overall effects of turbulent fluctuations to be predicted from the behavior of individual scales in the unburned gas mixture, as shown in Figure 9.

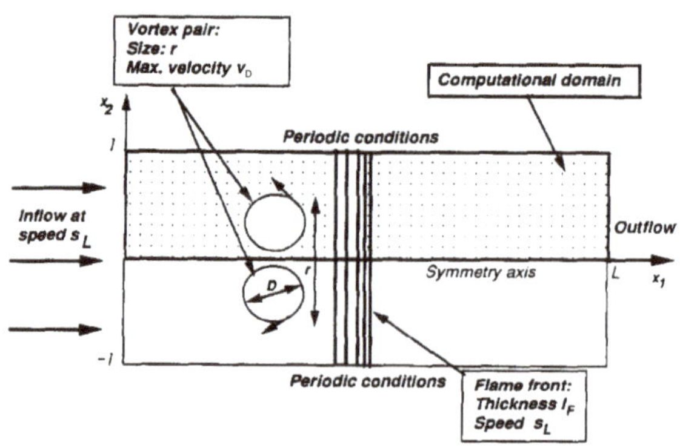


**Figure 9.** Configuration of flame vortex interaction [141,145].

By applying this model to a complete turbulent flow, it is assumed that the cumulative effect of all turbulent fluctuations can be inferred from the behavior of each scale. The limitation of this model is that it cannot account for vortex interactions. It is clear that this model has limitations when applied to PCSI engines, where strong turbulence intensity is distributed near the flame front of the hot turbulent jet and highly inhomogeneous turbulence exists. The production of flame surface density primarily results from the net flame stretch due to turbulence. This flame stretch is expressed as the large-scale characteristic strain ($\varepsilon/k$), adjusted by a function $C_t$, which considers the size of turbulence scales and viscous and transient effects [146]. $C_t$ depends on turbulence parameters and the properties of the laminar flame. Hence, the right-hand side of Equation (4) can be expressed as follows:

$$S_{production} = \alpha K_{eff}\Sigma \quad and \quad S_{destruction} = \beta\frac{\rho_{fu,fr}S_L}{\rho_{fu}}\Sigma^2 \tag{8}$$

Here,

$$K_{eff} = K_t = \frac{\varepsilon}{k}C_t \tag{9}$$

$K_t$ is a critical property since it influences the source term for the flame surface and, therefore, the mean turbulent reaction rate. The coefficient $\alpha$ is the stretching factor, and $\beta$ in Equation (8) is the arbitrary tuning constant used in ECFM.

However, it is important to better understand the contexts in which the ITNFS model can be effectively applied and where additional considerations or alternative models may be necessary. This is particularly due to the lack of consideration for nonlinear effects in the interaction of turbulence scales with the flame front, flame-generated turbulence, and the reignition of fresh gases crossing a locally quenched flame front [141].

Several previous studies [11,12,135,145,147] have reported the average turbulence–flame interaction experienced by engine flames in PCSI engines and recently developed highly downsized engines on the Borghi–Peters diagram, confirming the occurrence of multiple regimes during S.I. combustion. The results of these studies emphasize the necessity for combustion models to be predictive across all potential combustion regimes [135,141,147]. The high turbulence levels generated by the pre-chamber (PC), combined with the reduction in laminar flame speed due to dilution, significantly impact the combustion regime of PC-initiated combustion systems. These interactions are illustrated in the schematic Borghi diagram in Figures 10 and 11, where the Karlovitz (Ka) and Damköhler (Da) numbers are used to compare relevant timescales of turbulent combustion to determine the combustion regime. Consequently, the PC-initiated jet ignition combustion regime shifts into the thin reaction zone, bringing it closer to the stability limits.

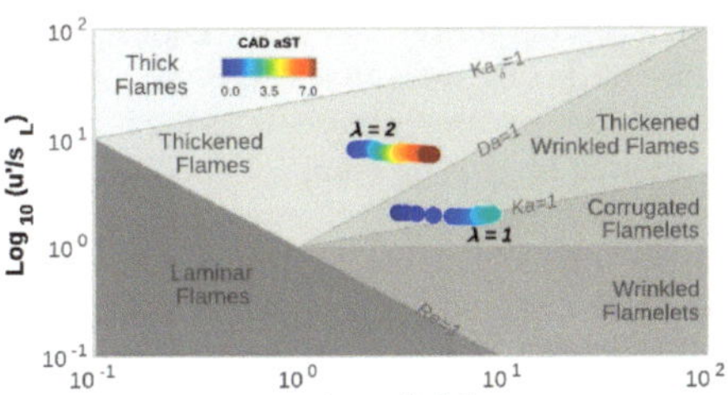


(a)

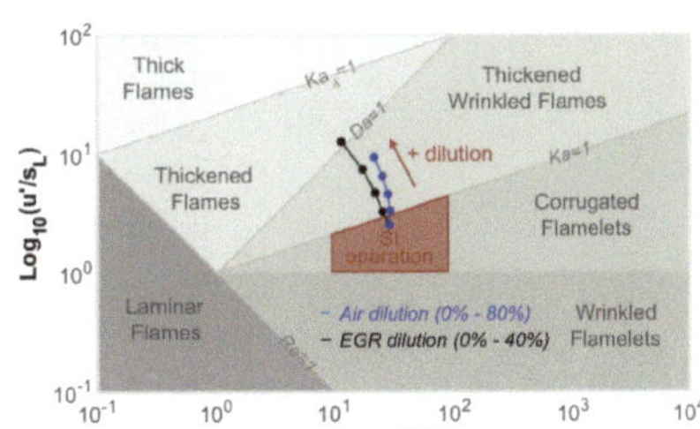


(b)

**Figure 10.** Evolution of the flame regime for (**a**) various air-fuel ratios and (**b**) dilution ratios as the pre-chamber combustion progresses [11,12].

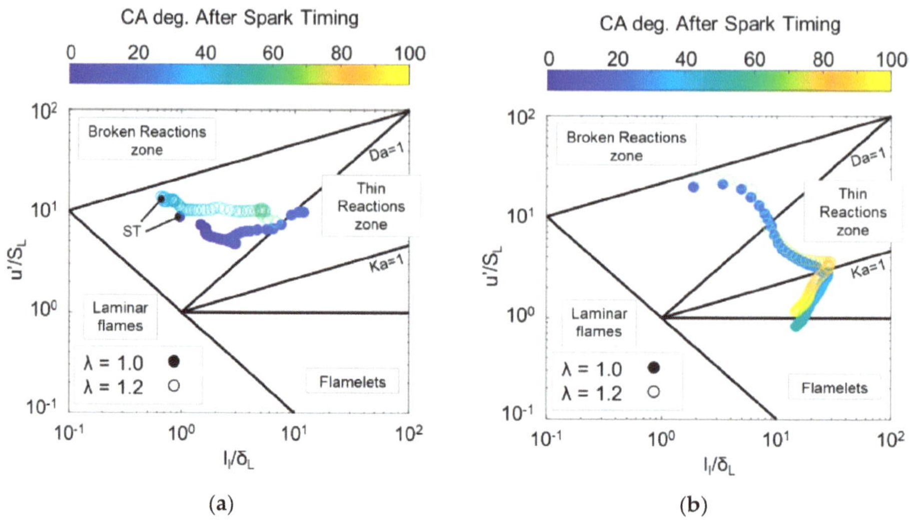


**Figure 11.** Evolution of the flame regions during pre-chamber (**a**) and main chamber (**b**) from spark timing to the hot gas ejection in the main chamber (**a**); from the start of main chamber combustion to MFB90 (**b**) [61].

The two figures below are based on the same passive PCSI engine with a port fuel injection system in a gasoline engine. Figure 10a shows the progression of the combustion region at 2000 rpm on the Borghi–Peters diagram for $\lambda = 1$ and $\lambda = 2$ cases [12]. Figure 10b illustrates the change in the flame structure during the combustion period at 4500 rpm by increasing the air/fuel ratio using air and EGR [11]. From these results, it is clear that lean combustion shifts the combustion characteristics to the thickened flame regime. As is well known, this occurs because the eddies become smaller than the flame thickness, allowing some eddies to penetrate the pre-heat zone of the flame. Another vital piece of information evident from Figure 10b is that EGR dilution exhibits higher sensitivity than air dilution [11,12].

Recently, the gasoline-fueled, passive, pre-chamber engine has been numerically modeled using the RNG k-ε turbulence model and MZ-WSR with the SAGE combustion model, and the combustion characteristics for $\lambda$ = 1.0 and 1.2 at 4000 rpm have been investigated [61]. Figure 11 shows the evolution of the turbulent regimes in the pre-chamber and main chamber for two operating conditions on the Borghi–Peters diagram. A noteworthy feature of this figure is that almost all pre-chamber combustion, even in lean cases, evolves in a thin reaction regime due to the high turbulence intensity produced by strong jet-to-jet interaction during compression stroke. As shown in Figure 11a, during the initial stages of combustion, the velocity scale ratio decreases due to the weakening of the initial turbulence intensity. Subsequently, it recovers as the residual gas decreases. The length scale ratio shows that the integral length scale ($l_I$) remains constant, and the laminar flame thickness ($\delta_L$) gradually decreases and stabilizes. Under lean conditions, the laminar flame thickness increases, and the laminar flame speed decreases, causing the curve to shift upward. From these results, it can be deduced that the MZ-WSR model is suitable for pre-chamber combustion. Figure 11b shows that compared to pre-chamber combustion, main chamber combustion occupies a broader area on the Borghi–Peters diagram, starting from the border of the broken reaction zone, passing through the thin reaction zone, and moving into the wrinkled flame zone. This occurs because the high turbulence intensity is distributed across the flame front when the hot turbulent jet is ejected. As the intensity of the jet and turbulence decreases, the combustion quickly transitions through the thin reaction zone and linearly moves into the corrugated reaction zone. Therefore, combustion in the main chamber experiences three combustion regimes due to the rapid changes in velocity and length scale ratios, indicating the presence of complex turbulence–flame interactions. This figure suggests that for PCSI engines, a multi-mode combustion model that can cover a wide range of the Borghi–Peters diagram needs to be developed. It is important to note that Figures 10 and 11 are based on the RANS type k-epsilon turbulence model using Boussinesq's isotropic eddy viscosity assumptions and flamelet-based combustion models, which may introduce some errors.

### *3.5. The Well-Tuned Versions of Combustion Models*

As previously explained, to accurately predict the combustion in a PCSI engine, developing a model capable of calculating multi-mode combustion is essential. Currently, utilizing a flamelet-based combustion model is a practical alternative. Therefore, recent research has focused on tuning the model coefficients $b_1$ and $b_3$ in Equation (2), which represent the large and small-scale turbulence enhancements in the G-equation model described by Equation (2) [64,65,91]. This tuning process is performed ad hoc.

In the case of combustion models using Flame Surface Density (FSD) or MZ-WSR, efforts have been made to adjust the flame stretch factor, $\alpha$ in Equation (8) for FSD and multiplier, $\gamma$ in Equation (7) ad hoc to align with experimentally obtained in-cylinder pressure traces and heat release rate profiles [148]. However, these treatments could not be adequate to achieve a precise correlation between simulations and experimental results in both the pre-chamber and the main combustion chamber [64,65,148]. As described above, such studies attempt to address the limitations of telling the turbulence–flame interaction using RANS-type turbulence models and flamelet-based combustion models by calibrating the model coefficients included in the equations that represent combustion speed or burn rate.

Kim et al. [64] tuned the multiplier of the MZ-WSR model within the range of 1.0 to 1.4. Additionally, they adjusted the coefficients $b_1$ and $b_3$ for the G-equation model between 1.0 and 2.5. for two air/fuel ratios at 1200 rpm. The engine used in this study is a 1.86-L passive PCSI PFI engine fueled by natural gas. They validated the improvement in prediction accuracy by comparing the in-cylinder pressure traces and heat release rate profiles with experimental data. Figure 5 compares the cylinder pressure and heat release rate obtained from experimental data and the simulation results using the tuned models. The experimental data includes results from 300 cycles (represented by light gray lines) and the averaged pressure (depicted by a bold black line). Figures 12 and 13 compare the simulation results to the corresponding experimental results without tuning the model coefficients. The results show that without tuning the model coefficients, it is impossible to predict the combustion characteristics of the PCSI engine accurately. Additionally, the discrepancies are more pronounced in the lean region, likely due to the inability to account for the transition to the thin reaction zone as the flame thickness increases and the laminar burning velocity decreases.

The tuning of model coefficients significantly improves the predictive accuracy of the G-equation model. However, the MZ-WSR model fails to enhance predictive accuracy with coefficient tuning, as it does not account for the effects of small-scale turbulence on the reaction rates.

Silva M et al. [65] conducted a detailed analysis on a 2.1-L active pre-chamber engine using natural gas as fuel, focusing on the impact of turbulent jets ejected from the pre-chamber on the burn rate in the main chamber. This study utilized the G-equation combustion and RANS-type turbulence models for two different orifice diameters. In this work, the turbulent flame speed equation was tuned with coefficients $b_1$ and $b_3$ set to 0.78 and 2.0, respectively. Figure 14 compares the pre-chamber, and central chamber pressure traces at 1200 rpm for the two orifice diameters between the calculated and experimental results. As observed, there is excellent agreement between the experimental and predicted data.

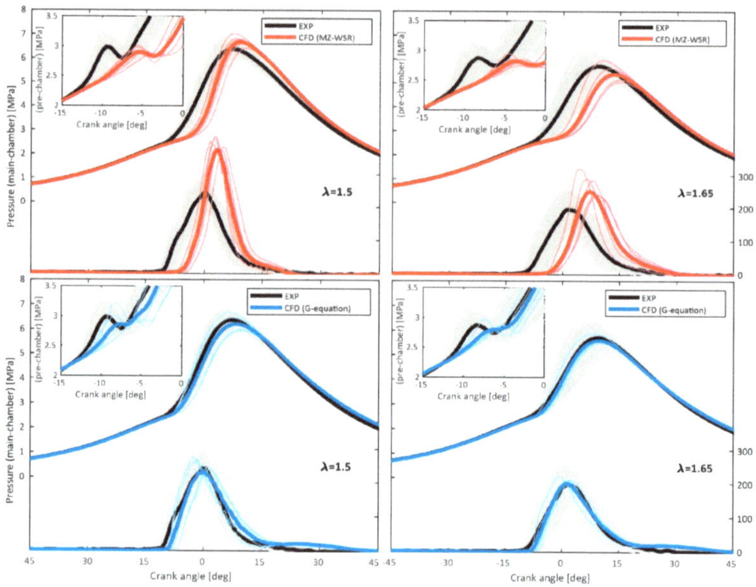

**Figure 12.** Comparison of pressure traces and heat release rate curves between experimental data and tuned simulation results for MZ-SWR and G-equation [64].

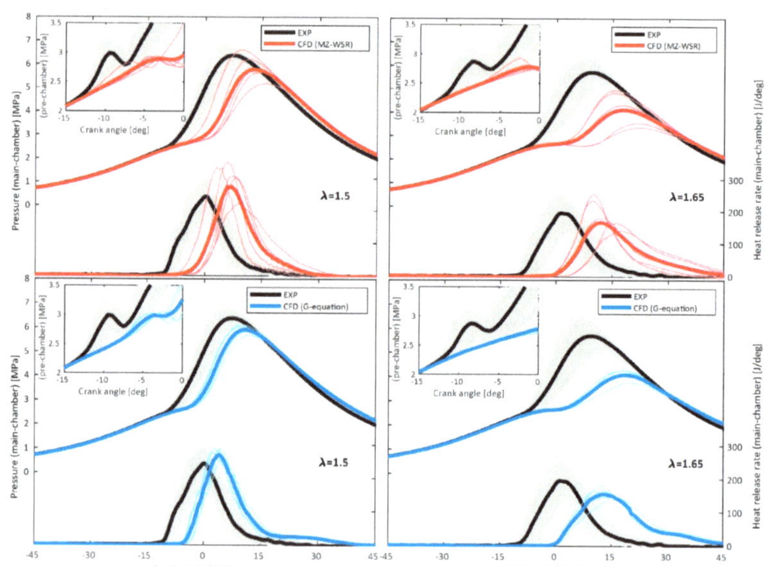

**Figure 13.** Comparison of Pressure traces and heat release rate curves between experimental data and untuned simulation results for MZ-SWR and G-equation [64].

In this computation, a highly sophisticated 0D/1D wave model [11,65] was employed to obtain initial conditions and temperatures at the walls, as well as temperature and pressure values at the inflow and outflow boundaries for the 3D CFD analysis. However, this model requires input conditions such as lift curves of intake and exhaust valves, pressure pulsations in the intake and exhaust pipes, and fuel lines. Additionally, experimentally obtained pressure traces are needed to analyze pre-chamber and main-chamber combustion phenomena.

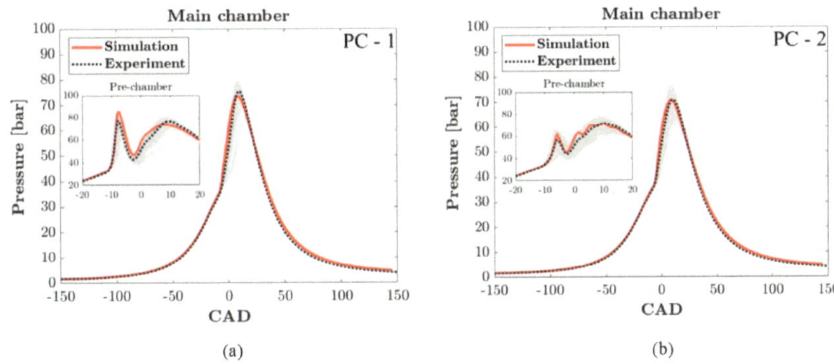


**Figure 14.** Experimental validation of pressure curves for throat diameters of (**a**) 3.5 mm and (**b**) 5.5 mm [65].

Therefore, performing precise 3D CFD analysis of the PCSI engine without preceding experiments is challenging. Moreover, tuning model constants in the combustion model based on experimental values is necessary, which varies depending on the engine displacement, fuel used, pre-chamber geometry, and fuel supply system.

To the authors' knowledge, only a few studies have utilized well-calibrated combustion models [45,76]. Although the physical phenomena of TJI combustion are not yet fully understood, model constants in the correlation or source term for various combustion models are adjusted to correct the laminar flame speed and align with experimental pressure traces [41,47,48]. However, the ability to accurately predict all phenomena involved in TJI combustion remains uncertain [45].

Figure 15 shows the recent share of (a) combustion models and fuels for PCSI simulations. The majority, 45.9% of studies, relied on the MZ-WSR model, while 32.4% opted for the G-equation model. Subsequently, the frequency of usage for the ECFM(3Z) model accounted for 16.2%, while less frequently used combustion models, such as the Weller model, were grouped as 'etc'. Additionally, the fuels used in the CFD simulation for the PCSI engine were categorized and shown in Figure 15b. As illustrated in the figure, 50% of PCSI engine studies utilized natural gas fuel, followed by gasoline fuel. Recently, research has emerged on using carbon-free fuels such as ammonia and hydrogen in PCSI engines [36,44,86]. These studies employ a dual fueling strategy, utilizing natural gas [84] and diesel fuel for the initial combustion stage, and have been grouped under dual fuel.

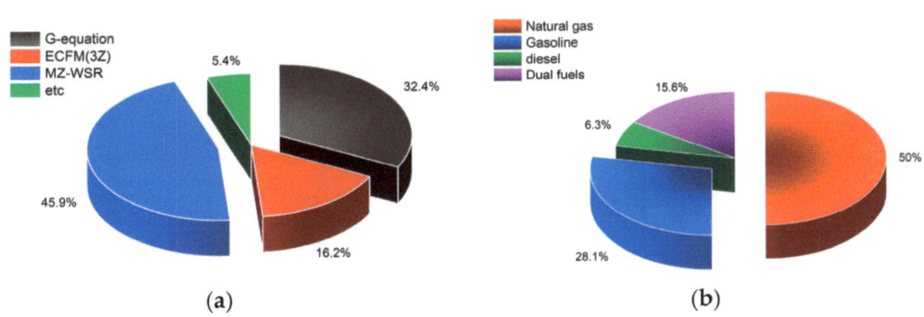


**Figure 15.** Categorical distribution of type of (**a**) combustion model and (**b**) used fuel for simulating PCSI engines.

### 3.6. Numerical Grid of Pre-Chamber Engine

The process of subdividing a physical domain into smaller subdomains is known as grid generation or spatial discretization, which divides the spatial domain into a mesh or grid. Each grid point, cell, or control volume represents a discrete location in the domain. In this context, meshing is crucial in defining the discrete elements where physical laws are applied. These meshes establish the cells or elements over which flow calculations are conducted. The primary goal of the simulation is to generate numerical values for critical variables, such as velocity, pressure, and temperature, at specific points within the mesh. As the grid size decreases, the solution to the discretized problem converges to the solution of the continuous problem. Thus, the grid significantly influences both the rate of convergence and the accuracy of the solution. Therefore, the grid generation step is a critical component of the simulation process [137], and generating a sufficiently fine and uniform grid for complex geometries is both challenging and time-intensive. It is difficult to ensure that the grid adequately captures all relevant scales of turbulence while keeping the computational requirements manageable.

#### 3.6.1. Mesh Generation for LES Turbulence Model

In the LES turbulent model, the mesh limits the size of the eddy that we can physically represent on the mesh [54,102]. LES resolves scales from the domain size L down to the filter size Δ, requiring substantial resolution of high wave number turbulent fluctuations. This necessitates high-order numerical schemes or acceptable grid resolution if low-order numerical schemes are employed [102,103,107]. Δ is the length scale or grid filter width and is determined as follows:

$$\Delta = (\mathrm{V})^{1/3} = l_0/5 \quad (10)$$

where the integral length scale, $l_0$, the size of the eddy, is directly related to the computational cell volume, V.

It is known that good LES should resolve at least 80% of the turbulent kinetic energy, which is dependent on mesh size [108]. As mentioned above, the remaining kinetic energy is modeled in the sub-grid model. Finally, the finer the mesh, the more turbulent kinetic energy is resolved. As shown in Figure 5, the sub-grid scale viscosity is also a function of the mesh size. Hence, we are solving different equations on each level of refined mesh. Consequently, a traditional mesh independence study, which aims to ensure that simulation results are not influenced by mesh density, becomes impractical. Instead, alternative validation techniques must be employed to ensure the reliability of LES results.

Accordingly, the LES turbulence model requires representing the integral length scale of each eddy with at least five grid cells, which demands an enormous number of grids. Additionally, it is not easy to ensure that Equation (10) conditions are satisfied meticulously. Due to the computational burden resulting from a large number of computational grids, studies using LES (Large Eddy Simulation) [32,52] have primarily focused on geometrically and computationally simple cases, such as engine-like structures like the RCEM or constant volume chambers with single-hole nozzles and low-pressure air [57]. However, recent studies [59,67] have conducted three-dimensional CFD analyses using LES models to understand the complex thermo-fluid dynamic processes throughout the entire working cycle of the engine. This trend is clearly illustrated in Figure 4.

#### 3.6.2. Mesh Generation for URANS Turbulence Model

It is well-known that huge grid sizes in RANS simulations can lead to under-resolved mean fields. This discrepancy significantly differs between the actual RANS and computationally resolved fields. Such under-resolution substantially impacts the accuracy of chemical simulations, often more so than the direct effects of turbulent fluctuations on the chemistry, which turbulent combustion interaction models are typically designed to account for [145]. Consequently, in scenarios of poor resolution, typical turbulent combustion models must compensate for under-resolution to maintain accuracy, even though they were not initially intended for this purpose. Therefore, increasing the resolution in under-resolved areas of the simulation can minimize the errors the unresolved field has on the chemistry. Once the resolution-related errors are mitigated, it can be observed that the remaining effects of turbulent interactions on combustion chemistry are of the same order of magnitude as the accuracy of the detailed kinetic mechanism being used. Additionally, it should be noted that in the case of an under-resolved flame front, the second derivative in the species conservation equation (representing diffusion) and the second derivative in the energy equation (representing conduction) would be under-estimated. This underestimation reduces the mixing, resulting in a lower calculated laminar flame speed.

All commercial CFD software provides grid generation tools in their pre-processors. When using the Converge CFD code, which is most commonly used for CFD analysis of PCSI engines, as shown in Figure 3, the Adaptive Mesh Refinement (AMR) algorithm is employed to increase the grid refinement level only in regions characterized by high velocities and large temperature gradients (e.g., flame fronts), without excessively slowing down the simulation with the uniformly refined grid. The mesh was re-generated at each iteration and dynamically refined. This approach resulted in the total number of cells in the computational domain ranging from 300,000 to 700,000 during the compression stroke to 1.23~3 [61] million to 9 million [88] at the end of the combustion phase. Studies using the Converge CFD code have set the base grid size in the pre-chamber to 1 mm [48,63,64]~4 mm [13,88] and, through a series of localized mesh refinements, achieved a minimum cell size of 0.125 [48,61,63]~0.5 [13,63,64] mm. By varying the grid size in the pre-chamber, the y+ value is monitored to ensure that the simulation results fall either within the recommended range (30 < y+ < 100) or within a range (20 < y+ < 30) that can be handled by a wall function [64]. Additional fixed mesh refinement is applied locally around the spark plug gap to capture the ignition phase [13,47,61,63,64] accurately. Additionally, the pre-chamber area, including the nozzles and the regions traversed by turbulent jets entering the main chamber, is locally refined to a resolution of 0.125 [13,47]~0.25 mm [64,88]. Figures 16 and 17 display the complete engine geometry and the computational grid layout along a plane intersecting the center of the pre-chamber. A cut-cell method using a Cartesian grid was applied to model the two different complex geometries with moving boundaries. As shown in Figure 17, the AMR algorithm can be used to optimize the grid distribution during the flame propagation phase in order to minimize computational errors caused by grid under-resolution in the ejection phase, where there are steep gradients in velocity, pressure, temperature, and concentration due to turbulent jets.

#### 3.6.3. Differencing Process

A differencing scheme is a specific method to approximate the derivatives in the discretized equations. These schemes define how the values at the discrete points in the grid approximate the derivatives of the flow variables. There are several differencing schemes for convection and diffusion terms. One frequently adopted for PCSI engine CFD simulation is the upwind differencing scheme [149], which uses values from upstream points to approximate the derivatives. It introduces numerical false diffusion [149] but is more stable for convection-dominated problems. There are several variations, such as first-order upwind and higher-order upwind schemes. Although higher-order differencing schemes such as the Quadratic Upwind Interpolation for Convective Kinematics (QUICK) scheme and Total Variation Diminishing (TVD) scheme [150] can achieve high-accuracy solutions, they often compromise solution stability. As a result, for the CFD analysis of PCSI engines, which requires the simultaneous resolution of complex multi-mode combustion coupled with strong turbulence and conjugate heat transfer, most studies do not employ differencing schemes with accuracy higher than second order. Finally, it is always a big challenge for CFD engineers to choose an appropriate differencing scheme that better balances between accuracy and stability. In automotive and mechanical engineering, the sensitivity of discretization schemes for convection terms in LES has been thoroughly explored, as documented [151].

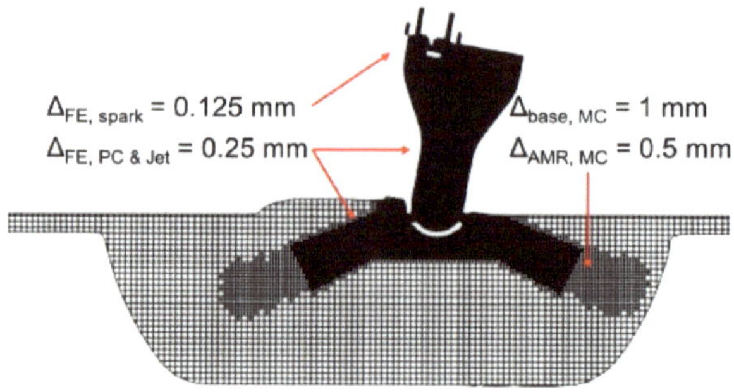


**Figure 16.** Cross section of computational grid configuration [63].

**Figure 17.** Local refinements and AMR strategy employed in PCSI Engine during the ejection phase [48].

The primary function of a Sub-Grid Scale (SGS) model in LES is to dissipate resolved turbulent fluctuations effectively. The SGS model is specifically designed to provide the appropriate level of dissipation. Therefore, in LES, we should employ central differencing schemes, as they do not introduce any numerical dissipation due to their symmetric treatment of fluxes, ensuring that no artificial viscosity is introduced into the system. Central differencing schemes are second-order accurate, balancing accuracy and computational cost well. However, they can be less stable than upwind schemes.

Central Differencing Scheme: Uses the average of values at surrounding points to approximate derivatives. It is second-order accurate but may introduce numerical dispersion and can be unstable in convection-dominated flows. Opting for the central difference scheme is generally favored to mitigate numerical inaccuracies; nonetheless, a common practice involves blending the outcomes of upwind schemes with those of the central scheme to enhance computational stability.

### *3.7. Time Discretization*

To obtain a numerical solution, partial differential equations must be discretized in both spatial and temporal domains. Partial differential equations must be discretized in spatial and temporal domains to obtain a numerical solution. Time-dependent variables, also known as transient terms, are mathematically represented as derivatives concerning time. However, from a physical standpoint, these terms require special handling. Transient terms describe the variation of a specific variable over time within an infinitesimally small control volume, adhering to conservation principles while preserving generality. When addressing the discrete treatment of transient factors, it is generally preferable to seek a time-varying solution, as this approach directly influences the accuracy of the numerical results. For predominantly steady flows, it is advisable first to determine a time-dependent approximation and transition to a steady-state approximation. Selecting the appropriate time step is crucial in simulating turbulent jet ignition and spark-ignition combustion in

PCSI combustion (IC) engines. Accurately capturing pressure fluctuations due to deflagration within the cylinder or other phenomena in two divided chamber IC engines depends heavily on the time step resolution used in the simulation. A variable time-step approach, utilizing the Courant–Friedrichs–Lewy (CFL) number, has been implemented in almost all previous literature [28–33,48,63,148]. The maximum CFL number differed across various domain regions, spanning from 1 to 5, and the time step was automatically adjusted within a range of 0.001 to 1 CAD [48] or $2.5 \times 10^{-5}$ to $1\times 10^{-8}$ s [88].

### 3.8. Appropriate Initial and Boundary Conditions

The most straightforward method for obtaining initial conditions (such as intake and exhaust temperatures, pressures, injection quantity, and ignition timing) and boundary conditions is to utilize experimental data. However, this approach cannot be employed until a prototype engine is constructed, so it is only applied in a few specialized research studies [29]. In addition, one of the primary difficulties in 3D CFD simulating PCSI combustion systems is the scarce availability of boundary and initial conditions within the main chamber and the pre-chamber. Typically, installing measurement instruments in the confined space of a pre-chamber is extremely difficult. Therefore, in most previous studies related to 3D CFD simulation, initial values of all required thermodynamic parameters (including composition) were directly transferred from the 0D/1D model(s). This section describes the 0D/1D simulations that provide the necessary initial and boundary conditions for the 3D CFD simulation of the PCSI engine based on the findings of previous studies.

#### 3.8.1. Inflow and Outflow Boundary Conditions of Intake and Exhaust

In the CFD analysis of a PCSI engine using RANS models, the most influential boundary conditions are the inflow and outflow conditions. It is necessary to obtain time-dependent variations in velocity, pressure, temperature, and chemical composition at the cross-sections of the engine's intake and exhaust pipes. The value of CFD simulations lies in their ability to provide precise foundational design data before the prototype is created. Therefore, an additional simulation is required to obtain the aforementioned boundary conditions before the prototype engine is built. Thus, time-varying pressure, temperature, and chemical composition profiles derived from a 0D/1D model built using commercially available 1D gas dynamic codes, such as GT-SUITE [152], GT-POWER [11,12,16,82–88], and WAVE [58], validated with experimental data, were implemented as the inflow and outflow boundary conditions. These same 1D CFD models were utilized to specify the chemical composition.

These one-dimensional gas dynamic commercial programs thermodynamically model the components comprising the engine and consider the intake and exhaust pipes, which connect each element, as one-dimensional reacting compressible. They utilize the Method of Characteristics (MOCs) [153] to calculate the pulsating flow characteristics during the intake, compression, combustion, and exhaust processes. This approach allows for fast and accurate calculations of these characteristics at the system level of the entire engine. The gas dynamic simulation consists of intake, exhaust system, and cylinder units. Geometrical data, flow losses, and heat transfer between the pipe wall and the gas are considered. The pre-chamber combustion engine model incorporates two different thermodynamic descriptions of the main chamber and pre-chamber operating cycle, integrated with the gas exchange processes through the intake and exhaust valves and orifices in the pre-chamber. The main chamber, pre-chamber, and intake and exhaust pipes are treated as a control volume, ensuring that energy and mass balances are maintained during the intake and exhaust phases, which utilize two distinct systems, each containing two thermodynamic zones, to independently model the combustion processes within the pre-chamber and main chamber. These systems are interconnected through nozzles, facilitating the exchange of mass and enthalpy between them. During the compression, combustion, and expansion phases of both chambers, the energy balance of the combustion and heat release rate is obtained from either experimental information [11] or a two-zone model-like approach [154,155], which requires the 3D-derived turbulent intensity profiles in both chambers because preliminary 3D CFD analyses or experiments should be carried out in motored conditions to tune the constants of the turbulence sub-model included in this model [8,9]. Another straightforward way of modeling the combustion of the pre-chamber is using the Wiebe function [11,153], which was obtained from the heat release rate from the experimental pressure trace [11].

A very recent, very delicate 1D model based on GT-Power is used to further refine initial and boundary conditions of pressure and temperature to the 3D CFDs model [11,12, 16,29,60,61,82–88]. However, this 1D model requires intake and exhaust valve profile curves, intake, exhaust, fuel lines, pre-chamber, and main chamber pressure traces. Combustions in the pre- and main chambers are expressed by heat release-based calibration (HRBC), where the heat release rates are computed from the measured pressure traces. The discharge coefficient through the pre-chamber nozzle is also calibrated to accurately replicate the pre-chamber pressure rise observed in the experimental data [45,65].

Figure 18 illustrates the single-cylinder active pre-chamber engine modeled using GT-POWER. This figure is a representative schematic diagram of an active PCSI engine modeled using a one-dimensional thermodynamic numerical analysis method (0D/1D model). As shown in the figure, the pre-chamber is modeled as a constant volume chamber, a check valve, and a nozzle. To obtain a more refined 1D model, calibration must be performed using experimentally obtained pressure traces from both the pre-chamber and main chamber to get accurate flow coefficients of the nozzle and check valve [45,65].

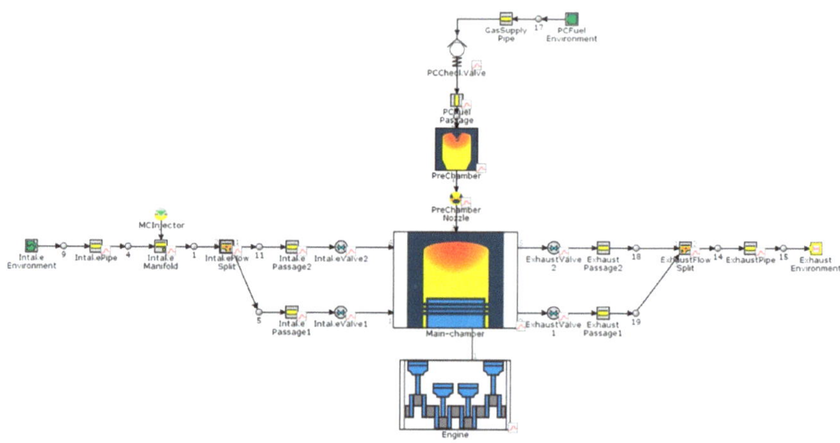


**Figure 18.** The 1D model is based on GT-Power for single-cylinder active PCSI engines [45].

Applying appropriate boundary conditions for LES is complex, especially since inflow conditions must accurately represent turbulent fluctuations. This is critical because LES requires supplying a dynamic velocity field that fluctuates across both temporal and spatial dimensions. However, studies using LES [32,33,59] do not provide detailed information regarding the boundary conditions.

### 3.8.2. Wall Heat Transfer Modeling

Given the challenges in experimentally measuring instantaneous gas-to-wall heat fluxes, 3D CFD simulations of in-cylinder processes have become indispensable. These simulations are crucial for assessing the overall heat transfer to the combustion chamber walls, its spatial distribution, and the heat losses through the boundaries of the computational domain. A conjugate heat transfer model is indispensable for precisely simulating the uneven temperature profiles in both the pre-chamber and principal chamber and effectively capturing their effects on the turbulent jet quenching and stretching phenomena. While numerous studies have emphasized the importance and necessity of developing models suitable for PCSI engines, no related research results have been reported to date.

In the 3D CFD simulation of IC engines, the heat flux to the walls significantly impacts the temperature inside the combustion chamber, thus greatly influencing the accuracy of predicting emissions such as NOx [156]. Therefore, numerous heat transfer models and wall heat flux correlations exist in the literature [157–161], many of which were developed and validated through experiments conducted in research laboratories over the past decades, typically under low-load and low-speed engine conditions. Due to engine downsizing, operating conditions in the low-speed high-load range have become increasingly important, resulting in a significant increase in the thermal loads on engine components facing the combustion chamber. Under high-load, low-speed operating conditions, the suppression and control of NOx emissions and knocking are critical design factors. As a result, extensive research has been actively conducted on the development and validation of wall heat transfer models with high predictive accuracy. Recent studies [156] have revealed that existing wall heat transfer models (Angelberger's [160] and Han and Reitz's [161]) tend to overestimate wall heat transfer, as evidenced by experimental engine thermal surveys and temperature measurements conducted on four currently produced engines. Consequently, existing heat transfer models require a new calibration. Recent studies have proposed alternative heat transfer models for optimizing wall heat transfer in highly downsized spark-ignition engines [156,162] and for engines using carbon-free fuels like hydrogen [163].

As inferred from previous research findings, all these commonly used wall heat transfer models are sufficient under certain conditions. However, their adequacy for PCSI engines has yet to be evaluated and remains uncertain. For PCSI engines, the validation and calibration of wall heat transfer models are crucial. The conjugate heat transfer between the pre-chamber surface and the gas greatly affects flame quenching, significantly impacting the flame speed prediction. The importance of developing wall heat transfer models suitable for PCSI engine modeling has been highlighted in previous research. Chinnathambi et al. [88] underscored that a specialized wall heat transfer treatment would be necessary for the pre-chamber wall and the nozzles to address the uneven temperature distribution and the quenching effects accurately.

The wall heat transfer model by O'Rourke and Amsden [74] and Angelberger [12,160] is the most frequently adopted wall heat transfer model for PCSI engines. This model is one of the most widely used for both isothermal and non-isothermal wall heat transfer. The O'Rourke and Amsden model assumes constant near-wall flow temperature and density, as well as a fixed Prandtl number in the boundary layer. In contrast, Angelberger's model allows for variations in near-wall flow temperature and density within the specified modeling range [164]. Surface temperatures are determined using a simplified predictive FEM model based on simulations that incorporate calibrated 0D/1D models [67].

Furthermore, utilizing the lumped model based on a thermal resistor network [153] with the experiments can yield the wall temperatures for the piston, liner, and cylinder head within the 3D CFDs model [29,165].

In the thermal resistor network model, the film coefficient, necessary to calculate heat loss between the gas and the wall, is obtained from empirical correlations such as Woschni's or Annand's correlations [153,163]. Recent research [166] suggested using a modified Woschni's correlation for pre-chamber heat transfer analysis to simulate PCSI combustion's characteristics precisely. It is important to note that the two heat transfer empirical formulas require calibration of constants, including the correlations, to accurately describe the conjugate heat transfer between the cylinder wall and in-cylinder gases based on factors such as fuel type, combustion method, bore, stroke, and turbocharging type [163,167]. However, a specific heat transfer model for pre-chamber walls is currently unavailable [166], and calibrating these empirical formulas for PSCI engines is time-consuming. Consequently, the uncertainties related to this aspect cannot be eliminated. Figure 18 illustrates a schematic diagram of the single-cylinder pre-chamber system modeled using a 0D/1D approach [59].

## 4. Experimental Validation

Experimental validation plays a vital role in CFD modeling as it determines the accuracy of a model in representing real-world scenarios for specific engineering applications. Conducting validation involves comparing CFD simulation outcomes with experimental data to ensure precision in engineering analyses. For PCSI engines, most numerical studies have used the in-cylinder pressure traces and heat release rate curves in the main chamber as validation data. As shown in Figure 19, multi-cycle LES calculations with the ECFM-3Z combustion model show different combustion paths with cycle variation on the in-cylinder pressure traces [67]. This trend is also observed in the heat release rate curves. This is because the HRR is derived from the pressure signal in the main chamber. It's important to recognize that the experimental HRR provides a simplified estimation of the energy released during combustion. Therefore, only the ignition onset, peak combustion rate, and duration should be rigorously considered for validation.

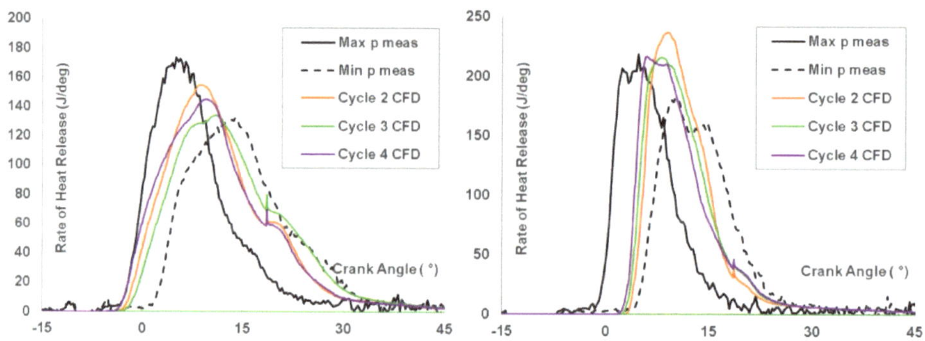


**Figure 19.** Comparison of in-cylinder heat release rate for small and big active PCSI engines between CFDs based on LES and Experiment [67].

Additionally, as seen in Figures 12 and 13, in the case of the G-equation, it was found that the model constants related to large and small-scale turbulence enhancement, which represent flame propagation speed, can significantly impact the pressure and heat release rate behavior inside the cylinder of a PCSI engine. However, this does not necessarily guarantee an accurate representation of turbulence–chemistry interaction. Therefore, before fine-tuning the physical models used in predictive PCSI engine modeling, it is crucial to identify and understand the primary sources of uncertainties. Many researchers have pointed out various uncertainties, including the lambda value inside the pre-chamber, excessively rich air/fuel ratios [168], and the correct calculation of laminar flame speed under ultra-lean conditions [91].

Syrovatka et al. [67] performed multi-cycle simulations on a single-cylinder active PCSI engine fueled by natural gas using the high-fidelity turbulence model LES and the ECFM-3Z combustion model. The results were compared to experimentally obtained heat release rate curves, as shown in the figure below. The operating conditions were set at a speed of 1800 rpm with an air excess ratio of 1.05. To analyze the cycle-by-cycle variation (CCV) effect, heat release rate curves for three cycles, excluding the first cycle calculated from the CFDs model, were displayed for two engines with different pre-chamber volumes. The maximum and minimum heat release rate curves obtained from the experimentally measured pressure traces over 120 cycles were shown for experimental validation. Figure 19 indicates that the LES modeling approach effectively captures the cycle-by-cycle variation of the PCSI engine, with all CFD traces falling within the limits of the experimental data. The figure also highlights the significant CCV in heat release rate curves due to lean combustion in the main chamber.

These results show that it is evident that analyzing model uncertainty by validating pressure and heat release rate curves averaged over 300 cycles using RANS-type turbulence models is very challenging. These findings suggest that the development of high-fidelity turbulence models capable of accounting for vortex-to-vortex interaction and non-uniform distribution of vortices, along with combustion models that can simulate multi-mode combustion, will be pivotal in establishing CFD analysis technology as a vital process in design engineering.

## 5. Critical Issues of CFD Simulation of a PCSI Engine

This study demonstrated that the ignition process in a pre-chamber spark-ignition (PCSI) engine is a complex phenomenon driven by the intricate thermo–chemical interactions governed by both chemical kinetics and fluid dynamics. The interaction between the pre-chamber jets and the main chamber charge plays a pivotal role in this process. Therefore, a thorough understanding of these kinetic and dynamic factors is crucial for a complete comprehension of pre-chamber behavior. To date, most of the fluid dynamic and chemical kinetics models employed in research have not been specifically developed for pre-chamber spark-ignition (PCSI) engines but rather have been predominantly designed for use in conventional spark-ignition (SI) engines. This indicates that, despite extensive research into the jet formation process and subsequent combustion in PCSI engines through three-dimensional CFD analysis, numerous issues remain that have yet to be fully understood. Figure 20 schematically illustrates the types of numerical models and their interrelationships used to represent the multi-physics nature occurring in a PCSI engine through CFDs. The figure provides an integrated overview of how these models are employed to simulate the complex interactions within the engine.

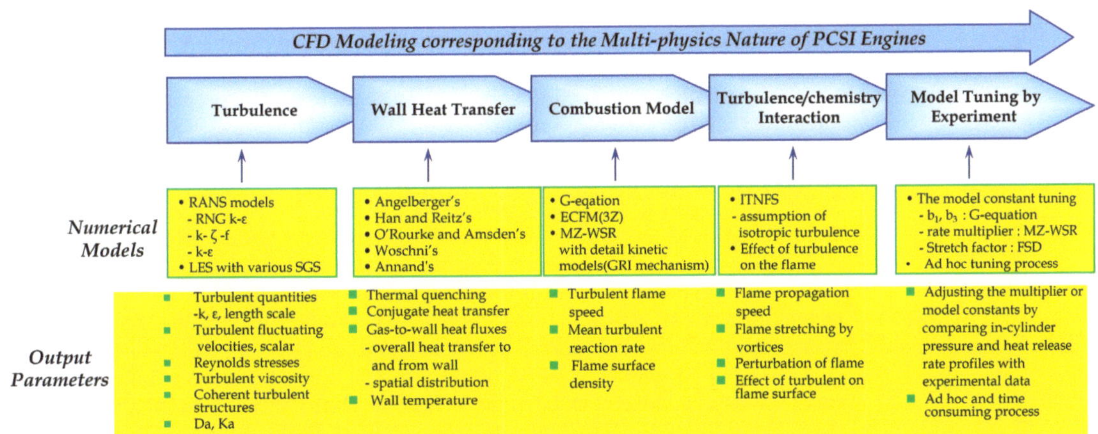


**Figure 20.** Schematical diagram of CFD modeling corresponding to the multi-physics nature of PCSI engine.

This study also provides a thorough and concise explanation of the limitations associated with various numerical models that have been widely applied to CFD modeling of PCSI engines. In this section, Figure 21 presents a schematic illustration aimed at clarifying the current predictive limitations and uncertainties of CFD models for PCSI engines. As depicted in the figure, uncertainties inherent in turbulence models propagate through to combustion processes, turbulence–flame interactions, and turbulent jet–wall interactions. A significant challenge in current CFD simulations are the uncertainty in predicting the anisotropic behavior of turbulent flames, which stems from the assumption of isotropic turbulence in widely used RANS-averaged turbulence models. Addressing this issue is critical for advancing the accuracy of CFD simulations.

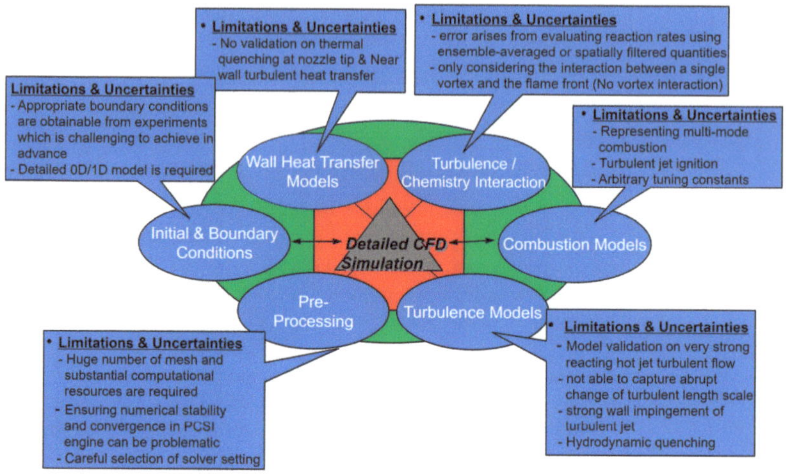


**Figure 21.** Schematic diagram of limitations and uncertainties of CFD models for the PCSI engine.

Additionally, the development of a multi-mode combustion model capable of accurately simulating turbulent premixed flames at medium and high Karlovitz numbers, particularly in PCSI engine combustion, is essential for enhancing the reliability of future CFD analyses. Furthermore, as discussed in Section 3.3, improving the ITNFS model, which currently simulates flame–vortex interactions independently without accounting for vortex-to-vortex interactions, is an urgent priority. Developing a model that incorporates flame interactions with multiple vortices could lead to more precise predictions of combustion rates and flame propagation speeds.

Another noteworthy limitation that should be highlighted is that when the mean field generated in a RANS simulation is insufficiently resolved, often due to the use of overly coarse grid sizes, a significant disparity can arise between the actual RANS field and the one represented on the computational grid. This under-resolution poses a major challenge to accurately simulating the chemical processes, with its negative impact likely surpassing the direct influence of turbulent fluctuations on the chemistry. When the flame front is insufficiently resolved, the second derivative terms in both the species conservation equation (pertaining to diffusion) and the energy equation (pertaining to conduction) are likely to be underestimated. This underestimation leads to diminished mixing, which, in turn, results in a lower calculated laminar flame speed. Finally, recent advancements in hardware and computational models have enhanced the quality of numerical analysis and made these tools more accessible to a broader range of researchers. However, as previously mentioned, 3D CFD simulations for PCSI engines remain highly time-consuming due to the large number of grids required, especially when compared to other simulation methods. Additionally, these simulations require boundary conditions that are often only obtainable through experimental methods or repetitive calculations using 0D/1D models, which represents a significant drawback to numerical simulations.

## 6. Conclusions and Future Research Directions

This paper comprehensively reviews how CFD technology has been applied and developed over the past 20 years, particularly in the context of emerging PCSI technology. It thoroughly analyzes the models used for 3D CFD simulations, highlighting their interactions and examining the limitations and errors associated with these complex physics-based numerical models. The paper also discusses the challenges faced in using CFD as a design tool, including difficulties in obtaining initial and boundary conditions, which currently hinder CFD's effectiveness in the initial design phase. Moreover, it explains why relying solely on 0D/1D analysis cannot fulfill the role of a predictive tool.

Despite its extensive use in industrial and academic research, as well as in optimizing pre-chamber designs, CFDs still exhibit significant limitations in their predictive capabilities. These challenges are particularly evident in turbulence modeling, affecting mixing and combustion, wall heat transfer modeling, and accurately depicting combustion processes. In the specific context of pre-chamber spark ignition, unresolved challenges remain, including flame and jet–wall interactions, ignition and combustion dynamics within the main chamber, and the ability of simplified combustion models in commercial CFD tools to accurately represent these complex processes.

To fully realize the potential of CFDs as more than just a predictive tool and develop it into an advanced design instrument, it is essential to enhance CFDs with precisely tailored, physics-based numerical models. Ultimately, for high-fidelity CFD modeling of pre-chamber combustion engines, the primary sources of uncertainty must be identified and corrected before proposing fine-tuning and improvements related to various turbulence and combustion models.

**Funding:** This work was supported by the Material and Parts Technology Development Project of the Korea Evaluation Institute of Industrial Technology, funded by the Korean government (Grant No. 20004900).

**Data Availability Statement:** Data are contained within the article.

**Conflicts of Interest:** The authors declare no conflicts of interest. The funders had no role in the design of the study, in the collection, analyses, or interpretation of data, in the writing of the manuscript, or in the decision to publish the results.

## References

1. Baumgartner, L.S.; Wohlgemuth, S.; Zirngibl, S.; Wachtmeister, G. *Investigation of a Methane Scavenged Prechamber for Increased Efficiency of a Lean-Burn Natural Gas Engine for Automotive Applications*; SAE Technical Paper 2015-01-0866; SAE International: Warrendale, PA, USA, 2015.
2. Zhu, S.; Akehurst, S.; Lewis, A.; Yuan, H. A review of the pre-chamber ignition system applied on future low-carbon spark ignition engines. *Renew. Sustain. Energy Rev.* **2022**, *154*, 111872. [CrossRef]
3. Novella, R.; Gomez-Soriano, J.; Martinez-Hernandiz, P.J.; Libert, C.; Rampanarivo, F. Improving the performance of the passive pre-chamber ignition concept for spark-ignition engines fueled with natural gas. *Fuel* **2021**, *290*, 119971. [CrossRef]
4. Wei, H.; Zhu, T.; Shu, G.; Tan, L.; Wang, Y. Gasoline engine exhaust gas recirculation—A review. *Appl. Energy* **2012**, *99*, 534–544. [CrossRef]
5. Bravo, Y.; Larrosa, C.; Lujan, J.; Climent, H.; Rivas, M. *Evaluation of EGR System Implementation in a GTDI Engine with Different Configurations: Assessment on Fouling and Corrosion Issues*; SAE Technical Paper 2016-01-1016; SAE International: Warrendale, PA, USA, 2016. [CrossRef]
6. Uyehara, O.A. *Prechamber for Lean Burn for Low Nox*; SAE Technical Paper; SAE International: Warrendale, PA, USA, 1995.
7. Toulson, E.; Watson, H.C.; Attard, W.P. *The Effects of Hot and Cool EGR with Hydrogen Assisted Jet Ignition*; SAE Technical Paper; SAE International: Warrendale, PA, USA, 2007.
8. Ronney, P.D. Laser versus conventional ignition of flames. *Opt. Eng.* **1994**, *33*, 510–521. [CrossRef]
9. Böker, D.; Brüggemann, D. Advancing lean combustion of hydrogen–air mixtures by laser-induced spark ignition. *Int. J. Hydrogen Energy* **2011**, *36*, 14759–14767. [CrossRef]
10. Getzlaff, J.; Pape, J.; Gruenig, C.; Kuhnert, D.; Latsch, R. *Investigations on Pre-Chamber Spark Plug with Pilot Injection*; SAE Technical Paper; SAE International: Warrendale, PA, USA, 2007.
11. Benajesa, J.; Novellaa, R.; Gomez-Sorianoa, J.; Martinez-Hernandiza, P.J.; Libertb, C.; Dabirib, M. Evaluation of the passive pre-chamber ignition concept for future high compression ratio turbocharged spark-ignition engines. *Appl. Energy* **2019**, *248*, 576–588. [CrossRef]
12. Benajesa, J.; Novellaa, R.; Gomez-Sorianoa, J.; Barberya, I.; Libertb, C.; Rampanarivob, F.; Dabiri, M. Computational assessment towards understanding the energy conversion and combustion process of lean mixtures in passive pre-chamber ignited engines. *Appl. Therm. Eng.* **2020**, *178*, 115501. [CrossRef]
13. Wang, M.; Leng, X.; He, Z.; Wei, S.; Chen, L.; Jin, Y. *A Numerical Study on the Effects of the Orifice Geometry between Pre- and Main Chamber for a Natural Gas Engine*; SAE Technical Paper 2017-01-2195; SAE International: Warrendale, PA, USA, 2017.
14. Onofrio, G.; Napolitano, P.; Tunestål, P.; Beatrice, C. Combustion sensitivity to the nozzle hole size in an active prechamber ultra-lean heavy-duty natural gas engine. *Energy* **2021**, *235*, 121298. [CrossRef]
15. Bozza, F.; De Bellis, V.; Malfi, E.; Teodosio, L.; Tufano, D. Optimal Calibration Strategy of a Hybrid Electric Vehicle Equipped with an Ultra-Lean Pre-Chamber SI Engine for the Minimization of $CO_2$ and Pollutant Emissions. *Energies* **2020**, *13*, 4008. [CrossRef]

16. Assanis, D.; Engineer, N.; Neuman, P.; Wooldridge, M. *Computational Development of a Dual Pre-Chamber Engine Concept for Lean Burn Combustion*; SAE Technical Paper 2016-01-2242; SAE International: Warrendale, PA, USA, 2016. [CrossRef]
17. Attard, W.P.; Toulson, E.; Huisjen, A.; Chen, X.; Zhu, G.; Schock, H. *Spark Ignition and Pre-Chamber Turbulent Jet Ignition Combustion Visualization*; SAE Technical Paper 2012-01-0823; SAE International: Warrendale, PA, USA, 2012.
18. Shah, A.; Tunestal, P.; Johansson, B. *Effect of Relative Mixture Strength on Performance of Divided Chamber 'Avalanche Activated Combustion' Ignition Technique in a Heavy Duty Natural Gas Engine*; SAE Technical Paper 2014-01-1327; SAE International: Warrendale, PA, USA, 2014.
19. Chao, Y.; Hu, K.; Wei, H.; Li, S.; Hu, Y.; Yu, J.; Scholten, I. Geely Jet Ignition System for 52.5% Indicated Thermal Efficiency. In Proceedings of the 43rd International Vienna Motor Symposium, Vienna, Austria, 27–29 April 2022.
20. Turkish, M.C. *3-Valve Stratified Charge Engines: Evolvement, Analysis and Progression*; SAE Technical Paper 741163; SAE International: Warrendale, PA, USA, 1974. [CrossRef]
21. Dale, J.D.; Oppenheim, A.K. Enhanced Ignition for I.C. Engines with Premixed Gases. *SAE Trans.* **1981**, *90*, 810146.
22. Toulson, E.; Schock, H.J.; Attard, W.P. *A Review of Pre-Chamber Initiated Jet Ignition Combustion Systems*; SAE Technical Paper 2010-01-2263; SAE International: Warrendale, PA, USA, 2010.
23. Oppenheim, A.K. *Combustion in Piston Engines: Technology, Evolution, Diagnosis and Control*; Springer: New York, NY, USA, 2004.
24. Date, T.; Yagi, S.; Ishizuya, A.; Fujii, I. *Research and Development of the Honda CVCC Engine*; SAE Technical Paper; SAE International: Warrendale, PA, USA, 1974.
25. Gentz, G.; Gholamisheeri, M.; Toulson, E. A study of a turbulent jet ignition system fueled with iso-octane: Pressure trace analysis and combustion visualization. *Appl. Energy* **2017**, *189*, 385–394. [CrossRef]
26. Toulson, E.; Huisjen, A.; Chen, X.; Squibb, C.; Zhu, G.; Schock, H. Visualization of propane and natural gas spark ignition and turbulent jet ignition combustion. *SAE Int. J. Engines* **2012**, *5*, 1821–1835. [CrossRef]
27. Di Sabatino, F.; Martinez-Hernandiz, P.J.; Novella, R.; Ekoto, I. Investigation of the effects of passive pre-chamber nozzle pattern and ignition system on engine performance and emissions. *Int. J. Engine Res.* **2023**, *24*, 2592–2613. [CrossRef]
28. Xu, G.; Wright, Y.M.; Schiliro, M.; Boulouchos, K. Characterization of combustion in a gas engine ignited using a small unscavenged pre-chamber. *Int. J. Engine Res.* **2020**, *21*, 1085–1106. [CrossRef]
29. Novella, R.; Gomez-Soriano, J.; Barbery, I.; Libert, C. Numerical analysis of the passive pre-chamber ignition concept for light-duty applications. *Appl. Therm. Eng.* **2022**, *213*, 118610. [CrossRef]
30. Wanker, R. Simulation Methods Covering Recent Technologies for GDI Engines. In Proceedings of the AVL International Simulation Conference, Graz, Austria, 22–24 October 2019.
31. Vedula, R.T.; Song, R.; Stuecken, T.; Zhu, G.G.; Schock, H. Thermal efficiency of a dual-mode turbulent jet ignition engine under lean and near-stoichiometric operation. *Int. J. Engine Res.* **2017**, *18*, 1055–1066. [CrossRef]
32. Mastorakos, E.; Allison, P.; Giusti, A.; De Oliveira, P.; Benekos, S.; Wright, Y.; Frouzakis, C.; Boulouchos, K. Fundamental aspects of jet ignition for natural gas engines. *SAE Int. J. Engines* **2017**, *10*, 2429–2438. [CrossRef]
33. Allison, P.; de Oliveira, M.; Giusti, A.; Mastorakos, E. Pre-chamber ignition mechanism: Experiments and simulations on turbulent jet flame structure. *Fuel* **2018**, *230*, 274–281. [CrossRef]
34. Bunce, M.; Blaxill, H. Methodology for combustion analysis of a spark ignition engine incorporating a pre-chamber combustor. In Proceedings of the SAE 2014 International Powertrain, Fuels & Lubricants Meeting, Birmingham, UK, 20–23 October 2014; SAE International: Warrendale, PA, USA, 2014.
35. Benajes, J.; Novella, R.; Gomez-Soriano, J.; Barbery, I.; Libert, C. Advantages of hydrogen addition in a passive pre-chamber ignited SI engine for passenger car applications. *Int. J. Energy Res.* **2021**, *45*, 13219–13237. [CrossRef]
36. Dong, D.; Wei, M.; Zhang, Z.; Wei, F.; Long, W.; Dong, P.; Tian, J.; Lu, M.; Wang, R.; Xiao, G. Enhanced ignition possibilities of ammonia by the prechamber fueled methanol: Rich, stoichiometric and lean combustion evaluations. *Sustain. Energy Technol. Assess.* **2024**, *64*, 103723. [CrossRef]
37. Liu, X.; Aljabri, H.; Panthi, N.; AlRamadan, A.S.; Cenker, E.; Alshammari, A.T.; Magnotti, G.; Im, H.G. Computational study of hydrogen engine combustion strategies: Dual-Fuel compression ignition with Port- and Direct-Injection, Pre-Chamber Combustion, and Spark-Ignition. *Fuel* **2023**, *350*, 128801. [CrossRef]
38. Liu, X.; Aljabri, H.; Panthi, N.; AlRamadan, A.S.; Cenker, E.; Alshammari, A.T.; Magnotti, G.; Im, H.G. Hydrogen pre-chamber combustion at lean-burn conditions on a heavy-duty diesel engine: A computational study. *Fuel* **2023**, *335*, 127042. [CrossRef]
39. Biswas, S.; Qiao, L. Ignition of ultra-lean premixed H2/air using multiple hot turbulent jets generated by pre-chamber combustion. *Appl. Therm. Eng.* **2018**, *132*, 102–114. [CrossRef]
40. Thelen, B.C.; Toulson, E. A computational study on the effect of the orifice size on the performance of a turbulent jet ignition system. *Proc. Inst. Mech. Eng. Part D J. Automob. Eng.* **2017**, *231*, 536–554. [CrossRef]
41. Biswas, S.; Qiao, L. Ignition of ultra-lean premixed hydrogen/air by an impinging hot jet. *Appl. Energy* **2018**, *228*, 954–964. [CrossRef]
42. Trombley, G.; Toulson, E. A fuel-focused review of pre-chamber initiated combustion. *Energy Convers. Manag.* **2023**, *298*, 117765. [CrossRef]
43. Alvarez, C.E.C.; Couto, G.E.; Roso, V.R.; Thiriet, A.B.; Valle, R.M. A review of prechamber ignition systems as lean combustion technology for SI engines. *Appl. Therm. Eng.* **2018**, *128*, 107–120. [CrossRef]

44. Zhou, L.; Zhong, L.; Liu, Z.; Wei, H. Toward highly-efficient combustion of ammonia–hydrogen engine: Prechamber turbulent jet ignition. *Fuel* **2023**, *352*, 129009. [CrossRef]
45. Hlaing, P.; Echeverri Marquez, M.; Cenker, E.; Im, H.G.; Johansson, B.; Turner, J.W.G. CFD-based methodology for the characterization pre-chamber combustion engines. *Fuel* **2022**, *313*, 123029. [CrossRef]
46. Shah, A.; Tunestal, P.; Johansson, B. *Effect of Pre-Chamber Volume and Nozzle Diameter on Pre-Chamber Ignition in Heavy Duty Natural Gas Engines*; SAE Technical Paper 2015-01-0867; SAE International: Warrendale, PA, USA, 2015.
47. Silva, M.; Sanal, S.; Hlaing, P.; Cenker, E.; Johansson, B.; Im, H.G. *Effects of Geometry on Passive Pre-Chamber Combustion Characteristics*; SAE Technical Paper 2020-01-0821; SAE International: Warrendale, PA, USA, 2020.
48. Distaso, E.; Amirante, R.; Cassone, E.; De Palma, P.; Sementa, P.; Tamburrano, P.; Vaglieco, B.M. Analysis of the combustion process in a lean-burning turbulent jet ignition engine fueled with methane. *Energy Convers. Manag.* **2020**, *223*, 113257. [CrossRef]
49. Gholamisheeri, M.; Thelen, B.; Toulson, E. *CFD Modeling and Experimental Analysis of a Homogeneously Charged Turbulent Jet Ignition System in a Rapid Compression Machine*; SAE Technical Paper 2017-01-0557; SAE International: Warrendale, PA, USA, 2017.
50. Thelen, B.C.; Gentz, G.; Toulson, E. *Computational Study of a Turbulent Jet Ignition System for Lean Burn Operation in a Rapid Compression Machine*; SAE Technical Paper 2015-01-0396; SAE International: Warrendale, PA, USA, 2015.
51. Gentz, G.; Thelen, B.; Litke, P.; Hoke, J.; Toulson, E. Combustion visualization, performance, and CFD modeling of a pre-chamber turbulent jet ignition system in a rapid compression machine. *SAE Int. J. Engines* **2015**, *8*, 538–546. [CrossRef]
52. Bolla, M.; Shapiro, E.; Tiney, N.; Kyrtatos, P.; Kotzagianni, M.; Boulouchos, K. *Numerical Simulations of Pre-Chamber Combustion in an Optically Accessible RCEM*; SAE Technical Paper 2019-01-0224; SAE International: Warrendale, PA, USA, 2019. [CrossRef]
53. Shapiro, E.; Ahmed, I.; Tiney, N. Advanced ignition modelling for pre-chamber combustion in lean burn gas engines. In Proceedings of the Ignition Systems for Gasoline Engines: Internationale Tagung Zündsysteme für Ottomotoren, Berlin, Germany, 6–7 December 2018; p. 104.
54. Bolla, M.; Shapiro, E.; Tiney, N.; Kyrtatos, P.; Kotzagianni, M.; Boulouchos, K. *Numerical Study of Turbulence and Fuel-Air Mixing within a Scavenged Pre-Chamber Using RANS and LES*; SAE Technical Paper 2019-01-0198; SAE International: Warrendale, PA, USA, 2019. [CrossRef]
55. Fei, Q.; Shah, A.; Zhi-wei, H.; Li-na, P.; Tunestal, P.; Xue-Song, B. Detailed numerical simulation of transient mixing and combustion of premixed methane/air mixtures in a pre-chamber/main-chamber system relevant to internal combustion engines. *Combust. Flame* **2018**, *188*, 357–366.
56. Sforza, L.; Lucchini, T.; Gianetti, G.; D'Errico, G.; Onofrio, G.; Beatrice, C.; Tunestal, P. *A 3D-CFD Methodology for Combustion Modeling in Active Prechamber SI Engines Operating with Natural Gas*; SAE Technical Paper 2022-01-0470; SAE International: Warrendale, PA, USA, 2022. [CrossRef]
57. Yin, Y.; Lei, Y.; Shen, H.; Yi, Y.; Zhao, T.; Qiu, T. Modeling Investigation on Transient Behaviors of Gaseous Ammonia Jet Flow with Direct Injection. *Fuel* **2024**, *358*, 129997–130011. [CrossRef]
58. Shapiro, E.; Tiney, N.; Kyrtatos, P.; Kotzagianni, M.; Bolla, M.; Boulouchos, K.; Tallu, G.; Lucas, G.; Weissner, M. *Experimental and Numerical Analysis of Pre-Chamber Combustion Systems for Lean Burn Gas Engines*; SAE Technical Paper 2019-01-0260; SAE International: Warrendale, PA, USA, 2019. [CrossRef]
59. Vavra, J.; Syrovatka, Z.; Vitek, O.; Macek, J.; Takats, M. *Development of a Pre-Chamber Ignition System for Light Duty Truck Engine*; SAE Technical Paper 2018-01-1147; SAE International: Warrendale, PA, USA, 2018. [CrossRef]
60. Zhou, H.; Meng, S.; Han, Z. Combustion characteristics and misfire mechanism of a passive pre-chamber direct-injection gasoline engine. *Fuel* **2023**, *352*, 129067. [CrossRef]
61. Piano, A.; Scalambro, A.; Millo, F.; Catapano, F.; Sementa, P.; Di Iorio, S.; Bianco, A. CFD-based methodology for the characterization of the combustion process of a passive pre-chamber gasoline engine. *Transp. Eng.* **2023**, *13*, 100200. [CrossRef]
62. Peethambaram, M.R.; Zhou, Q.; Waters, B.; Pendlebury, K.; Fu, H.; Haines, A.; Hale, D.; Hu, T.; Zhang, J.; Wu, X.; et al. Combustion Analysis of Active Pre-Chamber Design for Ultra-Lean Engine Operation. *SAE Int. J. Engines* **2024**, *17*, 705–720. [CrossRef]
63. Kim, J.; Scarcelli, R.; Som, S.; Shah, A.; Biruduganti, M.S.; Longman, D.E. Numerical Investigation of a Fueled Pre-Chamber Spark-Ignition Natural Gas Engine. *Int. J. Engine Res.* **2021**, *23*, 1475–1494. [CrossRef]
64. Kim, J.; Scarcelli, R.; Som, S.; Shah, A.; Biruduganti, M.S.; Longman, D.E. Assessment of turbulent combustion models for simulating prechamber ignition in a natural gas engine. *J. Eng. Gas Turbines Power* **2021**, *143*, 091004. [CrossRef]
65. Silva, M.; Liu, X.; Hlaing, P.; Sanal, S.; Cenker, E.; Chang, J.; Johansson, B.; Im, H.G. Computational assessment of effects of throat diameter on combustion and turbulence characteristics in a pre-chamber engine. *Appl. Therm. Eng.* **2022**, *212*, 118595. [CrossRef]
66. Zhao, P.; Ge, H.; Parameswaran, S. CFD-guided development of a pre-chamber ignition system for internal combustion engines. *Int. J. Powertrains* **2021**, *10*, 79–103. [CrossRef]
67. Syrovatka, Z.; Vitek, O.; Vavra, J.; Takats, M. *Scavenged Pre-Chamber Volume Effect on Gas Engine Performance and Emissions*; SAE Technical Paper 2019-01-0258; SAE International: Warrendale, PA, USA, 2019. [CrossRef]
68. Kyrtatos, P.; Bolla, M.; Benekos, S.; Bardis, K.; Xu, G.; Kotzagianni, M.; Wright, Y.M.; Giannakopoulos, G.; Frouzakis, C.E.; Boulouchos, K. Advanced Methods for Gas-Prechamber Combustion Research and Model Development. In Proceedings of the 16th Conference, The Working Process of the Internal Combustion Engine, Graz, Austria, 28–29 September 2017; pp. 167–183.
69. Wang, B.; Xie, F.; Hong, W.; Du, J.; Chen, H.; Su, Y. The effect of structural parameters of pre-chamber with turbulent jet ignition system on combustion characteristics of methanol-air pre-mixture. *Energy Convers. Manag.* **2022**, *274*, 116473. [CrossRef]
70. Convergent Science Inc. *Converge 2.3 Theory Manual*; Convergent Science Inc.: Madison, WI, USA, 2017.

71. Siemens PLM. *STAR-CD 12.02, Simcenter, Userguide*; Simens: Plano, TX, USA, 2020.
72. AVL. *AVL FIRE, Software Documentation, Version 2021 R2*; AVL: Graz, Austria, 2021.
73. Ricardo Software Ltd. *VECTIS CFD Release 2017.1 Theory Manual*; Ricardo PLC: London, UK, 2017.
74. Amsden, A.A. *KIVA-3V, Release2 A Block-Structured KIVA Program for Engines with Vertical or Canted Valve*; LA-UR-97-689; Los Alamos National Laboratory: Los Alamos, NM, USA, 1997.
75. The OpenFOAM Foundation. *OpenFOAM v2206*; The OpenFOAM Foundation: London, UK, 2022.
76. Xu, G.; Wright, Y.M.; Kyrtatos, P.; Bardis, K.; Schiliro, M.; Boulouchos, K. Experimental and Numerical Investigation of the Engine Operational Conditions' Influences on a Small Un-Scavenged Pre-Chamber's Behavior. *SAE Int. J. Engines* **2017**, *10*, 2414–2428. [CrossRef]
77. Wu, X.; Feng, Y.; Gao, Y.; Xia, C.; Zhu, Y.; Shreka, M.; Ming, P. Numerical simulation of lean premixed combustion characteristics and emissions of natural gas-ammonia dual-fuel marine engine with the pre-chamber ignition system. *Fuel* **2023**, *343*, 127990. [CrossRef]
78. Silva, M.; Mohan, B.; Badra, J.; Zhang, A.; Hlaing, P.; Cenker, E.; AlRamadan, A.S.; Im, H.G. DoE-ML guided optimization of an active pre-chamber geometry using CFD. *Int. J. Engine Res.* **2022**, *24*, 2936–2948. [CrossRef]
79. Krajnović, J.; Sjerić, M.; Tomić, R.; Kozarac, D. A novel concept of active pre-chamber engine with a single injector—The passive main chamber approach. *Appl. Therm. Eng.* **2024**, *250*, 123509. [CrossRef]
80. Kammel, G.; Mair, F.; Zelenka, J.; Lackner, M.; Wimmer, A.; Kogler, G.; Bärow, E. *Simulation Based Predesign and Experimental Validation of a Prechamber Ignited HPDI Gas Combustion Concept*; SAE Technical Paper 2019-01-0259; SAE International: Warrendale, PA, USA, 2019. [CrossRef]
81. Beran, R.; Wimmer, A. *Application of 3D-CFD Methods to Optimize a Gaseous Fuelled Engine with Respect to Charge Motion, Combustion and Knocking*; SAE Technical Paper 2000-01-0277; SAE International: Warrendale, PA, USA, 2000.
82. Shin, J.; Choi, J.; Seo, J.; Park, S. Pre-chamber combustion system for heavy-duty engines for operating dual fuel and diesel modes. *Energy Convers. Manag.* **2022**, *255*, 115365. [CrossRef]
83. Wang, H.; Wang, T.; Feng, Y.; Lu, Z.; Sun, K. Synergistic effect of swirl flow and prechamber jet on the combustion of a natural gas-diesel dual-fuel marine engine. *Fuel* **2022**, *325*, 124935. [CrossRef]
84. Wu, X.; Feng, Y.; Xu, G.; Zhu, Y.; Ming, P.; Dai, L. Numerical investigations on charge motion and combustion of natural gas-enhanced ammonia in marine pre-chamber lean-burn engine with dual-fuel combustion system. *Int. J. Hydrogen Energy* **2023**, *48*, 11476–11492. [CrossRef]
85. Dempsey, A.B.; Zeman, J.; Wall, M. A System to Enable Mixing Controlled Combustion with High Octane Fuels Using a Prechamber and High-Pressure Direct Injector. *Front. Mech. Eng.* **2021**, *7*, 637665. [CrossRef]
86. Zhu, J.; Liu, R.; Lin, H.; Jin, Z.; Qian, Y.; Zhou, D.; Yin, Y.; Li, Z.; Lu, X. Computational insights into flame development and emission formation in an ammonia engine with hydrogen-assisted pre-chamber turbulent jet ignition. *Energy Convers. Manag.* **2024**, *314*, 118706. [CrossRef]
87. Oliveira, W.P.; Mendonça, M.S.; Chavda, N.B.; Rodrigues Filho, F.A.; Baeta, J.G.C. Numerical and experimental analysis of the combustion in a Single-Cylinder research engine with passive TJI pre-chamber operating with hydrated ethanol. *Energy Convers. Manag.* **2024**, *310*, 118459. [CrossRef]
88. Chinnathambi, P.; Bunce, M.; Cruff, L. *RANS Based Multidimensional Modeling of an Ultra-Lean Burn PreChamber Combustion System with Auxiliary Liquid Gasoline Injection*; SAE Technical Paper 2015-01-0386; SAE International: Warrendale, PA, USA, 2015. [CrossRef]
89. Xu, L.; Li, G.; Yao, M.; Zheng, Z.; Wang, H. Numerical Investigation on the Jet Characteristics and Combustion Process of an Active Prechamber Combustion System Fueled with Natural Gas. *Energies* **2022**, *15*, 5356. [CrossRef]
90. Gholamisheeri, M.; Wichman, I.S.; Toulson, E. A study of the turbulent jet flow field in a methane fueled turbulent jet ignition (TJI) system. *Combust. Flame* **2017**, *183*, 194–206. [CrossRef]
91. Silva, M.; Liu, X.; Hlaing, P.; Cenker, E.; Turner, J.; Im, H.G. A Computational Assessment of Combustion Submodels for Predictive Simulations of Pre-Chamber Combustion Engines. In Proceedings of the ASME 2022 ICE Forward Conference, Indianapolis, IN, USA, 16–19 October 2022.
92. Thelen, B.; Toulson, E. *A Computational Study of the Effects of Spark Location on the Performance of a Turbulent Jet Ignition System*; SAE Technical Paper 2016-01-0608; SAE International: Warrendale, PA, USA, 2016. [CrossRef]
93. Jamrozik, A.; Tutak, W.; Kociszewski, A.; Sosnowski, M. Numerical simulation of two-stage combustion in SI engine with prechamber. *Appl. Math. Model.* **2013**, *37*, 2961–2982. [CrossRef]
94. Borghi, F.T.; Moreira, T.A.A.; Whanco, R.; Barros, J.E.M.; Valle, R.M. *Aerodynamic In-Cylinder Flow Simulation in an Internal Combustion Engine with Torch Ignition System*; SAE Technical Paper 2014-36-0298; SAE International: Warrendale, PA, USA, 2022.
95. Hernández, I.; Shapiro, E.; Tiney, N.; Kotzagianni, M.; Kyrtatos, P.; Boulouchos, K. Flame-wall interaction modelling for pre-chamber combustion in lean burn gas engines. In Proceedings of the 35th International CAE Conference and Exhibition (CAE 2018), Vicenza, Italy, 8–9 October 2018. [CrossRef]
96. Miccichè, S. Comparison of Optimization Methods for Prechamber Spark Plug Operations in Natural Gas Engines Using CFD-Simulation. Master's Thesis, Mechanical Engineering, Institut für Kolbenmaschinen (IFKM), Karlsruhe, Germany, 2019.

97. Nodi, A.; Sforza, L.; Lucchini, T.; Onorati, A.; Buttitta, M.; Marmorini, L. *CFD Modeling of Conventional and Pre-Chamber Ignition of a High-Performance Naturally Aspirated Engine*; SAE Technical Paper 2024-01-2102; SAE International: Warrendale, PA, USA, 2024. [CrossRef]
98. Addabbo, A. A CFD Methodology for the Design of Active Prechambers in SI Engines. Master's Thesis, Mechanical Engineering, Politecnico, Italy, 2023.
99. Posch, S.; Gößnitzer, C.; Rohrhofer, F.M.; Geiger, B.; Wimmer, A. Finding the Optimum Design of Large Gas Engines Prechambers using CFD and Bayesian Optimization. In *Scientific Computing 2023: Conference Proceedings*; Granigg, W., Ed.; Verlag der FH JOANNEUM Gesellschaft mbH: Graz, Austria, 2023; pp. 160–168.
100. Krajnovic, J.; Dilber, V.; Tomic, R.; Sjeric, M.; Ilincic, P.; Kozarac, D. *Numerical Simulations of Pre-Chamber Induced HCCI Combustion (PC-HCCI)*; SAE Technical Paper 2023-01-0274; SAE International: Warrendale, PA, USA, 2023. [CrossRef]
101. Pope, S.B. *Turbulent Flows*; Cornell University: New York, NY, USA, 2000; ISBN 9780521598866.
102. Hanjalic, K. Will RANS survive LES? A view of perspectives. *J. Fluid Eng.* **2005**, *127*, 831–839. [CrossRef]
103. Davidson, L. *Fluid Mechanics, Turbulent Flow and Turbulence Modeling*; Chalmers University of Technology: Goteborg, Sweden, 2024.
104. Chen, C.-J.; Jaw, S.-Y. *Fundamentals of Turbulence Modeling*; Taylor & Francis: New York, NY, USA, 1998; ISBN 1-56032-405-8.
105. Wilcox, D.C. *Turbulence Modeling for CFD*, 2nd ed.; DCW Industries Inc.: La Canada, CA, USA, 1994.
106. Yakhot, V.; Orszag, S.A.; Thangam, S.; Gatski, T.B.; Speziale, C.G. Development of turbulence models for shear flows by a double expansion technique. *Phys. Fluids A* **1992**, *4*, 1510–1520. [CrossRef]
107. Versteeg, H.K.; Malalasekera, W. *An Introduction to Computational Fluid Dynamics-the Finite Volume Method*, 2nd ed.; Pearson Education: Tamil Nadu, India, 2007.
108. The Smagorinsky Turbulence Model (Part 2). Fluid Mechanics 101, YouTube Channel. Available online: https://www.youtube.com/watch?v=GdXLyfRK188&t=992s (accessed on 1 June 2024).
109. Durbin, P.A.; Pettersson Reif, B.A. *Statistical Theory and Modeling for Turbulent Flows*; Wiley: Hoboken, NJ, USA, 2001; ISBN 0471497363.
110. Nzebuka, G.G.; Waheed, M.A. Thermal evolution in the direct chill casting of an Al-4 pct Cu alloy using the low-Reynolds number turbulence model. *Int. J. Therm. Sci.* **2020**, *147*, 106152. [CrossRef]
111. Hanjalić, K.; Popovac, M.; Hadžiabdić, M. A robust near-wall elliptic-relaxation eddy-viscosity turbulence model for CFD. *Int. J. Heat Fluid Flow* **2004**, *25*, 1047–1051. [CrossRef]
112. Durbin, P.A. Near-wall turbulence closure modeling without "damping functions". *Theor. Comput. Fluid Dyn.* **1991**, *3*, 1–13. [CrossRef]
113. Durbin, P.A. A Reynolds Stress Model for Near-wall Turbulence. *J. Fluid Mech.* **1993**, *249*, 465–498. [CrossRef]
114. Sunden, B.; Faghri, M. *Modelling and Simulation of Turbulent Heat Transfer*; WIT Press: Southampton, UK, 2005; ISBN 978-1-85312-956-8.
115. Wu, X.; Durbin, P.A. Numerical simulation of heat transfer in a transitional boundary layer with passing wakes. *J. Heat Transf.* **1999**, *122*, 248–257. [CrossRef]
116. Spall, R.E. An Assessment of k-w and v2-f Turbulence Models for Strongly Heated Internal Gas Flows. *Numer. Heat Transf. Part A* **2004**, *46*, 831–849. [CrossRef]
117. Argyropoulos, C.D.; Markatos, N.C. Recent advances on the numerical modelling of turbulent flows. *Appl. Math. Model.* **2015**, *39*, 693–732. [CrossRef]
118. Smyth, T.A.G. A review of computational fluid dynamics (CFD) airflow modelling over aeolian landforms. *Aeolian Res.* **2016**, *22*, 153–164. [CrossRef]
119. Kobayashi, H.; Hama, F.; Wu, X. Application of a Local SGS Model Based on Coherent Structures to Complex Geometries. *Int. J. Heat Fluid Flow* **2008**, *29*, 640–653. [CrossRef]
120. Ansys, Inc. *Fluent 18.0 User's Guide*; Ansys, Inc.: Canonsburg, PA, USA, 2017.
121. CHAM, PHOENICS v.1.0 User's Guide, October 2022. Available online: https://www.cham.co.uk/phoenics/d_polis/d_docs/tr316/tr316.pdf (accessed on 1 February 2024).
122. Germano, M.; Piomelli, U.; Moin, P.; Cabot, W.H. A Dynamic Subgrid-Scale Eddy Viscosity Model. *Phys. Fluids A* **1991**, *3*, 1760–1765. [CrossRef]
123. Lilly, D.K. A proposed modification of the Germano subgrid-scale closure method. *Phys. Fluids A* **1992**, *4*, 633–635. [CrossRef]
124. Nicoud, F.; Ducros, F. Subgrid-Scale Stress Modelling Based on the Square of the Velocity Gradient Tensor. *Flow Turbul. Combust.* **1999**, *62*, 183–200. [CrossRef]
125. Moukalled, F.; Mangani, L.; Darwish, M. *The Finite Volume Method in Computational Fluid Dynamics: An Advanced Introduction with OpenFOAM® and Matlab*; Springer: Berlin/Heidelberg, Germany, 2015; ISBN 978-3319168739.
126. Yoshizawa, A. Statistical Theory for Compressible Turbulent Shear Flows, with the Application to Subgrid Modeling. *Phys. Fluids* **1986**, *29*, 2152–2164. [CrossRef]
127. Kyrtatos, P.; Bardis, K.; Bolla, M.; Denisov, A.; Wright, Y.; Herrmann, K.; Boulouchos, K. Transferability of Insights from Fundamental Investigations into Practical Applications of Prechamber Combustion Systems. In Proceedings of the Ignition Systems for Gasoline Engines: Internationale Tagung Zündsysteme für Ottomotoren, Berlin, Germany, 6–7 December 2018; p. 442.
128. Zhou, L.; Liu, P.; Zhong, L.; Feng, Z.; Wei, H. Experimental observation of lean flammability limits using turbulent jet ignition with auxiliary hydrogen and methane in pre-chamber. *Fuel* **2021**, *305*, 121570. [CrossRef]

129. Xu, G.; Kotzagianni, M.; Kyrtatos, P.; Wright, Y.M.; Boulouchos, K. Experimental and numerical investigations of the unscavenged prechamber combustion in a rapid compression and expansion machine under engine-like conditions. *Combust. Flame* **2019**, *204*, 68–84. [CrossRef]
130. Rajasegar, R.; Niki, Y.; Garcia-Oliver, J.M.; Li, Z.; Musculus, M.P.B. Fundamental insights on ignition and combustion of natural gas in an active fueled prechamber spark-ignition system. *Combust. Flame* **2021**, *232*, 111561. [CrossRef]
131. Atis, C.; Chowdhury, S.S.; Ayele, Y.; Stuecken, T.; Schock, H.; Voice, A.K. *Ultra-Lean and High EGR Operation of Dual Mode, Turbulent Jet Ignition (DM-TJI) Engine with Active Pre-Chamber Scavenging*; SAE Technical Paper 2020-01-1117; SAE International: Warrendale, PA, USA, 2020.
132. Iacovano, C.; d'Adamo, A.; Cantore, G. Analysis and Simulation of Non-Flamelet Turbulent Combustion in a Research Optical Engine. *Energy Procedia* **2018**, *148*, 463–470. [CrossRef]
133. Dekena, M.; Peters, N. Combustion Modeling with the G-Equation. *Oil Gas Sci. Technol.* **1999**, *54*, 265–270.
134. Stiesch, G. *Modeling Engine Spray and Combustion Processes*; Springer: Berlin/Heidelberg, Germany, 2003.
135. Peters, N. *Turbulent Combustion. Cambridge Monographs on Mechanics*, 4th ed.; Cambridge University Press: Cambridge, UK, 2006; ISBN 9780521660822.
136. Chinnan, J.A. Simulation and Validation of In-Cylinder Combustion for a Heavy-Duty Otto Gas Engine Using 3D-CFD Technique. Master's Thesis, KTH Industrial Engineering and Management Machine Design, Stockholm, Sweden, 2018.
137. Ferziger, J.H.; Perić, M.; Street, R.L. *Computational Methods for Fluid Dynamics*, 3rd ed.; Springer: Berlin/Heidelberg, Germany, 2019. [CrossRef]
138. Ewald, J.; Peters, N. On unsteady premixed turbulent burning velocity prediction in internal combustion engines. *Proc. Combust. Inst.* **2007**, *31*, 3051–3058. [CrossRef]
139. Gulder, O.L. *Turbulent Premixed Flame Propagation Models for Different Combustion Regimes*; The Combustion Institute: Pittsburgh, PA, USA, 1990; pp. 743–750.
140. Bradley, D.; Lau, A.K.C.; Lawes, M. Flame stretch rate as a determinant of turbulent burning velocity. *Philos. Trans. R. Soc. Lond.* **1992**, *338*, 359–387. [CrossRef]
141. Meneveau, C.; Poinsot, T. Stretching and quenching of flamelets in Premixed turbulent combustion. *Combust. Flame* **1991**, *86*, 311–332. [CrossRef]
142. Lu, T.; Law, C.K. A Criterion Based on Computational Singular Perturbation for the Identification of Quasi Steady State Species: A Reduced Mechanism for Methane Oxidation with NO Chemistry. *Combust. Flame* **2008**, *154*, 761–774. [CrossRef]
143. Toulson, E. Applying Alternative Fuels in Place of Hydrogen to the Jet Ignition Process. Ph.D. Thesis, Department of Mechanical Engineering, The University of Melbourne, Parkville, VIC, Australia, 2009.
144. Veynante, D.; Vervisch, L. Turbulent Combustion Modeling. *Prog. Energy Combust. Sci.* **2002**, *28*, 193–266. [CrossRef]
145. Suillaud, E. Modelling of High Karlovitz Combustion in Spark-Ignition Engines. Ph.D. Thesis, Chemical and Process Engineering, Université Paris-Saclay, Gif-sur-Yvette, France, 2021.
146. Meneveau, C.; Sreenivasan, K.R. The Multifractal Nature of Turbulent Energy Dissipation. *J. Fluid Mech.* **1991**, *224*, 429–484. [CrossRef]
147. Salerno, F.; Bargende, M.; Kulzer, A.; Grill, M.; Burkardt, P.; Günther, M.; Pischinger, S.; Villforth, J. A Quasi-Dimensional Burn Rate Model for Pre-Chamber-Initiated Jet Ignition Combustion. *SAE Int. J. Adv. Curr. Pract. Mobil.* **2023**, *5*, 2258–2277. [CrossRef]
148. Kim, J.; Scarcelli, R.; Som, S.; Shah, A.; Biruduganti, M.S.; Longman, D.E. Evaluation of Combustion Models for CFD Simulation of Pre-Chamber Ignition in a Natural Gas Engine. In Proceedings of the 11th U.S. National Combustion Meeting, Pasadena, CA, USA, 24–27 March 2019.
149. Patankar, S.V. *Numerical Heat Transfer and Fluid Flow*; CRC Press: Boca Raton, FL, USA, 1980.
150. Toro, E.F. *Riemann Solvers and Numerical Methods for Fluid Dynamics: A Practical Introduction*, 3rd ed.; Springer: Berlin/Heidelberg, Germany, 2009; ISBN 978-3-540-25202-3.
151. Ikegaya, N.; Okaze, T.; Kikumoto, H.; Imano, M.; Ono, H.; Tominaga, Y. Effect of the numerical viscosity on reproduction of mean and turbulent flow fields in the case of a 1:1:2 single block model. *J. Wind Eng. Ind. Aerodyn.* **2019**, *191*, 279–296. [CrossRef]
152. GAMMA Technologies. GT-SUITE. 2024. Available online: https://www.gtisoft.com/gt-suite/ (accessed on 1 June 2024).
153. Benson, R.S.; Horlock, J.H.; Winterbone, D.E. *The Thermodynamics and Gas Dynamics of Internal-Combustion Engines*; Clarendon Press: Oxford, UK, 1982.
154. Bozza, F.; De Bellis, V.; Tufano, D.; Malfi, E.; Müller, C.; Habermann, K. *A Quasi-Dimensional Model of Pre-Chamber Spark-Ignition Engines*; SAE Technical Paper 2019-01-0470; SAE International: Warrendale, PA, USA, 2019. [CrossRef]
155. De Bellis, V.; Bozza, F.; Fontanesi, S.; Severi, E.; Berni, F. Development of a Phenomenological Turbulence Model through a Hierarchical 1D/3D Approach Applied to a VVA Turbocharged Engine. *SAE Int. J. Engines* **2016**, *9*, 506–519. [CrossRef]
156. Berni, F.; Cicalese, G.; Fontanesi, S. A modified thermal wall function for the estimation of gas-to-wall heat fluxes in CFD in-cylinder simulations of high performance spark-ignition engines. *Appl. Therm. Eng.* **2017**, *115*, 1045–1062. [CrossRef]
157. Reitz, R. *Assessment of Wall Heat Transfer Models for Premixed-Charge Engine Combustion Computations*; SAE Technical Paper 910267; SAE International: Warrendale, PA, USA, 1991. [CrossRef]
158. Jayatilleke, C.L.V. The influence of Prandtl number and surface roughness on the resistance of the laminar sublayer to momentum and heat transfer. In *Progress in Heat and Mass Transfer*; Pergamon Press: Oxford, UK, 1969; Volume 1.
159. Kays, W.M.; Crawford, M.E. *Convective Heat and Mass Transfer*, 3rd ed.; McGraw-Hill: New York, NY, USA, 1994.

160. Angelberger, C.; Poinsot, T.; Delhay, B. *Improving Near-Wall Combustion and Wall Heat Transfer Modeling in SI Engine Computations*; SAE Technical Paper 972881; SAE International: Warrendale, PA, USA, 1997. [CrossRef]
161. Han, Z.; Reitz, R.D. A temperature wall function formulation for variable density turbulent flows with application to engine convective heat transfer modeling. *Int. J. Heat Mass Transf.* **1997**, *40*, 613–625. [CrossRef]
162. Dou, X.; Yosri, M.; Talei, M.; Yang, Y. Impact of wall heat transfer modelling in large-eddy simulation of hydrogen knocking combustion. *Int. J. Hydrogen Energy* **2024**, *62*, 405–417. [CrossRef]
163. Heywood, J.B. *Internal Combustion Engine Fundamentals*, 2nd ed.; McGraw-Hill: New York, NY, USA, 2018.
164. Rakopoulos, C.D.; Kosmadakis, G.M.; Pariotis, E.G. Critical evaluation of current Heat transfer models used in CFD in-cylinder engine simulations and establishment of a comprehensive wall-function formulation. *Appl. Energy* **2010**, *87*, 1612–1630. [CrossRef]
165. Torregrosa, A.J.; Broatch, A.; Gil, A.; Gomez-Soriano, J. Numerical approach for assessing combustion noise in compression-ignited Diesel engines. *Appl. Acoust.* **2018**, *135*, 91–100. [CrossRef]
166. Hiraoka, K.; Nomura, K.; Yuuki, A.; Oda, Y.; Kameyama, T. *Phenomenological 0 Dimensional Combustion Model for Spark-Ignition Natural Gas Engine Equipped with Pre-Chamber*; SAE Technical Paper 2016-01-0556; SAE International: Warrendale, PA, USA, 2016.
167. Le Guen, S.; Maiboom, A.; Bougrine, S.; Tauzia, X. *Analysis of Systematic Calibration of Heat Transfer Models on a Turbocharged GDI Engine Operating Map*; SAE Technical Paper 2018-01-0787; SAE International: Warrendale, PA, USA, 2018. [CrossRef]
168. Silva, M.; Sanal, S.; Hlaing, P.; Cenker, E.; Bengt, J.; Hong, G.I. *A Computational Investigation of Fuel Enrichment in the Pre-Chamber on the Ignition of the Main Chamber Charge*; SAE Technical Paper 2021-01-0523; SAE International: Warrendale, PA, USA, 2021. [CrossRef]

**Disclaimer/Publisher's Note:** The statements, opinions and data contained in all publications are solely those of the individual author(s) and contributor(s) and not of MDPI and/or the editor(s). MDPI and/or the editor(s) disclaim responsibility for any injury to people or property resulting from any ideas, methods, instructions or products referred to in the content.

MDPI AG
Grosspeteranlage 5
4052 Basel
Switzerland
Tel.: +41 61 683 77 34

*Energies* Editorial Office
E-mail: energies@mdpi.com
www.mdpi.com/journal/energies

Disclaimer/Publisher's Note: The title and front matter of this reprint are at the discretion of the Guest Editors. The publisher is not responsible for their content or any associated concerns. The statements, opinions and data contained in all individual articles are solely those of the individual Editors and contributors and not of MDPI. MDPI disclaims responsibility for any injury to people or property resulting from any ideas, methods, instructions or products referred to in the content.

www.ingramcontent.com/pod-product-compliance
Lightning Source LLC
Chambersburg PA
CBHW041406210926
51850CB00013B/3194